An Environmental History
of
Great Britain

From 10,000 Years Ago to the Present

———

I. G. Simmons

EDINBURGH
University Press

© I. G. Simmons, 2001

Edinburgh University Press Ltd
22, George Square, Edinburgh

Typeset in Minion
by Pioneer Associates, Perthshire, and
printed and bound in Great Britain by
The Cromwell Press Ltd, Trowbridge, Wiltshire

A CIP Record for this book is available from
the British Library

ISBN 0 7486 1284 X (hardback)
ISBN 0 7486 1283 1 (paperback)

Contents

Preface

The genesis of this book is probably in my undergraduate days at University College, London. Historical geography, as taught by Clifford Darby, Terry Coppock and Hugh Prince, was very strong and we were always encouraged to look beyond the present towards origins and past developments. One evening at a social function I was introduced to H. J. Fleure, then retired from Aberystwyth and living in London. I remember him telling me that his grandfather had lived in the eighteenth century and I was so impressed with this continuity that I went away and read his book in the *New Naturalist* series, *A Natural History of Man in Britain* (London: Collins, 1951). So eventually when I wanted to write about the changes in the environment of Great Britain that had been brought about by people, I looked back to Fleure as a starting point. I want to move on from it in all kinds of ways, but I hope there is something of the same idea of human beings set in a biophysical framework, reacting to it and changing it at the same time. I do not see that the history of our environment in the last 10,000 years can be separated from our economic and social history, but of course the challenge is to select enough of the 'natural' and the social without eviscerating either. So if this book sits uneasily between ecological history and social and economic history that is why, and the challenge will remain to intertwine them more closely and effectively than I have done. Much of the best work conventionally described as environmental history has been done on the local and regional scales and I wanted to see if a coherent attempt of a political unit could be achieved: well 'maybe' is probably the right answer. But my thanks to members of a seminar in Toronto who spent an interesting few days wrestling with the problems of scale in this kind of historical study. My experience in teaching has been that interest in environmental history inevitably brings questions about the future, but I have largely avoided them here. My overall opinion would be that history does not have many lessons for the material future but certainly has some for the mythical setting of those material changes. So I leave prediction to the modellers and the prophets.

In some ways, an academic's books are like a composer's symphonies. Since musical equivalents to the edited collection are rare, then the single-authored book becomes the basis of the symphonic numbers. In that case, here we are at number 10 of those, though there are some co-edited pieces to add to the Opus total. For serious composers of the nineteenth century and after, 9 is the usual complement, though heroes like Sibelius stopped at 7. Mahler (not one of my symphonic heroes, I might say) makes it to 10 if later 'completion' is allowed and the great Dimitri Shostakovitch made it against all the odds to 15.

Enough: I am not in the same creative league as those people and this book will not last as their work has done. But the analogy does locate the book in a career and indeed in a life. *Changing the Face of the Earth* (my fourth book) is very empirical and so I veered off into the realm of ideas with *Interpreting Nature* (my sixth) and *Humanity and Environment* (the ninth). In the historical sphere, *Environmental History* (the seventh) is not very full of ideas but has in some quarters been interpreted as if it were. The present book is a different mix: it is largely empirical but is shot through with ideas about the non-material and the social.

The problems of selection have been enormous. Which processes and which 'events' (like legislation) to mention has been a task that inevitably results in leaving out matters which may well have been important. Likewise, I have tried to weave the story from a myriad and fragmentary set of books, journals, newspaper cuttings, the Internet, TV and radio and, of course, my own experiences. If you read the notes you will see allusions to these last mentioned and some of them have shaped my view of environmental history. I see no reason to apologise for this: I suspect it is true for everybody outside the strict realms of the natural sciences (and for some within, I bet) even if it is not always acknowledged. The result is a skein: the same colours wind their way in and out but different ones dominate the surface at different points.

How to use this book

There are two ways to look at the book. You can look at the whole spectrum at any one of the periods of time by which the text is largely divided: all the way from climate to population growth for the eighteenth century for example. Alternatively, you can read all the bits on woodlands in each chapter one after the other to get a sense of how that environmental system has changed from the Palaeolithic to the 'present', where the latter is the date of the last alterations permitted by the publisher. I recommend the former, as I always hope my books will be read rather than reproduced in note form. At the same time I hope there are enough data to make the book useful for reference as well as narrative. The scholarly apparatus is tucked away as endnotes, so it is easily ignored. It is barely adequate, of course, since quite a lot of material is simply the accumulation of over 60 years of living, which defies any citation method.

All references to sources will be found in the endnotes which, as I have said, sometimes contain material such as personal observations or marginal matters. I have defined technical terms in the Glossary and marked their first instances in the text in **bold** type. I have also included an appendix containing brief biographies of some of those who appear in the text where such persons' names are printed in SMALL CAPS where first mentioned. The illustrations are numbered by chapter and not consecutively through the book. I have listed in a Select Bibliography books that I found especially interesting or relevant. Most quantities are given in metric units but I have tended to retain older systems when they have been used in historical sources, especially where these provided rounded numbers, and I have given a conversion nearby.

ACKNOWLEDGEMENTS

My thanks must include a spectrum of people who have helped in various ways. My understanding of pre-agricultural communities has been helped by Tim Schadla-Hall and the late Don Spratt, for example, and the generosity of both the Vale of Pickering Trust and the Leverhulme Trust has helped with elucidating the environmental connections of that era. Medieval England has always come to life in the company of my colleague Brian Roberts, and a number of guides to industrial history has increased my understanding: Ray Hudson has been a leader among them. Dick Watson kindly read over the material on the literary scene of the nineteenth century and made some enlightening comments. Having the opportunity to teach about the whole sweep of environmental change to successive groups of school teachers from California has helped to sort out some problems of emphasis and presentation: thank you, Slade Backer, for bringing them here. In the directly practical sphere, a period of summer residence at St John's College, Oxford, by courtesy of the hospitality of the President and Fellows of the College, provided a space away from everyday concerns and was extremely helpful. The British Academy gave me one of their small grants which allowed me to employ Vicki Innes to help with the later stages of the production of a complex set of texts, captions, drawings and pictures and I want to thank her very much for her effort and patience. The environment of my department at Durham has as always been conducive to academic work of all kinds and, among its facilities, the Cartographic and Design Unit has been unfailingly helpful: Chris Orton has realised all the maps and diagrams in this book. Friends like Judith and Rob Catty provided supportive perspectives and practical help at difficult times. Among her many other contributions to my life, my wife Carol has often done the driving along narrow lanes while I tried to find just the right place to see some feature and try to get a picture: all my thanks as always.

AN OUTLINE UPDATE

Since the submission of the manuscript to the publishers, a lot seems to have happened. The consolidation of the Environment Agency (inaugurated in 1996) has been important in the institutional framework, while on the ground there have been street protests over motor fuel prices, strong resistance to a ban on hunting with dogs, exceptional flooding levels, the closing of cod fishing in part of the North Sea and an epidemic of Foot-and-Mouth disease. The USA has repudiated the Kyōtō protocols. The UK government has twice affirmed its commitment to 'greener' policies but without being very specific and certainly not by being radical. What seems to link these together is a sense of limits being reached, whether in terms of industrial agriculture and fisheries or lack of adaptation to changing patterns of land use and climate. All of which might suggest that history is only a partial guide to the future.

I. G. SIMMONS
Durham, August 2000

Permissions

Grateful acknowledgement is made to the following sources for permission to reproduce material previously published elsewhere. Every effort has been made to trace the copyright holders, but if any have inadvertently been overlooked, the publisher will be pleased to make the necessary arrangements at the first opportunity.

All uncredited photographs in the book are © Ian Simmons.

Batsford, London (3.2, 3.3); Blackwell, Oxford (4.10); Blackwell Scientific Publications, Oxford (6.3, 7.2, 8.13); British Archaeological Reports British Series 95 (4.5); British Wind Energy Association © BWEA (8.27); Cambridge University Collection of Air Photographs (App. 1–8); Cambridge University Press (3.4, 4.3, 5.2, 6.2, 6.7); CPRE and Countryside Commission, London (8.16); Dr Ben Horton, University of Durham (2.1); Edinburgh University Press (2.4); English Heritage, London (3.2); T. Fawcett, London, BM catalogue no. E.143 (14) (4.2); Fulton, London (8.9); IAHS Press, Wallingford (8.10); Geographical Association, Sheffield (7.8, 8.18, 8.19, 8.26); HarperCollins Publishers Ltd. (5.1, 8.4, 8.12); HMSO, Department of the Environment (2.3, 7.1, 7.9, 8.1, 8.5, 8.14, 8.15, 8.17, 8.20, 8.22, 8.25). Crown Copyright is reproduced with the permission of the Controller of Her Majesty's Stationery Office; © John Wiley & Sons Ltd. (8.3, 8.8, 8.11, 8.23); Landscape History (4.1); Leicester University Press (4.6); F. Macavoy (4.9); Northumbrian Water (8.6); Novello, London (7.10); Oxford University Press and Wolfson College Lectures 1985 © Wolfson College, Oxford 1985 (3.1); Oxford University Press (4.11, 8.24); M. L. Parry (4.7, 4.8); Pearson Education Limited (1.1, 2.2, 7.4, 7.5, 7.6, 8.28); Penguin Books Limited, Harmondsworth and the Department of Economic and Social Affairs (1.2); Professor B. K. Roberts (4.4); Routledge, London and New York (5.3, 6.4, 6.5, 6.6, 7.7, 8.2); David Simmons (Plate 5.2); XYZ Digital Map Company and Wildgoose (App. 9).

CHAPTER ONE

Introduction

THE PURPOSE AND SCOPE OF THE BOOK

When the British write about their history and their places, they often use as an epigraph the lines from T. S. Eliot's 'Little Gidding',

> We shall not cease from exploration
> And the end of all our exploring
> Will be to arrive where we started
> And know the place for the first time.

Although much impressed by this piece of Eliot's poetry, I have never quoted these lines until now, when they seem especially apposite. For this book deals with an often-told tale: that of the history of Great Britain. But it does so from a different perspective from most other histories, both popular and academic. It is an *environmental history* which takes as its central theme the interaction between nature and culture in these islands which lie off the western fringe of the Eurasian continent. It is therefore positioned between an 'objective' scientific history and a 'moral history' which is all cultural construction. Other phrases from *Four Quartets* come to mind for this book: the idea of 'timeless moments' to describe the act of actually writing history; 'shafts of sunlight' to remind us that it is never possible to illuminate the whole and that some of the story has to remain in shadow; and 'experience and meaning' to remind academics, public and policy-makers alike that they must go beyond the 'facts' to think about questions of significance.[1]

The ways in which we can write about the interaction of humans and nature are not comprised solely of the data of the environmental scientist: the pollen analytical profile and the rainfall gauge have their places but they are not the totality. Yet neither is the contrasting study of the power relations among the social groups of eleventh-century England. Both may in fact be relevant to one understanding of how farming was affected by natural conditions such as soil and weather and how cultivation in turn altered or mitigated some of those circumstances. When people wrote about their surroundings or depicted them, it fed back into what they did or avoided doing: subsistence activities have always been important parts of life but never its totality.

The equal accord given to different kinds of data and various kinds of framework for discussion is paralleled in the view that the natural environment does not determine the way in which people live.[2] Within the context of the climate, soils and coastal

1

bounds of Britain, it would have been possible at various times to have run a very different set of economies. No law of physics made it imperative that industrialism developed in the eighteenth and nineteenth centuries, for example. The cultures which developed after the eighteenth century have shown no taste for the kind of densely populated, apartment-dwelling urban life characteristic of for example Paris, which has environmental consequences different from a city surrounded by suburbs of low-density, semi-detached housing. So this book is not an attempt to show that nature governs all our actions and that we are bereft of choices. Neither is it an assertion of total freedom from the natural world, for more than ever we perceive ourselves to be embedded in an ecological web at all spatial scales including to the global. Further, it acknowledges that the cities and the towns are important elements of an environmental history since they are absorbers and emitters of materials, and sites of the gestation of ideas.

The discipline of environmental history attempts, therefore, to undertake studies of environments in a way which highlights the interfaces between humans as agents, acting in the light of all their manifold human characteristics (both social and individual) and the non-human world in all its complexities and dynamics. So far, many of the best known studies have been of, for example, an individual river or of a particular region at a key time in its history.[3] Studies of a whole political unit are relatively rare, and surveys of long periods of time (in this case the whole of the period since the last fading of an episode of glaciation) even rarer. An exploration is therefore quite a reasonable word to have near the beginning; quite what word or phrase may be appropriate at the end will have to emerge from both writer and reader.

The best studies in environmental history also have one more feature. They carry through an environmental process involving both nature and culture from its beginning to its end. So a mention of for example the atmospheric contamination caused in the late nineteenth to mid-twentieth centuries by the many iron and steel works in Rotherham (South Yorkshire) is not an isolated 'air pollution problem' but part of a process starting with the mining of iron ore and coal and ending with the fall-out of sulphurous compounds onto moorlands, housing and human lungs, and taking in the construction of water impoundments on the way. Since, however, words have to be placed sequentially it is rarely possible to deal with the simultaneity of the ramifications; given British history it is true as well that some of the 'outer' parts of processes have taken place far beyond our coastlines. Hence, simplification in time and space is an inevitable part of the account which is given here. The aim is, however, to talk of an environmental history which acknowledges the full resource flow rather than simply a *landscape history* which reconstructs the visible landscape in front of us or our ancestors.

'Great Britain' is the term used for the area covered. That is, it covers the mainland and immediately adjacent islands of England, Scotland and Wales; Ireland is left out entirely. The reasons for this selection are in part pragmatic and in part historically defensible since it deals with the area of the core political unity after 1707. This does include areas with strong cultural differences, such as the Outer Hebrides and Shetland, and zones of strong linguistic influence in Wales and even the separate Parliament of

Man but, in general, the unifying influence of the Crown and the spatial coherence of a mainland with some generally small off-shore islands makes for a unit which has a perceptual coherence in many people's minds.

The time period covered can be more easily divided since any horizon is arbitrary in some fashion or other. The aim, however, is to weld both scientific information and socio-historical data (whether from the statistical database, the landscape, the archive or the dig) together and so the obvious place to start seems to be the beginning of the present geological era, the **Holocene**. This being only 10,000 years ago (abbreviated hereafter as 10 kya) and these islands having already a discernable human population, it seems a good horizon. The 10 kya date is the time when the last glaciers disappeared from the British Isles and the present period of relatively temperate climate began. Though there has been climatic change during the last 10 ky, it has been of a much lower amplitude than that of the great Glacial-Interglacial cycles which preceded it. (The difference in mean July temperatures between 10 kya and the warmest part of the Holocene is about 4°C; between the coldest glacial and the warmest interglacial about 20°C.) Contrary to everyday thinking, there is no scientific evidence to suggest that the latest glacial period, ending at 10 kya, is necessarily the last. Another 5000 years or less could see the climate turning Arctic once again and the ice sheets advancing, as they did at the maximum, to a line joining the Thames and Severn estuaries. Sports entrepreneurs may perhaps need to extend their long-term planning interests beyond ice hockey to the commercial possibilities of bobsleigh competitions down Cheddar Gorge.

The other end of the period is 'the present day'. This is a moveable occasion, which in fact depends very largely upon when an author stops fiddling around with a manuscript and sends it to the publisher. The publisher is then reluctant to bear the costs of many subsequent alterations and so 'the present day' usually means 'some months before this book is published'. Having spent so much time as it were running up to the present, an author can rarely resist treating those years as a springboard from which to leap into prediction and the discussion of the future or of alternative futures. On the whole, this has been resisted here, except in the most general of terms.

Nature and culture in the last 10,000 years

At the outset, we have to consider two features of our history: first, the basic endowment of our position on the globe and the consequences of that for climate, landforms and vegetation history, and then the more plastic matter of our culture and its history.

The world of nature

In the world of environmental studies, position is always important. The islands of which Great Britain is a part-occupier are within the temperate zone of the northern hemisphere, situated off the western perimeter of the Eurasian continent (Plate 1.1). The Isles of Scilly lie at 50° 10′ N and the northern limit of Shetland is at about 61°N. If we were off the eastern margin of North America, then we should extend from Fogo in Newfoundland to the northern tip of the Labrador peninsula; on the west coast the

PLATE 1.1   *Island Britain. The British Isles are a series of islands and they stretch as far as St Kilda, which is 80 km west of the Outer Hebrides. It exhibits the combination of sea, high ground and bay-head settlement characteristic of the far north-west. St Kilda is Britain's most 'natural' World Heritage Site, largely on account of its seabird populations.*
Photograph: David Simmons, 1999.

stretch is from Vancouver Island to the northern limit of the Alaska panhandle, with the Aleutian Islands stretched across the latitudes of England. The climatic consequences of being on the eastern shore of the North Atlantic are well known. Importantly, the North Atlantic drift, originating in the Caribbean and the Mexican Gulf, keeps up sea temperatures. Atmospheric depressions originating in the temperature contrasts of those sub-tropical areas bring warm and moist air at any season. So the cool and damp but equable climate of Britain is to some extent an anomaly and the contrast with the continent to the east is quickly felt by travellers in most seasons. Within Britain, there are north–south gradients of temperature, ameliorated by sea-borne warmth on the west coasts, and an east–west precipitation gradient that makes part of East Anglia as dry as areas of the Great Plains of the USA. Though the relief is never extreme (Ben Nevis is the highest mountain at 1343 m), the temperature lapse rates are steep: 0.5°C for every 100 m of altitude is typical. In the coastal lowlands of Wales, the growing season extends from mid-March to mid-December but barely from mid-April to late November in the upland areas. Here also, precipitation may increase by 2.5 to 5.0 mm per metre of altitude.[4] Only in Scotland, however, is there any possibility of the kind of winter sports that require serious investment. Some snow, however, can be found covering more than 50 per cent of the ground at Moor House in the Northern Pennines (556 m) on an average of 70 mornings per year.[5] A nearby weather station at 847 m recorded England's highest wind gust of 116 knots in January 1968. This was a westerly, which is not surprising since west and south-west winds are dominant.

In spite of Britain's small extent, some local or regional features of climate are found. The Helm Wind, for example, rushes down the steep west-facing scarp of the North Pennines under certain conditions of east or north-north-east airflow into the Vale of Eden at force 7 (it is akin to the Bora of the Adriatic), signalled by a stationary cloud over the hill crest. In summer a cold North Sea means that low cloud and fog (sea-fret or haar) comes onshore between Fife and the Humber and keeps the coastal temperatures at 15–17 °C while the rest of the country flourishes in sunny weather with temperatures at least in the mid-20s.

The basic shape of Britain owes much to its solid geology, with more superficial modifications induced by the last million years. A very broad generalisation has the oldest and hardest rocks in the north and west, with the younger and often softer strata folded across the south and east. Any of these zones may be interrupted by volcanic intrusions of a younger date, though the lowlands of the south and east are largely free of them. These rocks control the basic topography but most have been subjected to the ice ages of the **Pleistocene** period in the last million years. Ice sheets laid great mantles of clays, sands and gravels across many types of rock, and these sheets often comprise the key material for soil formation. Glaciers carved out deep valleys in areas like Snowdonia, the Lake District and the Highlands of Scotland and left transverse barriers of debris and the shores of former ponded-up lakes (Plate 1.2). Then, after the withdrawal of the ice, worldwide melting caused sea levels to rise and so first the Irish Sea was formed and then the English Channel and North Sea. The final insulation came between 8000 and 7000 BP, with the last land bridge running away from Lincolnshire in the north-easterly direction to the Zuider Zee[6] (see Fig. 2.1).

Most of the systems in which wild plants and animals live have been subject to alteration by humans: one of the main themes of this book will be the historical depth of such changes. So to talk of the 'natural environment' in terms of plants and animals is to attempt a reconstruction of some stable interval in the past. But perhaps there have been no periods of equilibrium between climate, soils and **biota** that could attract the title of '*the* natural environment of Britain', for the last 10,000 years at any rate (Fig. 1.1). All we can talk of are periods well after the main post-glacial warming (dealt with in Chapter 2) when human influence on the biota was apparently small, or make imaginary maps of what the vegetation and animal communities would be like now if humans were absent and the resulting 4000-odd years of changes (which ecologists call **succession**) were truncated into a single day. Much of Great Britain south of the Scottish Highlands would be covered in oak (*Quercus*) forest, with varying admixtures of lime (*Tilia*), beech (*Fagus*) and hazel (*Corylus*); the present moorlands would be mostly wooded but some high open areas would have thick peat, others grassland. Scottish forests would have pine (*Pinus*) and birch (*Betula*) dominating their ecology but mountain vegetation would occupy the highest tops. Even the Isles would have some woodland, usually of birch or hazel. Red (*Cervus elaphus*) and roe (*Capreolus capreolus*) deer would everywhere be the largest herbivores, with their

PLATE 1.2 *Upland England. Kentmere in the Lake District in 1988 showed a zonation of environments from open, heathy fell summits, down through grassy slopes which were obviously once wooded, to enclosed and improved grassland with farms. Apart from the rocks, this sequence has been created by humans in the last 8000 years.*

numbers being kept in check by wolves (*Canis lupus*), and the bear (*Ursus arctos*) and beaver (*Castor fiber*) alike would diversify the fauna to which we are accustomed. There would be fewer animals of edge habitats, however: smaller numbers of foxes (*Vulpes vulpes*) and songbirds in all probability.

'Nature' as a human construction

In academic papers and books on human-environment relations, it is common to see *nature* set as 'nature'. The reasons for this are two-fold. The first derives from the many uses of the word and so alerts us to the fact that here one particular usage is being employed, that is, the whole of the material components of the universe including, for instance, gravity, water, plants and of course their interactions such as the **hydrological cycle**. Humans are part of this in so far as their biology is concerned: they have to eat and drink. So 'nature' here stands for the whole of our material surroundings as an outcome of billions of years of cosmic evolution and a few million years of organic evolution. Within this huge spread of time and space, the word 'environment' is usually taken to mean (again, in this context) those parts of nature which are (1) non-human and (2) interactive with the human species. So, the local river which receives our sewage is part of our environment, as is the vegetation of the African savanna which absorbs some of the carbon dioxide emitted from our heating

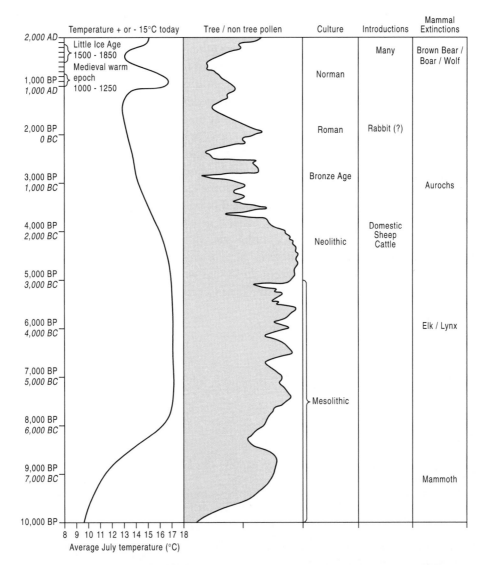

FIGURE 1.1 *The course of climatic change during the Holocene period. The Hypsithermal period of Mesolithic and early Neolithic time is the dominant fluctuation though the Medieval Warm Epoch and the Little Ice Age are far from insignificant. It should not be inferred that the changes in biota are necessarily related only to climatic change.*
Adapted from M. Bell and M. J. C. Walker (1992), Late Quaternary Environmental Change, *Harlow: Longman, p. 55, Figure 3.3, with numerous additions.*

boiler. But, in material terms at least, the constellation of Orion is not part of 'environment', but it is part of 'nature'. For the purposes of writing, 'environment' comprises the atmosphere, landforms, water, biota and land cover, with driving and explanatory influences including climate at a global scale, human populations, technology, economy and cultural perceptions. There is a second reason for printing 'nature'. This relates to the finite character of our human perceptions. Our senses can only receive a limited proportion of the stimuli present in the world (we can neither hear bats squeak nor see infra-red light, for example) and the instruments that aid in extending our range are necessarily selective of what we want to find. Further, the human brain (even aided by the computer) is limited in what it can apprehend, comprehend and remember. Our minds do, though, carry a great deal beyond the material. They are clearly the seats of emotions, which are rarely irrelevant to matters like our surroundings. They are also the repository of social behaviour, in which we submit our individuality to the greater sanctions of the group. Further, humans are probably unique among animals in having an extended concern for the future.

In sum, 'nature' and 'environment' are to some extent culturally constructed. This is obvious in visual terms when we see a field of wheat: we know that land has been cleared, that humans have sown the seeds and spread the fertiliser and will come to harvest the grain (Plate 1.3). It is less obvious when we see an ecosystem which is composed of apparently wild creatures like an oakwood, but historical study will reveal it as a human-influenced system. In fact, as will be demonstrated in the course of this book, there are very few if any nooks of Great Britain which remain unaffected by human activity.

PLATE 1.3 *The margins of lowland England, looking west from Wenlock Edge in Shropshire, 1984. The further slopes have the enclosures and hedgerows with timber characteristic of much of England's farmlands but the nearer land has undergone the field amalgamations produced by agricultural intensification. This pattern may yet change again in a 'post-productivist' countryside.*

It goes beyond that, however. Cultural construction means that we perceive nature through lenses which our culture can make. The Western world-view within which we now operate sees the world as a series of atomistic resources to which science and technology give access. These resources supply societies which are mostly in a state of economic and demographic growth. The systems of these communities are viewed as mechanisms whose states can be accurately predicted and controlled.[7] Our local version is complex. It contains needs which are material and essential, like food and water, apparent needs which are not metabolically essential but desirable, like money and central heating, and highly longed-for luxuries like holidays and private vehicles. This tradition also contains non-material strands like the value of open countryside, the sight of predatory birds and the preservation of historic buildings. There is no doubt that the lenses through which we see nature and culture can offer us a representation of environment which is different from what science tells us is actually 'out there'. Take the example of the English and Welsh moorlands: ecological science (itself a part of Western culture) tells us that they are a kind of wet desert, depleted of species and soil nutrients by centuries of burning and grazing. The popular outlook, however, is to see the wild openness as a contrast to the cities and the closely cultivated lowlands, and demands their preservation as they are, with no further tree-planting or other forms of 'reclamation'. But science itself must begin and end with human beings in all their complexity. It is always provisional and often political; fewer people nowadays believe that it somehow reveals an objective truth.

The importance of culture is so great that some commentators have followed the philosophical system of **idealism** and averred that there is nothing 'out there': that it is all a mental (indeed linguistic) construct and thus 'nature' is in our heads and nowhere else. While not denying the immense importance of culture, this book will accept that there is a world of entities independent of our apprehension, but that all we can know of them is filtered through our human capabilities. The universe is not only queerer than we suppose, but queerer than we *can* suppose, said the biologist J. B. S. Haldane in 1927. To acknowledge that autonomy, neither nature nor environment will normally be enclosed in quote marks hereafter.

Human cultures

The term **culture** is used here in its full anthropological and sociological sense as referring to the whole repertoire of learned behaviour which comprises inherited ideas, beliefs, values, knowledge and technology, and which is transmitted and reinforced by members of the group. It is not synonymous with notions like 'high culture' or 'she is a cultured person', which refer only to a particular sector of culture. Neither does it exclude such portions of the whole.

It is obvious that culture changes. At the start of our period of 10,000 years all the humans present in Britain were hunter-gatherer-fishers whose technology was comprised of implements of wood, bone and stone and whose prevalent non-material culture was very possibly dominated by an all-pervasive myth that provided explanations of the nature of the cosmos and the spawning of salmon alike. Now, our technology embraces the fissioning of atoms on the one hand and on the other a belief in several religions and in none. All the characteristics of a culture (the sources of food, access to energy, power structures, religions, popular amusements) have changed and it is difficult to say whether periods of different economic structures have been the occasion of, or caused by, changes in culture. 'Both' is the likely answer. But although cultures change, and can alter quickly, history seems never to be absent entirely. Though we erect systems of periodisation and even assert the presence of discontinuities in history, some traces of the past always persist into later periods however defined. Naturally enough, their presence gets fainter as time goes on.

The basic position of British culture is not in doubt. It is primarily one of the European societies which underwent a transition to an industrial economy in the eighteenth to nineteenth centuries and has remained in that basic mode ever since. Britain is now a participant in the type of global economy made possible by electronic communication and by trade, but outside links have always been strong, for parts of Great Britain were once components of the Roman and the Norman empires. The country thereafter built an empire of its own, at its apogee in the nineteenth century, and joined the European Union in the 1970s. There has never been a phase comparable to that of Tokugawa Japan between 1603 and 1842, when the Shōgun sealed off the nation and even tried to prevent Japanese citizens from leaving their native islands. So although this volume is devoted to the lands of Great Britain and the adjacent seas (Plate 1.4), the cultural influences have never all been home-grown.

It would be an error to assume that Great Britain has always exhibited a homogeneous culture, for within the general aegis of the label 'British' there have been variations, sometimes attracting the term 'culture' in their own right, at other times being dubbed 'sub-cultures'. Hugh Kerney has written of a history of four nations within the British Isles, forming a 'Britannic melting pot'.[8] He includes Ireland but, even without it, there is strength in his conception of a post-Roman arena in which several Celtic and Germanic cultures competed for territory and power. From the ninth to the eleventh century large areas fell under Scandinavian influence; in the mid-eleventh century the Normans took control of England and Wales and tried to extend their power northwards and westwards; the sixteenth century finally saw the political dominance of English Protestantism. Some variant of Western Christianity formed a cultural link through all the regions of Great Britain from the sixth century AD until very recently. Since the industrial revolution, the cities of Britain have always been multi-ethnic, with immigrants from Ireland, eastern Europe, the West Indies and southern Asia. They certainly emphasise that social diversity has existed within a generally recognisable British medium although by no means all these cultural changes have brought specific environmental alterations in their wake.

Every variant has nevertheless had to acknowledge that it is placed on a set of islands centred on 50°N where any month in the year can be the wettest month in that year, and where there are some long shallow estuaries as well as a few deep inlets; and where cereals do not ripen reliably above 300 m ASL; and where many land surfaces have coal underneath them, sometimes to be gained by hand-dug pits but in other places only by shafts deep under the land and then out under the sea. No amount of cultural imagination can change those realities, though it may variously

PLATE 1.4   *The estuary of the Ore in Suffolk in 1996. On the landward side, salt-marsh has been reclaimed first for grazing and then for arable farming; the shingle spit of Orford Ness is a nature reserve although it has had many years of military use. The supply of sediment to a spit can be strongly affected by coastal development.*

view them as dangerous or picturesque, or as economic or uneconomic. The strength of the cultural is indicated even here by the use of the word 'picturesque', for it suggests that we may evaluate a landscape in terms of what sort of picture it might make and this has indeed been the case, most notably in the eighteenth century. So the reflexivity of the whole process where environments influence culture and then the culture in turn affects our assessment of the environment's potential and qualities, which then produces actual manipulations of 'nature', is a tightly bound set of processes and influences.

FRAMEWORKS FOR THE READER

No book can survive without being divided up. No matter how holistic the treatment nor unitary the history, there is a limit to the amount of undivided text the reader can take. This is especially true if the text is being used for reference purposes; it may be even less valid for searchable electronic texts. Here, though, a clear and orienting framework is needed since several different types of environment and 10,000 years of time form a matrix to be differentiated out for the reader. The history of Great Britain is usually sorted into periods such as prehistory, the Roman period, the Anglo-Saxons, the Middle Ages, the Tudors and the Stuarts, the eighteenth century, and so on. Such divisions are not irrelevant to environmental history, for the gaining and wielding of power has always been germane whether it is authority over other people or hegemony over nature. In many instances it is the capability of a ruling class to harness the labour of a number of workers which actually produces an environmental manipulation. The legitimacy of power derives from a political structure and so is of interest to our present theme. For example, the Norman aristocracy loved hunting and so appropriated great tracts of terrain as chases, warrens and forests, with the Royal **Forests** as the special preserve of the king. The special laws applying to the Forests often meant that they were unavailable for colonisation by ordinary folk and so stayed as woodland or moor when neighbouring lands were cleared for agriculture; some still form the heart of uncultivated areas like the New Forest or Epping Forest.

Political events may not define environmental eras. Nevertheless the notion of power remains attractive, to the point where we may mark off a periodisation of the last 10,000 years according to the material power which whole societies were able to control, that is, their access to the energy sources of their environments and hence the basic type of economy they were able to pursue. Table 1.1 sets out the basic lineaments of such a division of these last 10,000 years, with approximate dates, the main energy sources and economic activities, and subsidiary activities. Within such basic divisions, the gatekeepers of power – in all senses – remain important.

All these activities require that humans have access to energy, like any other animal. With humans, one great difference falls between the consumption of an early Holocene hunter who used energy to stay alive, forage and reproduce, and of twentieth-century humans whose demands are dominated by cultural factors. Today in Great Britain we still need the energy for those basic metabolic processes but have added to them a thousand-fold. Since there are many more of us than in about 9000 BC, it follows that the wellsprings of energy have either been extraordinary in their fecundity or that humans have gained access to more concentrated sources. This latter is, of course, the case. The hunter-gatherers used relatively recently fixed solar energy which was concentrated as wood and other burnable materials, and as live or just-dead plants and animals. On land, animals have a tendency to run about and plants often grow in mixed communities. The usable energy, therefore, is dispersed about the landscape and so it is a feature of hunter-gatherers world-wide that they needed a large area of territory as a resource base and that they often moved through this terrain in a yearly cycle. Some archaeologists have quoted a kind of 'average' figure of 26 km^2 per person for such an economy. A population of hunter-gatherers with sufficient genetic diversity for healthy reproduction might total 500 individuals. So their bailiwick would be 13,000 km^2, which is a square 114 x 114 km or, if arrayed along a river basin or basins to a distance of 100 km from the sea would be 130 km wide. Then 500 people would be present in the whole of the counties of Durham and Northumberland. Any environmental modification undertaken by these people would then very likely be directed at increasing the density of energy available to them. The sea was likely a better source of food than the land in some seasons and so seasonal movement was probable.

TABLE 1.1 Economy and energy in the last 10,000 years

	Energy use per person/year (MJ)	Main sources of energy	Major economic activities	Other economic activities
Hunting-gathering-fishing 10 kya–3500 BC		Solar energy (SE) as wild plants and animals Controlled fire	Food collection with some environmental manipulation	Unknown but social and ceremonial life very likely
Solar-powered agriculture 3500 BC–AD 1800		As above plus SE as domesticated biota; SE as wind, water and tidal power	Food production; industrial production fuelled by wood; minerals; trade	Pleasure for a few: hunting, gardens
Industrialism 1800–1950		As above but SE greatly supplemented by fossil fuels, especially coal, then oil and natural gas (NG)	Production of goods; also food. Services in major urban zones, e.g. the City	Pleasure for minorities: shooting, fox-hunting; for the many: coastal recreation
Post-industrialism 1950–present		Oil and NG dominate with some nuclear power and electricity from hi-tech solar sources	Production of goods but services become more important	Consumerism, including more outdoor recreation

NOTES: These are only the main lineaments of the economy. No particular environmental consequences should be inferred at this stage – more detail will be added at appropriate places in the text.
Dates are mostly approximate

The energy sources of agriculturalists show the importance of more concentrated sources (more calories per nearby hectare, as it were) since they focus on **domesticated** plants and animals as the major sources of food. The energy can be controlled in the sense that the plants are grown in concentrated stands (that is in fields) and the animals corralled by being herded or stalled. Technological advances allowed less human effort per unit of output as exemplified by the windmill, the watermill and even the tide-mill. Charcoal is less bulky and lighter to transport than wood, even though it consumes more wood per unit of heat generated. Provided that fertility could be kept high in the fields, then energy surpluses could be generated so that social stratification might take place. This allowed energy consumers to live different lives from energy producers: kings, bishops, Isaac Newton and Captain James Cook alike did not have to produce their own food, except perhaps for Newton's serendipitous skill with apples. Energy surpluses allow accumulation: of material possessions, of towns, of time for invention and leisure, and of capital to finance environmental manipulation at home and abroad.

One of the developments thus financed was a move to a more concentrated energy source than charcoal for purposes such as smelting iron. Whether or not it was the depletion of forest resources that caused the shift is disputed but the outcome is certain. Lump for lump, coal contains far more energy than wood and so can be transported further and more cheaply; a mine might yield energy surpluses of thirty-fold and upwards. Inventions like the railway locomotive could carry their own fuel supply, as could the internal combustion engine when distilled liquid fuels became available. The story of the nineteenth century, though, is the use of a new energy source to gain access to all kinds of resources on hitherto unrealised scales: a chemicals revolution took place as a consequence. One result was the production of fertilisers and so freedom from the **Malthusian** spectre of insufficient food was brought about. This release was aided by the import of food from the oceans and from the colonies, neither of which would have been so productive without the steamship, which was the sea-going railway train, so to speak. Its task was to bring in the produce and in the case of the empire to carry out the troops to make sure the colonial people realised their proper place in this scheme of things.

A desire to lessen dependence on fossil fuels led to the adoption of the hitherto military technology of atomic fission for civilian use in Great Britain. Its presence after the 1950s, together with a much enhanced role for oil and natural gas, suggests a new era, especially when we found we had our own supplies of these resources. Consumption per head of energy rose swiftly after 1950 with the shift to much more technology in the home, more widespread private vehicle use and a general rise in material standards. Concern with energy futures has now arisen: the supply of fossil fuels at prices that can be afforded is not assured once the native resources are drawn down and both the economic and health aspects of nuclear power are questionable. One response has been to intensify the use of energy so that output per calorie (though energy is now usually measured in joules) rises; another has been to employ new technology to look again at the dispersed sources of energy such as the sun (for example, via photovoltaic compounds), the wind (via revived windmill technologies), waves, tides and **biomass**. Landscapes of both town and country reflect that much of the energy we use daily comes in the form of electricity.

LINKS BETWEEN ECONOMY AND ENVIRONMENT

Another complexity that must also enter into later discussion is the type of spatial linkage between an economic activity and the environment. In the solar agriculture phase, making metal implements meant that charcoal and ore (and skilled craftsmen) had to be on one site and near their materials; in the industrial period the railways meant that coal could be brought to the ore; now, the ore and the fuel could well come from abroad so that only the product and the wastes are likely to relate to the British environment. A few people walking briskly up Cairn Gorm in the 1890s exerted far less environmental impact than several hundred (with a fair proportion of recalcitrant teenage stragglers) in the 1990s. 'Pollution-free' vehicles now heralded for the new millennium are generally powered by electricity, thus shifting the waste gases from the cities to the sites of electric power generation rather than actually eliminating them.

Population growth

One measure of the outcome of interaction between culture and environment is the size and distribution of the human population. The growth of the population over much of our period of interest (necessarily estimated until the censuses began in the nineteenth century) is given as Figure 1.2 with some numerical estimates in Table 1.2 and it can be seen that there are several stages at which the curve changes direction.

TABLE 1.2   Population estimates for England up to 1801

Date	Period or source	Estimate
10,000 BC	Upper Palaeolithic (HGF)	2,000
7500 BC	Later Mesolithic (HGF)	4,000
3000 BC	Early Neolithic (Agric)	20,000
500 BC	Iron Age	250,000
AD 43–400	Roman occupation	1–2,000,000
590	Saxons and British	440,000
1086	Domesday	1,750,000
1340	1377 Poll Tax and Plague	4–6,000,000
1430	Inquests Post Mortem	2,100,000
1603	Ecclesiastical Census	4,100,000
1650	Hearth Tax	5,000,000
1750	Registers	6,300,000
1801	First Census (Industrialist)	9,200,000

Source: D. Coleman and J. Salt, *The British Population: Patterns, Trends and Processes*. Oxford: Oxford University Press, 1992, p. 5, Table 1.1.

Notably, there is a decline after the Roman withdrawal; the Black Death reduced the population by perhaps one-third so that the level of the thirteenth century was not reached again until the seventeenth century. There has been a ten-fold increase between 1690 and 1990, for the population doubled between 1780 and 1831 and exceeded 20 million by 1851. This doubled again to 40 million by 1911 and is now more or less level in the high-50s: the UK in 1994 was 58.09 million. Projections for the next 25 years add barely one million to the overall total. The population of England has always been the dominant element.[9] The lowlands have always been more densely populated than the cooler, wetter uplands, though the then forested uplands were probably much more inhabited in the hunter-gatherer and early parts of agricultural phases than subsequently. In common with much of Europe, the outstanding phenomenon is the extent to which the people have gathered in cities, especially after the

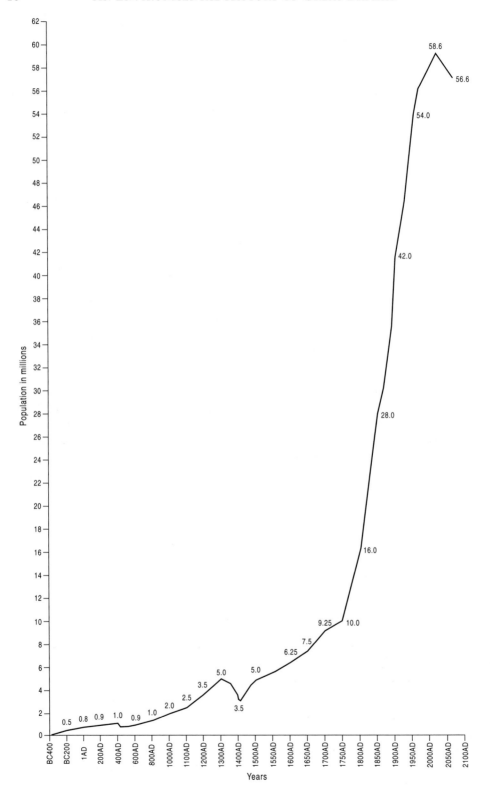

sixteenth century: whereas in 1700 one-tenth were developing early versions of Estuarine English, in the mid-nineteenth century about 30 per cent of them lived in cities with one-third of the entire population living in London.

Gross population data, like information on economies, are not direct guides to environmental relations such as resource demands and ecological impact. A group of relatively poor people living mostly on plant foods has different linkages from a well-off group with a high meat consumption and two cars per household. Inhabitants of cities exert relatively little environmental pressure on-site once the buildings have covered the land, but they reach out for food, water and energy supplies into a peripheral area which today may be world-wide. It is impossible to ignore the fact that we are a people mostly living at a high density and virtually all with strong material demands.

Non-material aspects of culture

Economic and demographic data, essential though they are, do not give much indication of the 'softer' aspects of culture which nevertheless may be important in any set of environmental relationships. For instance, there has never been the appetite, it seems, for killing all species of birds on the scale of Italy or Malta.[10] The horse has never been a regular item of diet as it is only 50 km away in France. There are as well differences within the realms of Great Britain: some regions are now well-known for their high incidences of cruelty to both wild and domesticated animals. In short, there are native ways of thinking and of formulating attitudes which do not derive in any obvious way from economy or population density. We can cite the way in which the countryside of Britain was portrayed in wartime, for example, or the 'Englishness' of certain types of music from the 1930s, in contributing to a temper of the times which in turn might influence planning legislation with environmental consequences. Literary examples of such sensibilities are even more common and some of them will appear at relevant points in the following narrative sections.

Let there be no doubt that all these views on culture and nature are very much in the Western post-enlightenment tradition. The important role of science, the key function of rationality and above all the implicit separation of the species *Homo sapiens* from the rest of nature and our self-elevation to a dominant position are all Western phenomena. They are inescapable in Britain.

COMMITTED HISTORY?

The theorists of history have comprehensively undermined the idea of objective and detached historiography. Yet to abandon the aspirations of a scientific approach completely is to trivialise something which is vital to our culture. To suggest that all

FIGURE 1.2 (opposite) *The course of population growth in Great Britain from 400 BC to AD 2050. Adapted from: Colin McEvedy and Richard Jones (1978),* Atlas of World Population History, *Harmondsworth: Penguin, with material from United Nations Population Division, Department of Economic and Social Affairs, accessed at* www.popin.org.pop1998/3.htm.

knowledge is always, all at the same time, relative to the social system that produced it, is a travesty of what humans can achieve. Yet all knowledge must be open to revision: otherwise there is only dogma. So it is perhaps incumbent on an author of a book such as this to lay out a few principles of selection: those of attitude towards what is picked out for mention among all the possible examples, for instance.

Primacy is accorded to the four main types of economy delineated above. Environmental resources are at the heart of human economies and they in turn receive the waste products from them. In addition, some environments are valued simply for what they are: wild places are the obvious example. More often, the various parts of the resource–environment interaction will inevitably suggest a segmentation of the modern economy and environment into a series of parallel systems which sometimes interact synergistically but which are generally managed separately. The results of apartness in both spatial and perceptual terms will therefore weave in and out of the story.

Overall, though, the book is not committed to a Whiggish view in which all is cumulatively for the best nor to a Stygian outlook in which town and country alike are descending into the well-meaning but ugly mélange of which Angus Wilson writes so convincingly in his novel *Late Call* which is set in one of the New Towns of the 1950s.[11] More, the present volume embraces historical processes and events in which there have been signs of hope and signs of tribulation, both rendered uncertain by the contingent occurrence. There may be long-term findings which can be teased out of so many years and so much change. There is, for example, a tension between processes which produce coalescence and those which bring about fragmentation. Examples of the former might be membership of a trade bloc, or being at the heart of an empire, or contributing to global increases in certain atmospheric gases. Examples of the latter might include private land ownership, the designation of land and water for 'conservation' and the separation of home and workplace. These apparent opposites form strands that continually interweave and are unlikely ever to be resolved. Some have their roots in technology, some in obvious social and economic developments, most in fundamental lineaments of our culture. Very few if any are solely due to natural changes.

The organisation of the material

Chapters 2–8 form the core of the book. These consist of a series of accounts of **environmental systems** in sections arranged by the major economic types identified in Table 1.1. Some of these systems belong largely to the realm of nature and connect Great Britain very widely indeed: the atmosphere is a good example. Others are mostly human in their components (military use of land and sea for instance) but relate to environmental opportunities and to environmental impact. A list of the categories of environmental systems used in this work is given as Table 1.3. Since the aim is to write about most of these, there is no space to give an exhaustive treatment of changes in species. The categories are not static through time: during the middle Holocene when hunter-gatherers occupied the land, then the dominant environmental system was that of mixed deciduous forest. The moors and heaths, wild though they look, have mostly been formed by human hands (Fig. 1.3). Later phases of agriculturalists, right up to the end of the medieval period, differentiated out most of their land uses from

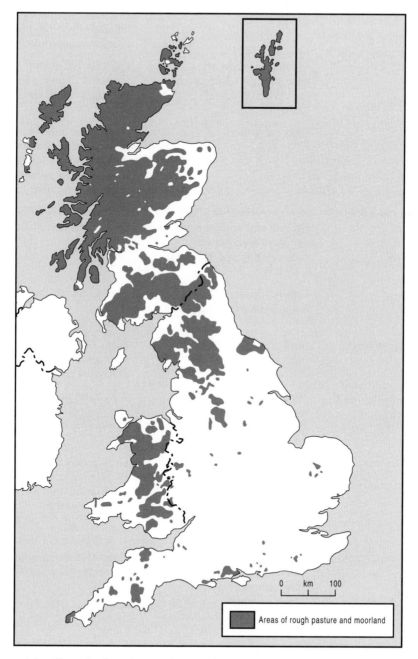

FIGURE 1.3   *The upland semi-natural areas of mountain and moorland, together with some lowland heaths, in the 1970s. At this scale, only the general distribution has any meaning but the difference between the north and west and the rest of Great Britain is striking.*

TABLE 1.3 Environmental systems

System	Regional distribution	Comments on system links
Landforms	Ubiquitous	Product of geology, Glaciation weathering, etc., but also affected by human activity since prehistoric times
Fresh water	Ubiquitous except on limestone and Chalk uplands (e.g., White Peak and Chilterns)	Quantity and quality changed by humans from early agricultural era
Atmosphere	Truly everywhere	Changed by humans from industrial times onwards, except at microscale where creation of, e.g., hedges alters some features
Woodland and forests	Never very dense in SI and right on tops of mountains and moorlands but very widespread in mid-Holocene, c. 8000 BC	The matrix out of which most humanised ES have been created
Heaths, moors mountains	Only heaths in LE, otherwise found in each regional unit	Most affected by humans via control of fire and/or animal numbers
Cultivated grasslands	In all units, though less common in SI and UE	From 3500 BC onwards with continuing importance as basis of rural ES
Rural settlement	Everywhere but very different in spatial arrangements and density in LE from rural W, for example	A human creation but not simply a peasant response to economy and environment
Urban settlement	May mean different things, but here taken as large urban-industrial complexes and as such not found north and west of central lowlands of SM	Grew greatly in nineteenth century but relatively large cities found from seventeenth century onwards
Industry	Not always in conurbations: isolated steelworks founded in nineteenth century; ironstone mining in remote UE in that century and other metal ores before that	Impact has been strong in relatively wild areas of SM and UE
Transport	Trade from very early agricultural times (or even before) must have ensured its relevance everywhere	ES always important in location of routes until advent of air transport in twentieth century

Leisure and pleasure	Happens everywhere	Dedication of ES at first an aristocratic privilege but widens through nineteenth and twentieth centuries
Military	Found in all regions, some preference for wild-looking terrain	Impact from medieval SM is recorded
Dunes, estuaries and salt-marshes	By definition, in all regions	Often subject to 'reclamation' from Roman times onwards
Wetlands	All regions: bogs, fens	Often a target for conversion to other systems
Open sea	Surrounds the whole	A larder of resources, permanently affected by humans from the nineteenth century onwards
Energy systems	Found everywhere	Regional differences in energy intensity are found

Regional abbreviations: ES = Environmental Systems; LE = Lowland England; SI = Scottish Islands; SM = Scottish Mainland; UE = Upland England; W = Wales.

PLATE 1.5   *The layers of industry: near Blaenavon, Glamorgan. This is a landscape which has been mostly industrialised since the nineteenth century, with iron working altering the landforms of the far hillside (note the absence of agricultural enclosures of any kind: sheep tend to roam the streets of small towns in valleys like this) and the new developments intended to create employment once the coal mines (near one of which we were standing in 1998) were closed. Some amenity planting of conifers has created a little biodiversity.*

that matrix. Forest became wood pasture under the influence of cattle grazing, which in turn became ley grassland under more intensive regimes. In the New Forest, woodland was differentiated out into heath, managed woods and improved areas of grass, or 'lawns'. Rural settlements in Lancashire became urban expanses with heavy industry which has since retreated and been reclaimed for recreation, housing and shopping (Plate 1.5). Today there is a picture of an intricate series of fragments belonging to all kinds of uses and a few to none. All of these have the particular energy flows which assign them to an era of Table 1.1.

Imperfect though this procedure must be, it nevertheless gives us a framework in which to lay out the results of some studies which can be used to construct this account. As will become apparent, most of the studies have been made for other purposes. Some are palaeoecological; some are pieces of historical geography or economic history; few were labelled environmental history at the outset. This may make it more difficult from time to time to describe the whole system in the way mentioned above as desirable. But our reach should always exceed our grasp.

CHAPTER TWO

Hunter-gatherers and fisherfolk
10,000 to 5500 BP

INTRODUCTION: THE BEGINNING

Ten thousand years ago, Great Britain was very different from today. A small amount of ice remained in north-west Scotland, but otherwise the glaciers and ice sheets had gone, leaving a land which was Arctic in nature but nevertheless with a small human population. Between then and the coming of agriculture at about 5500 **BP** (3500 bc)* there were many environmental changes. Most of these were natural responses to shifts in climate and other consequent events but a few, as we shall see, were brought about by humans.

The evidence for the state of the islands and the environmental changes is nearly all from the natural sciences.[1] There is no documentary evidence and we rely on the evidence of archaeology, **palaeobotany** and zoology, **palaeoclimatology**, **geomorphology** and **palaeohydrology**. In sum, those sciences which together comprise Quaternary Studies and which when interacting make up palaeoecology. These sources of data, rich though they are, tell us only about material life, so that any mention of myth, religion, attitudes to nature or similar ideas has to be inferred from material remains or back-projected from modern ethnographic studies. 'Shafts of sunlight' there may be but the unilluminated patches are dominant.

THE CURRENT INTERGLACIAL

Certain natural processes pursue their course through an interglacial period even if there is a strong human presence. They may be affected by human action but the forces which condition their natural progressions are still present. In the case of the Holocene interglacial, some six-tenths of the time so far have in any case been dominated by natural rather than human pressures.

---

* For the period before AD 1, the chronology will be described in terms of years BP (i.e., radiocarbon years Before Present), where Present is AD 1950 in the conventions of radiocarbon dating. The straight conversion of this to years Before Christ is denoted as years bc. If a calibration based on tree rings is applied to the radiocarbon dates or the date is derived from other sources such as Roman records then the convention cal BC (calendar years BC) is used.

Atmospheric changes: the climate

The changes in the temperature regime of the last 10,000 years were greatest during the hunter-gatherer phase (Fig. 1.1). The first few thousand years of the Holocene were not only crucial in their own right but also brought about alterations in other systems in their wake. Indeed 'wake' is the right word since other systems usually lagged behind the climatic shifts. The rate of temperature change in the early Holocene in Great Britain may have been 1.7 to 2.8°C per century during these years and so by *c*.9500 BP (7500 bc), both winter and summer temperatures were similar to those of the present day. Mean January temperatures in 10,000 BP were 1°C but in 9000 BP had risen to 5°C; for July the equivalent figures are 16°C and 18°C. (Present levels are 4°C and 17°C.) Between 8000 and 4500 BP (6000–2500 bc) there was a 'climatic optimum' period or **Hypsithermal** when temperatures were 1–2°C higher than recently. After 5000 BP there was some falling-off, often in the form of short-lived oscillations, especially after 2500 BP (500 bc), and of the order of 1–2°C in magnitude. These take us into the period of agriculture.

Superimposed on such progressions, random short-term fluctuations may take place, as when a volcano emits large quantities of dust into the upper atmosphere and it brings about cooling (perhaps of 1°C) for a year or two. The same may be true of the sudden collapse of a major ice sheet, releasing large quantities of cold water and icebergs into an ocean. Any accounts of climatic change at the scale of thousands of years must acknowledge that there would sometimes have been years and perhaps decades of what appeared to be atypical climate and weather.

Fresh waters

The glacial stages of the Pleistocene released large quantities of outwash materials into the drainage. As the climate warms up, the broad and braided river channels choked with sediment give way to deeper and meandering channels with far less silt. Such channels incise themselves into the sands and gravels of the colder phases. Phases of increased precipitation mean more floods and possibly a return to sediment deposition, in phases of **aggradation**. Precipitation was subject to variation. The early Hypsithermal may have had rainfall totals of about 110 per cent of today's and the period from 6500 BP onwards seems to be one of rising water-tables and of peat growth. Glaciation leaves bodies of still water behind. Large but shallow pro-glacial lakes are one form, small but deep kettleholes another. As climate gets warmer, lakes become colonised with rings of vegetation whose dead material gradually fills up the water body, with the vegetation zones moving inwards until the lake may eventually disappear. On top of the sediments which filled the lake area woodland may develop, or if it is still too wet, a peat bog (often with a domed surface, called a 'raised bog') may grow.

Landforms

The most notable changes to these are discussed below in relation to sea level. Away from the coasts, the hunter-gatherer era is one of the stabilisation of landforms as they become vegetated. Screes on mountain and hillsides, and spreads of gravel along river valleys would have gradually become smaller features of the landscapes; as woodland colonised steep valley sides, the soils became stabilised (Plate 2.1).

Seas and coasts

The melting of ice sheets and glaciers brings about a number of other changes. The first is a lowering of sea temperatures. The extra volume of water causes sea levels to rise, and higher temperatures add to the effect by thermal expansion of the volume of water. The release of the land from the weight of an ice sheet, however, allows the land surface to emerge and this may offset some of the increases in water volume. The actual position of sea level is therefore always relative to movements in both water and land.

In Great Britain, major changes in the environments of the hunter-gatherers took place after sea-level had been lowered by about 130 m at the height of the last glaciation, some 18,000 years ago. The wasting of this ice began about 16,000 years ago and sea-level rose rapidly overall until about 7000 BC, when the rate of rise slowed up. This resulted in the separation of all of Great Britain from Ireland shortly before the Holocene, and thereafter the separation from the European mainland, with the North Sea basin attaining its present shape at *c*.7800 BP (Fig. 2.1) Great Britain in fact sits

PLATE 2.1 *Loch Einich, c.1961. Many parts of upland Britain had a bare aspect before the period of forest colonisation. Steep unstable slopes and much more open water, with woodland encroaching from the lower ground would have been found even in less mountainous districts than the Cairngorms.*

10000 BP

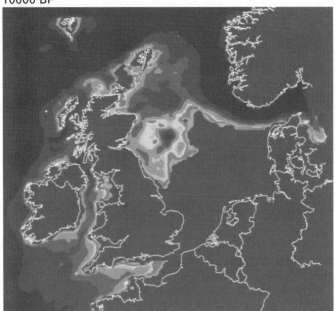

7500 BP

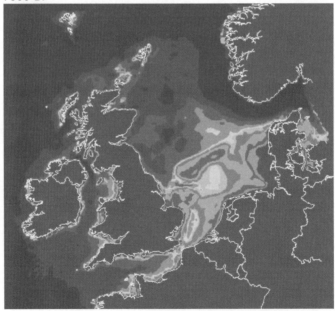

Elevation (m)
relative to MSL

-100 -50 -30 -20 -15 -10 -5 0 +5 +10

m MSL

across an important hinge-line, with land to the south of a line from the Severn to the Humber warping downwards at up to 2 mm/yr at present and northwards rising upwards at a maximum of 2 mm/yr.[2] The orderly change of coastal landforms and vegetation in response to these long-term shifts may be interrupted by especially vigorous storms or by **tsunami** caused in British waters by the massive slumping of sediment down the continental slope. The response of the coastal zone is not necessarily linear: for example if the sea-level rises at a rate of 5 mm/yr or above, then there is widespread coastline retreat within estuaries.[3] Given that the rate of rise around Great Britain during the hunter-gatherer time was probably 1 cm/yr, then many of the areas important to hunter-gatherer societies for resources may have been far from equilibrium. Another source of variability in their coastal environment was uneven rates of rise of the sea level: general maps like Figure 2.1 conceal short-term advances and retreats.

In Great Britain, one example of rapid change is given by the events in the Fenland of East Anglia. The hunter-gatherer populations of this area lived on a flat but well-drained plain until the seventh millennium bc. Rising sea levels after 6850 bc caused water to back up in the rivers and allow peat accumulation in the deeper river valleys. Indeed, low-lying areas like Wood Walton and Holme Fens in Cambridgeshire, now at −5 m OD (ordnance datum), were peat-clad by 7500 bc. Further rises in sea-level then brought about silt accumulation south of the Wash and a narrow band of peat accumulated in front of the rising sea-level as well as in river valleys. This continued until 5400 bc. So at a rate of about 10 m/yr, the dry plain was replaced by fenland conditions; the resulting peat deposits have obscured many of the traces of hunter-gatherer times until recently when they have been eroded away and have revealed the former land surface.[4]

The open sea is important as a resource base for hunter-gatherer cultures. Most archaeologists agree that dugout canoes of the type found in Mesolithic Denmark and Ireland would have been confined to rivers and lakes but that skin boats (which leave few if any traces in the archaeological record) would have been much more sea-worthy in, for example, the Irish Sea and the Inner Hebrides. Skill in constructing such boats would have given people access to such food sources as the birds, chicks and eggs of offshore islands and otherwise inaccessible cliffs, to the shellfish beds of similar islands and reefs, and to offshore populations of fish, some of which are known to be seasonal. Saithe (*Pollachius virens*) is the most likely example; any cod (*Gadus morhua*) caught would have had to come from quite dangerously open waters.

Vegetation and soils, animals

At 10,000 BP, the British Isles' vegetation was on the cusp of rapid change. Since

FIGURE 2.1 (opposite)   *The insulation of Britain during the Holocene. At 10,000 BP, most of the North Sea basin is still land, with a big embayment reaching southwards towards the Humber. At 7500 BP (c.5000 BC) there are some islands of dry land left off The Wash and on the Dogger Bank but since the Dover Strait was formed early in the Holocene Britain is now essentially a set of islands. Maps courtesy of Dr Ben Horton, University of Durham.*

*c.*14,000 BP a tundra-like vegetation had prevailed everywhere that was not ice-bound (Fig. 2.2). Under the influence of an ameliorating climate, this was changing rapidly, and the reconstructions of vegetation in 10,000 BP show four zones. Ireland, Scotland, north Wales and northern England were covered in tundra vegetation. This resembled in many ways the vegetation of northern Eurasia or Canada today, with many herbs and low shrubs. Where there was shelter then thickets of willow (*Salix* spp) and dwarf birch (*Betula nana*) flourished. In midland England, south Wales, the south-west peninsula and East Anglia, there was a belt of birch woodland with some Scots pine (*Pinus sylvestris*); the south-east of England was already occupied by deciduous woodland with birch and hazel, shortly to be colonised by elm (*Ulmus*), with oak (*Quercus*) and lime (*Tilia*) to follow. Oak expanded at a rate of 350 to 500 m/yr in England, though only by 50 m/yr in the less favourable conditions of Scotland. By 9000 BP, all of Great Britain was in a deciduous forest province. There were variations in this forest: some Scottish mountains and exposed islands never developed closed forest. The Highlands were mostly pine-birch woodland but the far north and Outer Hebrides birch-hazel. Above the tree line, mountains displayed a rich flora of cold-tolerant ('arctic-alpine') herbs. The immigration of tree species into Great Britain continued even into the agricultural era, with beech and hornbeam arriving only after the first farmers. Natural means of dispersal did not carry them beyond southern England. Along with the development of vegetation, soils changed as well. The accumulation of raw humus on the tops of tundra soils or the shallow incorporation of organic matter in a very thin soil layer (**rendzina** soils) gave way to deeper soils with a more even distribution of organic matter down the profile, usually known as **brown earths**.

So by the time of the onset of the Hypsithermal, or climatic optimum, *c.*8000 BP, the main vegetation type of much of Great Britain was oak forest. In lowland England, this was diversified by large quantities of lime; elsewhere this tree was replaced by hazel (*Corylus*). Riparian zones held wet woodland dominated by alder (*Alnus glutinosa*). Ground floras are not well understood since they leave little pollen in the deposits but the presence of natural openings and wet places will have allowed patches of bramble (*Rubus fruticosus*), rushes and sedges to grow. On uplands of low slope which had not been colonised by trees, peat had started to grow in small basins and then to spread out onto hitherto drier ground.

Along with the changing vegetation came different animal communities. The rate of establishment of the Holocene climatic patterns can in fact be traced most effectively from insect remains in peats and lake muds, since these creatures responded more quickly than plants to temperature fluctuations and in any case could migrate more rapidly. We can imagine, too, that climatic changes must have brought about different patterns of bird migration, with the present pattern of summer and winter visitors becoming established at *c.*8000 BP. Of greatest interest perhaps are the populations of large mammals. Two characteristic mammals of the tundra are reindeer (*Rangifer tarandus*) and wild horse. They vanished as the woodlands took over: the birch-willow phase is especially favourable to moose (*Alces alces*),[5] few of which survived into the oak forest. The aurochs (*Bos primigenius*) or wild ox seems to have made the transition but became extinct in Great Britain during the Bronze Age. So the dominant mammals of the deciduous forest are the red (*Cervus elaphus*) and roe (*Capreolus capreolus*) deer and the wild pig (*Sus scrofa*). Like the wild pig, the beaver (*Castor fiber*) is now extinct in Great Britain but before its demise in medieval times

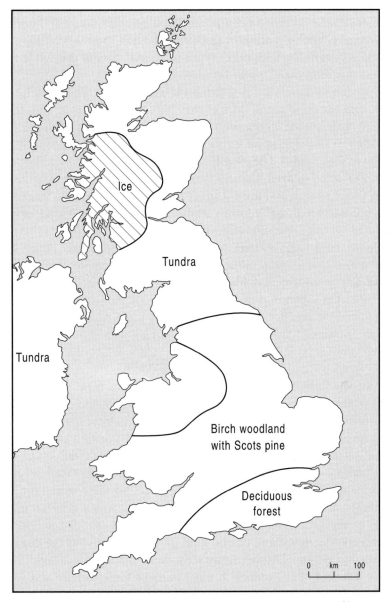

FIGURE 2.2   *Vegetation of Great Britain in 10,000 BP. There is more variety of biome type than later in the Holocene when deciduous forest came to dominate. The ice was more or less all gone within a thousand years and the birch-pine woodland was nearly all found north of the Highland boundary line.*
Source: *Redrawn from M. Bell and M. J. C. Walker,* Late Quaternary Environmental Change, *Harlow: Longman, 1992, p. 98, Figure 4.13.*

(in the sixteenth century in Scotland) it could have been a powerful influence on woodland environments. It is not certain that European beaver built dams though they would have felled small trees near water. The main predator on the large herbivores was the wolf (*Canis lupus*). Gaps of any kind in the deciduous woodlands would have borne populations of birds like the robin (*Erithacus rubecula*), wren (*Troglodytes troglodytes*), willow tit (*Parus atricapillus*) and woodpeckers. Some of the wild animals of today were possibly less common since they flourish best where there is a lot of woodland edge interfacing with other types of habitat. So the fox, while not absent, would be less common than in later times.

The key process in the clothing of Britain is immigration. The trees, however, had to migrate from the continental refuges they had occupied during that long winter. So their arrival depended upon such factors as the location of the refuge on the continent, the rate of dispersal of viable seeds, the suitability of the receiving habitat and obstacles to their spread. The early Holocene tree species had no major sea-crossing by way of obstacle between the continent and Great Britain, whereas the later immigrants such as beech and hornbeam (*Carpinus betulus*) had to be carried across the Channel or the North Sea by some means: birds such as jays (*Garrulus glandarius*) and rooks (*Corvus frugilegus*) are a possibility.[6] But insulation means that the British flora and fauna tend to have fewer species than their continental equivalents: there is only one native species of coniferous forest tree for example: the Scots Pine. The spruces and firs of continental Europe have only come to Britain if brought by humans. The earlier separation of Ireland meant that, for example, the wild pig was never present and the absence of snakes is probably more related to sea-level rise than the impressive powers of St Patrick.

The end of the beginning

The whole of the hunter-gatherer period takes place in times of some instability, though at various scales. Overall, there were periods of major climatic change with interruptions from minor oscillations toward wetter or drier, warmer or colder conditions. Sea-level fluctuated: it proceeded in one overall direction but there were times of retreat as well as advance. The forests had openings whose creation was not readily foretold and whose subsequent colonisation had variations in detail even if renewal of the dominant species was an eventual outcome. So there were no millennia of stasis or equilibrium in the environmental systems with which humans interacted.

The map of vegetation in the mid-Holocene gives us one datum line for future changes, namely that woodland of one kind or another is by far the dominant type of environmental system. Though some areas were too high or too wet or geomorphologically unstable, the tendency in most parts of Great Britain was for forest to develop wherever the conditions allowed it. If we were to have a concept of 'natural ecosystems' then it would be one in which there was preponderance of woodlands. While allowing therefore that there was no equilibrium between climate, soils and vegetation in the middle of the Holocene, we can nevertheless accept that many ecological systems thereafter were carved out of a forest matrix and that later still, successor environments might still show some traces of that origin.

PEOPLE AND CULTURES

The first humans in Great Britain were present in earlier interglacial periods and so the period with which we are concerned does not represent an initial occupation of the land. However, the glacial maximum at around 18,000 years ago whose periglacial zone extended to all of Great Britain provides a clean enough slate in terms of both nature and culture for us to start here. Archaeological research identifies three cultures before the coming of agriculture, each with some distinctive stone tools and often other characteristics of location or apparent lifestyle as well.[7]

Late Upper Palaeolithic

The Late Upper Palaeolithic is conventionally dated from 13,000 to 10,000 years ago, so it is just about finished when our interest begins. Its terminal phase, though, is a convenient datum line: some 126 sites were known by the early 1990s, all of which fall south of a line joining the Tees estuary and Morecambe Bay. The Late Upper Palaeolithic culture is sometimes referred to as the Creswellian after a site at Creswell Crags in Derbyshire, occupied from 13,000 to 11,000 years ago (Plate 2.2). Their stone technology is diagnosed by a range of points made on blades and usually regarded as spear points or arrow tips.

PLATE 2.2 *The Upper Palaeolithic leaves few visible traces: here at Cresswell Crags, Derbyshire, in the mid-1990s, the limestone gorge has a number of caves and rock shelters which housed a Paleolithic group of hunter-gatherers. The gorge was then dry-bottomed, but in the mid-1990s the lake was created by a land owner for scenic and duck-shooting purposes. A small museum nearby puts the site and its finds into context.*

Early Mesolithic

Early Mesolithic culture starts at about 10,000 BP and lasts for about 1500 years before giving way to its successor. The flint tools include both big implements and microliths not more than 50 mm long. The microliths of the Early Mesolithic are often broad-bladed compared with later examples. Where, rarely, preservation conditions permit, worked bone, antler and wood are seen to be important parts of the material culture. The most intriguing of these were found at Star Carr in the Vale of Pickering (east Yorkshire), where a number of antler frontlets were recovered. These consist of parts of red deer skulls with the antlers attached. The antlers are thinned to reduce their weight, as are some of the skulls. Some examples have pairs of holes in the skull as if cords were attached to them. No consensus exists as to whether these were disguises used in stalking deer (something similar was used on the High Plains of North America by indigenous people at the time of European contact) or part of the ritual gear of a 'shaman'. Since there were 21 of them, then the case for meat rather than mysticism is perhaps rather stronger, though the separation of the two may have been less wide than our current formal distinction between supermarket and shrine. The distribution of the sites of this period is about the same as for the preceding group, with an apparent withdrawal from west Wales. This may be a consequence of continuity and overlap between Palaeolithic and Mesolithic cultures as shown by the range of radio-carbon dates, a progressive decrease in the size of stone tools, the perpetuation of some technological usages like the groove and splinter method of working antler, and above all the continuity in the occupation of some sites (Plate 2.3).

Later Mesolithic

The later Mesolithic is found after 8500 BP and thus includes the first British culture to develop in physical isolation from the rest of Europe. Not until the opening of the Channel Tunnel in 1994 could humans once again walk dry-shod between Britain and our eastern neighbours. The tool-kit is dominated by microliths, often developed upon narrow blades; organic finds are very rare, though analogies with Swedish material, for instance, suggest that wood was worked for arrow shafts. At first, this culture is confined by the Tees–Morecambe Bay line with an extension northwards up the western side of Scotland. By the time of its demise, however, sites referable to the culture are found throughout Great Britain.

ENVIRONMENT AND CULTURE

Energy relations of hunter-gatherers

The ecological relations of hunter-gatherer people in north-west Europe have been explored by a number of palaeoecologists from the point of view of how such human groups obtained enough energy to survive and reproduce.[8] The first and key point to make is that nearly all the access to energy needed by such people was for food. In this connection, fire is useful: it can be used to break down plant and animal tissues and thus render edible materials otherwise too tough for human dentition and digestion.

PLATE 2.3   *Vale of Pickering, East Yorkshire, 1998. Sites from the Early Mesolithic are rarely visible today. Beneath these fields (whose peat substratum gives them a dark colour) lie a whole series of lake basins and small islands where the organic remains of Early Mesolithic times are well preserved. The peat is wasting under current agricultural practices and so plant and animal remains from that time are becoming less easy to recover. Small gravel islands like the one on which the farm buildings stand are 'rising' from the flat landscape quite quickly – on a scale of years rather than decades.*

Beyond that, it is possible to use it in some ecosystems to manipulate both plants and animals to increase the likely yield to human groups. One important element in the equation of energy expended/energy gained is the concentration of foods. A large mammal such as an aurochs at 1400 kg live weight is a better kill than a few dozen squirrels. Traps are one obvious response (though no land animal traps have been found from the British Mesolithic, only fish traps) and the use of a controllable animal to seek out game species and perhaps to flush them out in the direction of hunters is another: apparently domesticated dogs were present at Star Carr and Thatcham (Berkshire) in Mesolithic contexts. Energy relations therefore centred on food, though the importance of fire for warmth and as a symbol of social relations was probably high: consider only the way in which most men encountering an open fire in a room will approach it and warm their backsides at it.

Professional consensus seems to be that plant material has been undervalued by archaeologists because its preservation is less good than that of, for example, animal bone, and that animals of all kinds, including fish and molluscs, were very valuable additions to a strongly vegetarian diet. One attraction of meat would centre on the pleasure of eating it compared with that of, say, starchy roots like those of bullrushes (*Typha* spp)or even perhaps bracken rhizomes. Ecological approaches have suggested that the main nutritional problem of Mesolithic times would have been a 'lean season' in late winter/early spring when wild plant foods would, in the latitudes of Great Britain, have been very sparse and when wild animals would themselves be at their leanest. The resources of the sea might well have been the most important at that precise time, when sedentary organisms like mussels (*Mytilus edulis*) and oysters (*Ostrea edulis*) provided plentiful calories without having to be chased or trapped. The appeal of dense beds of molluscs persists in those restaurants (rather few in Britain) which serve great mounds of seaweed, ice and shellfish in a *plateau de fruits de mer*.

Mussel-beds apart, the nature of hunter-gatherer food means that the animal component comes in relatively small portions, quite widely separated. Given that animal foods were actively sought, then ways of increasing the energy surplus of hunting and fishing (as distinct from the gathering of plant materials) would have been part of the behavioural repertoire of Mesolithic groups. For example an intimate knowledge of the animals' behaviour: where they were likely to be at a given time of day or year, would be essential.

Even though the knowledge of ecosystems possessed by hunter-gatherers must have been tuned to the kind of fine degree which we cannot even imagine, adaptation to the conditions of these islands at that time was inevitable, even though attempts were made to make improvements in the energy equations at the margin. One cultural unknown in the Mesolithic period is the extent to which separation from the continent (complete by about 8000 BP/6000 bc) produced an insular culture which developed its own practices in ignorance of what was happening in the low countries, Germany or France, or to what extent there was a continuity based on ancestral links when it was all one landmass. Given, though, that boats (of canoe size) certainly existed and that the later knowledge of agriculture found its way across the Channel or the North Sea, then it seems unwise to indulge in assertions of an early development of the xenophobia which later afflicted the culture of, especially, the English.

Post-glacial change, post-glacial cultures

The Early Mesolithic cultures lived during a rapid transition from tundra through to the earlier deciduous forests; the later Mesolithic people were inhabitants of more stable mixed deciduous forests in which lime, oak and birch were variously dominant regionally. Sites from both cultural groups are found at or near the coasts.

The effects of changing environments are seen most strongly in the recovery of animal bones which are taken to be the remnants of food resources. The tundra environment of Palaeolithic times shows up in finds from Cresswell Crags of mountain hare, horse, reindeer and ibex (*Capra ibex*), from *c.*12,400 BP; grouse and ptarmigan (*Lagopus mutus*) seem to have diversified the diet. At Gough's Cave in the Cheddar Gorge of Somerset (*c.*12,500 BP), horse predominates but red deer, wild cattle, saiga antelope, mountain hare (*Lepus timidus*) and brown bear (*Ursus arctos*) are also found. In each case, pollen analysis of appropriate deposits showed an open landscape, with trees confined to sheltered valleys: woody plants were usually dominated by willow though juniper (*Juniperus communis*) and birch might also figure in the pollen counts. An interesting feature of the Palaeolithic flint tool assemblage at Cheddar is the provenance of 95 per cent of it in the Upper Cretaceous rocks of Wiltshire and a small amount from the Portland area now on the south coast: periodic migration or some form of trade took place.

By contrast, the Early Mesolithic site at Star Carr in eastern Yorkshire was placed in birch woodland, with hazel and pine not far off. It was on a lake shore and so preservation of organic remains has been good; material for radiocarbon dating is abundant and *c.*10,500 BP (8550 bc) is a common date for the occupation. The fauna is dominated by the bones of red deer, but wild cattle, moose,[9] roe deer and wild pig also contributed to the menu. In terms of weight of meat, it is possible that the wild cattle were the most important, followed by moose and with the red deer in third place. Bird remains are relatively poor but several fish-eating species were present; there are, however, no fish remains. Extraction of seasonal information from the animal bones is complex, but a summer occupation seems certain even if other seasons might also have seen a temporary occupation of the sites. Charcoal analysis shows that the reed-beds were burned in two phases beginning in 8750 BC and 8600 BC respectively.[10]

TABLE 2.1   Hunter-gatherers' ways of improving their energy balance

Practice	*Found in Great Britain?*
Knowledge of daily/yearly location of resources	Inferred
Movement of groups of people seasonally to tap resources	Very probable
Planting of wild foods to increase their density	Not known
Fire to improve conditions for edible plants	Yes – for hazel (*Corylus avellana*)
Fire to improve conditions for game	Yes – for hazel (leaf material) and grasses
Fire to manipulate vegetation to improve hunters' sight lines	No evidence
Fire to flush out game	No evidence
Fire to increase edibility of foods	No direct evidence
Use of domestic dog	Yes
Traps and snares	Evidence for fish, but not land mammals
Keep human population at relatively low level	No direct evidence

The first column is derived from worldwide hunter-gatherer studies; the second from a survey of archaeological work on the Mesolithic period in Great Britain.

At Thatcham in Berkshire another waterside site overlooks a tributary of the River Kennet.[11] The site seems to have been occupied from about 11,500 to 10,500 BP. During this interval, birch woodland became diversified by the immigration of hazel and pine. Red deer, roe deer and pig were important food sources, although the range was wide: fox (*Vulpes vulpes*) and cat (*Felis silvestris*) bones were found, and again summer occupation seems the most likely. A possible structure was found: it would have been a shelter of bent and tied saplings covered (presumably) by skins and having a floor area of about 35 m². The extent of the bone and stone remains and a spread of radiocarbon dates suggests that this site was occupied over a long period.

If we compare the archaeological and palaeoecological evidence from sites such as Thatcham and Star Carr with Table 2.1, then the Early Mesolithic saw the development of most of the known British stock of environmental adaptations. But they do not show us the extent to which a rich resource like the sea could so strongly influence the subsistence activities of a human group. For examples of this we have to look to the later Mesolithic and to Scotland.

The gull's way and the whale's way

Two sites in Scotland provide evidence of the usefulness of knowing about the sea and its potential harvests. On the coast of Fife, a site now 4 km from the sea has been dated to about 6300 BP. It was sited in an open oak woodland but also adjacent to salt-marsh and tidal mudflats. The chief form of subsistence debris is a **midden**, which is composed mostly of shellfish remains. These are dominated by two species, cockles (*Cardium edule*) and the Baltic tellin (*Tellina bathica*), but there are 38 other species of mollusc as well. Mammals yielded 22 species and birds 11 species; there are numerous fish bones of which only a small proportion could be identified; these were mostly cod. Taking all the evidence together, excavators think of it as being occupied for short periods but that any season might have been propitious for one or other of the resources.[12]

On the Hebridean island of Oronsay, shell middens are the source of detailed evidence[13] about Mesolithic settlement in the period 6100 BP to 5400 BP (4150 bc–450 bc). Now some 10 m above high tide, they were formed close to the water's edge; where the onslaught of the Highland midge might be somewhat reduced in ferocity. The middens are composed of fish remains and limpet (*Patella* spp) shells, and the greatest quantity of food would have come from the inshore fish, saithe; the limpets may have been bait. Hunting marine mammals was also carried out: bones of seal (*Phoca vitulina*), dolphin (*Delphinus delphus*) and rorqual (*Balaenoptera acutorostrata*) turn up, but it was clearly secondary to the fishing. Land mammals were a poor third and it looks as if the birch–hazel woods did not support these species: their bones were brought to the island for use in toolmaking. A seasonal element to the occupation is indicated by studies which show that the saithe were mostly caught in the autumn; this is supported by the teeth structures of grey seal pups and the occurrence of hazel nuts.

Early Mesolithic

In this and the following section, we consider whether the environmental systems of Great Britain were changed by human activity during the hunter-gatherer era. We are certain that natural changes took place but here we want to know if human agency produced similar trends or metamorphoses that were different.

Evidence of human-environment interaction in the Early Mesolithic is quite strong: the sites at Thatcham and Star Carr show us the killing and butchery of a number of species of animals. None of these investigations, however, adds up to a convincing case that, for instance, animal populations were made extinct locally or regionally by human agency in this period. The analysis of pollen from peats and lake deposits of the early Holocene does, however, show occasional traces of the opening or repression of pine or birch woodland and of the burning of heathy plants. None of this can be unequivocally tied to human activity, but neither is it sufficiently synchronous to be tied to climatic oscillations. Examples include the recession of juniper-pine-birch woodland in the presence of charcoal on northern Dartmoor before 8758 BP; at about the same time, channel peats on the North York Moors show inwash stripes of silt which appear to result from soil erosion. On the Wolds just to the south, birch woodland underwent human manipulation which allowed the survival of species-rich chalk grassland, and, in South Wales, an upland birch forest was apparently kept open by human activity, resulting in heathy vegetation.[14] For each of these examples, there are others where no such interference is postulated, so it seems to be a phenomenon which is restricted in scale. But there are sufficient hints of a process in which human control of fire in the landscape happened often enough (and perhaps at critical times) to produce a vegetation different from the pre-existing condition. The evidence is extended by the discovery that the Mesolithic inhabitants of the Vale of Pickering burned the abundant reed-beds which fringed the lake around which they lived.[15]

Later Mesolithic

The actual evidence for the later Mesolithic is confined almost entirely to stone tools and toolmaking waste and to palaeoecological data. There are virtually no human bones, for example. Conditions of preservation have meant that there is a greater weight of evidence from the uplands of England and Wales; this certainly means that Mesolithic communities were present there in the period 8500–5500 BP. It does not necessarily mean they were absent from the lowlands. Evidence for forest recession at that time is quite plentiful: the pollen of forest trees is partially replaced by species of open ground; there are many deposits with charcoal in them, and woodland is overtaken by bog vegetation. It is often difficult to disentangle natural processes from human actions and much of the discussion revolves around the probabilities of the two processes.

The recession of woodland in upland England and Wales

We shall first examine some examples of later Mesolithic forests which are replaced with other kinds of vegetation, in contexts which mean that the possibility of human agency must be examined. The main conclusion to be reached is that no type of woodland was necessarily immune from disturbance at that time. The main forest type to exhibit the process is the mixed deciduous woodland dominated in the pollen spectra by that of oak. Disturbance and recovery often afford a foothold for species like ash (*Fraxinus*) as well as letting hazel flower more abundantly. Unmanaged woodland has a great deal of dead wood, which carries a distinctive insect fauna.

The upper edge of the woodland was the main scene of disturbance. Such an **ecotone** is of course always the most susceptible to climatic change, but this zone is also prone to the kinds of disturbance in which fire is implicated. The clearest indications of the processes at work come from central Pennine sites.[16] At 6000 BP, a pollen-influx diagram for Robinson's Moss proposes that the upper level of the 'lowland forest' was at 425 m and that of the 'upland forest' at 460 m OD. This date, however, coincides with a second temporary retraction of the limit of the upland forest which is associated with evidence of burning. This is seen by Tallis and Switsur as just one of a series of fires in which burning 'probably prevented the upward spread of the component tree taxa . . . right from the time when upward forest expansion was just commencing in the early Flandrian'.[17]

The principal woody species at the tree-line at 6000 BP was probably hazel. In spite of the fires the tree-line moved up slowly between 6800 BP and 5500 BP, which probably signifies a continuing response to climatic change. Given that exposure, rock type and slope are all additional variables, it is to be expected that the impacts of burning at the upper edge of woodland are highly variable from place to place. Between 8000 and 6000 BP on Dartmoor, hazel is dominant in the woodland community but its values appear to have been negatively influenced by fire after 7500 BP.[18]

In many investigations, estimates of areas involved in disturbance rely simply upon the degree to which certain pollen frequencies alter. The outstanding example of a more reliable data-set comes from Waun-Fignen-Felen on the Black Mountain of South Wales, where analysis of multiple profiles within a small basin has allowed the construction of a series of diagrams reconstructing the vegetation at 8000, 7500, 6500, 5700, 4700 and 3700 BP. The actual edges of the vegetation mosaic are not mapped, but at 8000 BP, for example, the mixed woodland shows a burned-over opening which is beginning to accumulate **mor** humus; a part of this area is a grassy clearing being recolonised by birch and hazel. The opening abuts a small shallow lake, contains an 'Early Mesolithic' flint knapping site and is *c.*200 x 200 m in size. This site forms a nucleus for the spread of blanket bog to the north-west but on other sides continues to be set in mixed woodland until 4700 BP when mixed woodland appears to form islands in a sea of blanket bog (the lake being now covered with acid peats), a reverse of the position at 6500 BP (Plate 2.4). Local topography is clearly important in determining the actual sites of vegetation change but it is apparent that fire is implicated in the basal layers of the blanket peat and played a role in its inception.

Forest recession brings about vegetation change. What had been the ground layer of a woodland might become a dry grassy glade with a variable amount of bracken and/or heather. Herbs which are intolerant of shade but do not require deep soils of high base status are found, including cow-wheat (*Melampyrum*). Another possibility would have been a dry heathy glade with accumulating *mor* humus and a dominance of heather (*Calluna vulgaris*). If the environment was wetter then there might be a damp grassy glade, with wet-tolerant grasses and herbs. The common nettle (*Urtica dioica*) might be found as a natural component of base-rich areas of this kind. Along this spectrum of dampness the next stage might well have been a glade in which there was a carpet of wet-tolerant, peat-forming plants such as *Sphagnum* moss or a sedge like cotton-grass (*Eriophorum* spp). There might also have been openings with pools of open water in them. These might have as causes partial deforestation, for example,

PLATE 2.4 *A limestone upland on the Black Mountain, Dyfed. Palaeoecological research has shown how Mesolithic populations manipulated the woodland around a small lake just beyond the skyline and caused the first of a series of environmental impacts which resulted in the largely treeless landscape of the mid-1990s. For the mid-Holocene, we have to imagine these slopes covered with mixed deciduous woodland.*

or wallowing by large mammals. Water plants and acid mire plants are all likely colonists.

The growth of blanket peat in the uplands

In many places, the earlier landforms are blanketed out with a layer of peat commonly 2 to 4 m deep and found in two main topographic situations:

1. water-receiving sites which are the recipients of natural drainage, such as dry basins, basins with small lakes, channels and poorly drained cols between stream-heads;
2. water-shedding sites, of which the convex slopes of low angle which form the majority of the upland terrain today are the major category. Complex slopes with convex segments are also included. Such terrain underlies much of the blanket bog characteristic of higher ground today but this type of peat has also overgrown sites of type (1) as at Waun-Fignen-Felen in South Wales.

Any time between 9000 and 1000 BP might have seen an episode of onset of such growth: it therefore took place in the presence of humans. If inception happened shortly after a major climatic shift to wetter conditions, such as is postulated for the British Isles around 7500–7000 BP, then the role of climatic factors must be suspected, as permissive if not necessarily decisive.[19]

The evidence for fire is also germane to any study of the origin of these types of peat. After burning there may be a period of nutrient enrichment detectable in some peat profiles by remains of the moss *Drepanocladus*, a species of flushed mires. Another ecological effect of fire is to block the pores in the surface layers of soil with fine charcoal and thus to reduce its permeability to water and hence to increase the amount of water on the soil surface.[20] The accumulation of water is enhanced in woodlands by the removal (by any means) of trees. Deciduous trees act (1) as a shelter layer, intercepting precipitation and re-evaporating it from the canopy and trunk, and (2) as water pumps, removing water from the soils via their root systems and transpiration mechanisms. Experiments have shown that runoff increases by as much as 40 per cent after clear felling of deciduous forest. The pathways to peat accumulation are to some extent more complicated than allowed in this brief description: Figure 2.3 shows some of the complexities but prior to them all is the removal of woodland and/or the presence of fire leading to waterlogging. Once that process has started then the sequence of biochemical mechanisms will usually lead to peat formation.

It is important to ask whether the process is reversible. The possibility is shown when wood layers are found in peat stratigraphy. These indicate that the peat was thin and dry enough to permit recolonisation by trees: birch and pine are common species, for example, both at Lady Clough Moor in the south Pennines and birch at Bonfield Gill Head on the North York Moors. Generally, recolonisation by trees coincides with lower levels of charcoal in the peat, suggesting that fewer fires as well as (or even rather than) any climatic shifts was a factor in their re-growth.

To the populations of the later Mesolithic, though, heather moors accumulating *mor* humus and becoming seasonally waterlogged (with underlying soils undergoing gleying) and invaded by wet-tolerant sedges and *Sphagnum*, cotton-sedge mires which were wet year-round, *Sphagnum* bogs, open hazel and birch scrub with a variety of wet-tolerant ground flora species and a high proportion of dead trees,[21] as well as wet mires in water-collecting sites, must have been a familiar part of their environment.

Causes of fire: lightning strike?

The ability of fire to run through heath and moorland habitats is not in doubt. Nor is there any uncertainty about the fire-prone nature of largely coniferous forests. The fire characteristics of deciduous forest are more difficult to elucidate, since most analogous woodlands in continental Europe are intensively managed and those in North America are, likewise, carefully guarded against fire or are largely secondary woodland rather than mature communities.

Observation of today's deciduous woodlands, especially in upland England and Wales, tends to regard them as likely to burst into flame as a load of wet football socks. But a lightning strike which hit after a noticeably dry period in spring or autumn might kill a tree and also ignite the fine fuels of the forest floor or, for example, a thick

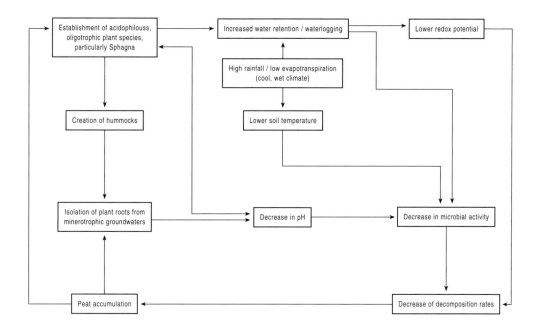

FIGURE 2.3   *The main routes through which peat accumulation takes place once tree cover has disappeared through either climatic change or human activity.*
B. D. Wheeler and S. C. Shaw, Restoration of Damaged Peatlands, *London: HMSO, Department of the Environment, 1995, p. 12, Figure 16.*

but dry tangle of bramble. Any resulting fire would, however, be confined in area and would at most kill some saplings and perhaps a few birch or aspens. There would certainly be no widespread crown fire.

Other natural factors: senescence and windthrow

Discussion of this undoubted source of disturbance (to which could be added the loss of trees from ice-glaze during certain winter conditions) tends to be conflated with fire in much of the literature. North American levels of about twelve storms in 300 years in an area prone to remnant hurricanes is not a high frequency, and seems likely to have been exceeded in the cyclonic climate of the British Isles. Thus the forests of the mid-Holocene would have been subject to such an influence, with the possibility of medium-sized areas being felled as well as single trees. Windstorms have the potential to wreak very extreme disturbance if accompanied by intense rainfall: soils can then be lost to erosion and landslides may remove the entire ecosystem over a small- to medium-sized area.

Beaver

The habits of the beaver (*Castor fiber*) are well known in the sense that in North America they fell trees to make dams which then inundate areas of woodland, killing the stands adjoining the pond which they have created. They eat aspen (*Populus tremula*), birch and willow as well as aquatic plants; one effect might be to create a kind of natural coppice from stump-sprouting. The size and duration of beaver clearings is very variable and it is difficult to estimate the area of forest that might be subject to forest disturbance at any one time; filled-in pools would create grassy clearings even after the beaver had left.[22] However, the European beaver only makes dams where the soils are unsuitable for burrowing and so we are uncertain of the effect they had on streams during prehistoric times.[23]

Grazing

At this time in the Holocene there is no suggestion that domestic animals were part of the Mesolithic culture and so the question is whether the wild fauna could create and/or maintain openings in forests.[24] If aurochs in particular spent the year in groups of at least from 4 to 6 then some impact on vegetation can be imagined. Smith and Cloutman affirmed this kind of possibility at South Wales:

> The [flora revealed by pollen analysis] gives the impression of a damp, grassy, somewhat disturbed area such as might have been quite intensively grazed by wild herbivores. In these circumstances it could have been kept open since early Flandrian times. Alternatively a clearing may have been created by prevention of woodland regeneration.[25]

THE POSSIBLE ROLE OF HUMANS

In the above account, there are places where the natural processes appear to fail to give satisfactory explanations of the phenomena inferred from the palaeoecological evidence. It is time, therefore, to turn to human societies as possible forcers of change in ecosystems. It is clear from the archaeological discussion that the later Mesolithic people had no very advanced technology with which to manipulate their environment. The visible tools are not massive, for example, nor are they highly honed as with the Neolithic polished stone axes which have been shown to be so effective in cutting trees. It has to be assumed that fire could be used and controlled away from the hearth and that devices such as pits and nets could be constructed. So the tool kit in essence consisted of fire, the means to capture and/or kill animals up to the size of *Bos primigenius*, coupled with ways of gathering plant material.

Tree-felling?

It can be asked whether the tool kit of the later Mesolithic communities allowed them to fell trees of a substantial size. Assuming that the oaks were mature trees, then a tree typically 36 m high with a single straight trunk that had little branching below 25 m would have to have been tackled. Roughly the same would be true of lime, oak and pine trees; most specimens of birch, ash and alder would have had thinner bark and more slender trunks than the oaks and limes. It seems unlikely that any stone tool found in later Mesolithic contexts would have made much impression on mature forest trees, and would not have been very effective against smaller individuals. Where relatively small tools might have been effective was in ring-barking, which would in due course mean their death. The trees would be left standing but there would be no summer canopy to deter shade-intolerant species of the forest floor or shrub layer.

The analytical evidence does not suggest that fire was used in an attempt to clear woodland in any direct fashion, but it is possible to think of a fire which burns off grasses that do not recover much in the next year but then have from 5 to 6 years of better growth and which also keeps bracken (*Pteridium*) down for about 3 years before it comes back. *Corylus* pollen peaks both at and between charcoal peaks and so is probably not affected by the burning. It seems unlikely that leaf-litter and cured herbaceous material would be dry enough to burn in spring. Hence, a late summer or autumn incidence of fire is most likely. Hazel browse and nuts are important products of management but not the only ones. Grasses are quickly killed off by the fire but come back soon and stay high. They will diminish if there is no more fire, partly because this may well be the end of the disturbance phase, and shading by woody species takes place. Grasses will under certain circumstances compete successfully with heather: the accumulation of nutrients (especially nitrogen) is one circumstance. Another is trampling and grazing by large mammals, which will lead towards the replacement of heather by grasses such as *Deschampsia flexuosa*. Heather will also disappear under conditions of continuous waterlogging, though not of merely seasonal quagginess.

The wild animal resource

The evidence of microliths suggests that wild animals were hunted in the uplands. Was this a simple chase or were there attempts to improve the hunters' chances of making a kill? The management of open areas (and especially those near water) to attract wild animals may have had a wider context. Humans might seek to attract game to sites where killing was relatively easy. Such a site, in this context, would be open and attractive to wild animals by virtue of providing food and/or water and possess good cover for the hunters from more than one side, to take account of varying wind directions.

To provide such conditions, human groups might try to manipulate the vegetation so as to improve the attractiveness of the fodder for game species, in terms of species available and quantity; this might attempt to compensate for seasonal deficiencies or extend a seasonal resource. They might also intervene in the flow of water should it be possible to improve its appeal to the animals; or manipulate the surrounding vegetation to provide cover and/or good sight-lines (which sounds contradictory) for the benefit of the hunters.

Human-induced manipulation of vegetation to attract the larger species would have been focused on improving the grass and herb content of the forest floor. (Heather is eaten by deer, but bracken is not.) Three possible practices immediately come to mind:

1. maintain any existing openings free from the encroachment of woody vegetation or indeed any species not eaten by the game animals;
2. create openings with the same purpose in view;
3. attract animals with fodder: leaves are an obvious source and seasons lacking browse resources for chiefly browsing animals (red and roe deer especially) a possible purpose.

The sequences of burning with higher grass values and lower bracken would benefit wild animals and indeed, given some apparent preferences for grassy fodder, are likely to have been a more important means of manipulation than foddering. All these practices fit the sequences of fire and vegetation change detected in the palaeoecological data.

The Wild Plant Resource

Apart from the frequent finds of hazel nuts in Mesolithic contexts, very little evidence of what might have been the basis of plant resources has accumulated for the British Isles: finds of small-seeded plants such as *Chenopodium* and the remains of wetland genera like *Menyanthes*, *Phragmites* and *Nymphaea*, whose tubers are edible, have added to the list a little and remains of pear have been found in Ireland. No over-stretching of our speculative faculties is needed to consider that edible fruits, nuts, seeds and possibly tubers, were in fact eaten and that some tools (for example, antler mattocks) might relate to such procurement.[26] Oil-rich seeds must have had special attractions.

It would be interesting to know whether the vegetation was actively managed to enhance the productivity of such plants and whether other forms of management might incidentally produce such higher levels. Several processes might produce the same type of result. If the seeds of *Chenopodium* were useful to eat, then attracting wild animals who broke the soil and allowed the ruderal to grow would serve both processes. Encouragement of hazel for winter browse for deer (for example, by opening the canopy of the forest dominants) might increase the number of shoots bearing nuts. Openings mean lots of 'edge', which helps the production and gathering of blackberries. Grassy clearings for animal feed might produce small-seeded grasses which may have had a culinary role. In most British discussions, little mention is made of acorns as a possible food, though in many cultures worldwide they are so used, both with and without processing. British acorns would need processing before they were comestible: they are unlikely to be 'snack' foods to be eaten on the move away from a base. Unlike hazel shells, the acorn cups do not turn up in archaeological contexts, but their absence could be due to the way they were processed rather than their total absence.

Here, as in other sections, it seems prudent not to rule out any such possibilities but to note that the evidence is still tenuous. But there seems little doubt that a landscape with a mosaic of forest, scrub and edge, together with open water and wetland, would attract and carry the highest densities of most edible plants as well as game animals.

The Scottish Islands

A number of Scottish islands yield the signs of forest demise associated either with human artifacts or with palaeoecological evidence charcoal during the Mesolithic. On South Uist, for example, there are two phases of removal of birch-hazel woodland, at 8040 and 7870 BP: the pollen of heather and grasses and the presence of charcoal all increase. On Shetland, a reduction in fern spores is sometimes attributed to the import of red deer as a food resource.[27] Analogous evidence has been found on Arran.[28] These signs must be placed in the context of work that shows the vegetation of the Western Isles of Scotland to have been neither treeless nor uniform at this stage of the post-glacial period. The impact of humans seems to have begun in the Mesolithic period and thereafter combined with climate to allow blanket bog to expand over areas formerly occupied by birch, hazel and pine woodland. After 4300 BP, Scotland became much more oceanic and the outer isles were treeless after about 2500 BP.[29]

A SUMMARY OF ENVIRONMENTAL CHANGE INVOLVING HUMANS

The natural world was still a profound influence in the life of every group of hunter-gatherers. Hence, it is reasonable to postulate a sequence of events leading to ecological change in which both natural and human-directed processes have important roles.

A possible progression might be:

1. A mixed oak forest on the uplands, covering almost all the terrain, though broken where there was open water and mire accumulation in water-receiving sites. The upper edge of woody vegetation was marked by a hazel scrub. The forest underwent normal processes of death and renewal, which included gaps of various sizes caused by windthrow. The lower edge near streams underwent a sharp transition to an alder wood with other deciduous species, including elm.

2. Openings in the forest with a ground cover of grasses and herbs attracted mammal herbivores on a seasonal basis. This had two consequences: (1) the concentration of animals made regeneration of the woodland less likely; and (2) humans noticed the concentration of animals and wished to maintain and/or enhance it.

3. Humans thus took over some of the existing openings and maintained them, using fire (Plate 2.5) to keep a grassy sward rather than allow bracken to cover the ground in the early years. Heather grew as well and was burned, though it too attracted grazing animals. Attempts to extend the virtues of a grass-herb sward to the alder woodland were also made.

4. During this time climatic change and soil maturation brought about the accumulation of *mor* humus over podsolic and gleyed soils.

5. Increasing human populations (or more resource-hungry groups) wanted to try to create extra openings in the woodlands or at their ecotones. They did this by

killing trees using ring-barking and by opening the canopy by breaking off leafy branches, which were also useful in attracting animals to feed.

6. Some of the openings underwent rises in water-table and became invaded by rushes; thereafter peat accumulation began to get under way even on ground which was water-shedding rather than water-collecting in a micro-topographic sense.

7. None of the human-induced processes totally replaced natural events: natural openings continued to be formed by natural processes.

This is a set of temporal statements of a very general kind. In an attempt to reconstruct the environmental conditions of which later Mesolithic people were a part, it is feasible to postulate a model of their seasonal round and its relationships with resources.

Towards a model of resource-nature relationships

A number of models can be made, using different combinations of terrain and population size but a most-likely story can be selected as a better fit than the others to (1)

PLATE 2.5 *Under the heather at Ouse Gill Head on the North York Moors, a layer of flints and charcoal gave the first clues that later Mesolithic people had been instrumental in altering their environment and that control of fire had been a key tool. Work in deeper peats has enabled us to tell a more detailed story but an essentially similar one. This photograph was taken in 1974 in the presence of Raymond Hayes and Don Spratt, immense contributors to our knowledge of the archaeology of North Yorkshire.*

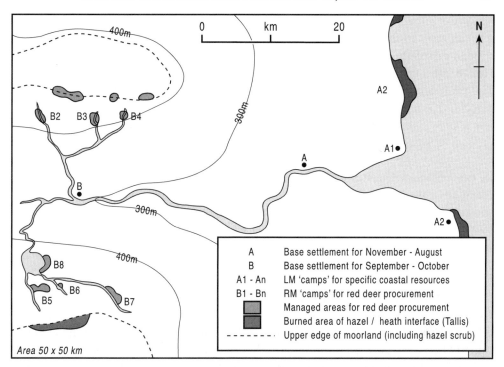

FIGURE 2.4 *A postulated yearly round for a later Mesolithic hunter-gatherer group. The main evidence for environmental alteration comes from the areas marked B2 . . . B7 and above them on the hills. A major area of ignorance is the environmental impact of these people in the 'A' areas.*

I. G. Simmons, The Environmental Impact of Later Mesolithic Cultures, *Edinburgh: Edinburgh University Press, 1996, p. 215, Figure 5.2.*

the direct palaeoecological and archaeological evidence, and (2) analogous circumstances in time and space (Fig. 2.4). It postulates a largely sedentary group which spends much of the year near the coastline (near enough for sea-level change to have swallowed some of the evidence) but which moves inland up the river valleys as a unit at the end of the summer. Once out of the terrain that can be hunted over on a daily basis from the main settlement, smaller sub-groups of hunters are spun off and some of these make for upland hunting grounds for red deer. In order to increase the chances of success, areas in the woodlands and at the forest/scrub edge have been fired in order to encourage plants known to be favoured by the deer. For summer-autumn hunting, grasses are more important than browse plants. Fire may also act as a territorial marker; in effect it appropriates the land to a particular group. The camp sites are not likely to be the same exact places as the sites whose vegetation is manipulated to attract the deer, so each 'visit' spawns at least two openings and two fires. Even repeated visits to the same small area might not use the same campsites. Hence, a small hunter-gatherer population can produce a relatively large detectable environmental impact, especially in ecotonal vegetation. The provision of evidence about the impact of hunter-gatherer people in the lowlands and at the coast is more difficult than in the uplands since the necessary deposits are fewer. But examples of similar palaeoecological events have been found at Lytham (Lancashire), Westward Ho! (north Devon), Hartlepool Bay (Cleveland) and Shippea Hill in the Fenland.[30] The probabilities are that a peripatetic group managed its environment in as many places as it could but that the surviving evidence is filtered.

INTO THE AGRICULTURAL ERA

In no sense can the hunter-gatherer period be presented as one of an unchanging set of environmental systems inhabited by quiescent people who lived solely of the usufruct of nature. Instead we have a story of changing natural environments occupied by people who strove to make their lives less risky by engaging in environmental manipulation.

Legacies

The later Mesolithic was the time when a key element of the continuing landscape was fixed in place: that of human-created openness. Before that period, open landscapes were probably the result of entirely natural processes except in narrow zones around settlements. The later Mesolithic saw the creation and maintenance of human-induced open areas at a landscape scale. But probably not everywhere: the short time period of 4650–4200 cal BP when clearing and tilling took place, was the time of the 'first significant human impact on the Cheviots'.[31] In many parts of Great Britain, nevertheless, a mosaic of land cover types was created which proved a receptive environment for the new economy based on cultivation and domesticated animals.

It was also the first time when it can definitely be inferred that there existed purposive human behaviour towards environmental manipulation; the evidence for earlier periods is still too thin. Nevertheless in terms of today's knowledge, this is a pivotal moment comparable with the discovery of the full potential of steam power, for example: the redirection of the energy flow patterns of the Earth in terms of human-defined, that is, cultural, desires. Such a realisation does bring home the great depth of our ignorance of hunter-gatherer times when we think about the non-material. We know nothing of the myths which probably governed people's lives and which explained their environment to them in terms both of its origin and its uses. The songs, poetry and religious beliefs of more recent hunter-gatherers in, for example, Australia and North America are rich and diverse, but of our hunter-gatherer's non-material past we know virtually nothing. We can still guess at it and tell stories, like the landscape appreciation qualities reconstructed by Christopher Tilley[32] for parts of the Welsh borderland in the later Mesolithic, and this is a necessary accompaniment to the flints, the pollen and the charcoal. In a similar fashion, we may note the sacred nature of a stream between two lakes in the Nenet culture of the Archangel region of Russia: this Arctic culture selected the site as a frontier through which strangers could not pass without a sacrifice for purification. The locality and the environment are very like Star Carr.[33] At a more abstract level, Bradley has suggested that in Europe as a whole, the Mesolithic people lived in a world in which human identity was not felt to lie outside nature, in contrast to the dominance over it which the building of monuments and the domestication of biota conferred.[34]

Innovations: links forward

The later Mesolithic is a turning point in yet another respect. It marks the full human occupation of the mixed deciduous forest which was spatially the most important element of the vegetation cover of England and Wales during the Holocene. As such, it was the beginning of the differentiation of a series of land uses out of the matrix of the forest; each subsequent environmental impact was the result of another valuation of the resources of the land and its flows of energy, water and living materials, and the diversion of these flows to human purposes and societies. So from a single land use in, say 6000 BP, there has been a progression to many and multiple land uses.

The boxed material which follows points briefly to a historical event or process which is still relevant to policy today. The boxes relate forward from situations in hunter-gatherer times through to 1950. After this latter date, everything is presumed to be relevant.

The open nature of the moorlands of England and Wales was in part a product of the later Mesolithic period, when human manipulation of openings and tree-lines prevented the forest from occupying the full area which the climate and soils would have permitted. Later periods of use for grazing and agriculture in both prehistoric and later times have consolidated that picture. The growth of blanket peat was in many places a response to the removal of tree cover. These uplands are now low in floral and faunal diversity, prone to erosion, and not very productive. Yet they are regarded as the most important areas in the nation's landscape protection system, for they are the cores of most of the National Parks, with even Pembrokeshire containing one such area, the Preseli Mountains. So there is a conflict between ecological history, which tells us that these are depauperate areas long over-exploited, and cultural history, which values them highly. Government policy since 1949 has been to emphasise the latter at the expense of the former, almost totally. A group in the Southern Uplands (The Borders Forest Trust) plans to restore an entire watershed to 'wildwood' condition in the Carrifran valley.

One interesting fact is that the later Mesolithic was a turning point in the sense that it was then that the turn to an ecologically downgraded condition began and that this state is, paradoxically, recognised by the nation as the best landscapes it has to offer its citizens and visitors. Where this impinges upon the story as told here is that a landscape archaeology which attempts to preserve the concatenation of field system, monument, open pasture and blanket bog is contributing to the maintenance of an ecologically degraded landscape and should, presumably, be done with a consciousness of that actuality. In practical terms, however, there seems no movement to reclaim the uplands for deciduous woodland and so *de facto* conflict is at present unlikely on these grounds. If EU sheep subsidies fall yet more and the uplands become yet more marginal to productive agriculture then this could possibly become a live question.

Evaluation

In stating that the hunter-gatherer period was some six-tenths (60 per cent, that is) of our last 10,000 years of history, there was the risk that it would be interpreted simply as background which had to be as it were cleared away before the real business could begin. This discussion has shown that to be untrue. Many lineaments of both the natural and the human-created were established which have relevance today and a continuity with the past has been established. If we wish to single out a key feature then it must be the beginnings of deliberate management of the environment during the Mesolithic.

CHAPTER THREE

Shafts of sunlight: agriculturalists

INTRODUCTION: A VERY LONG DURATION

To allocate the 6000 chronological years of this and the next two chapters to one era has a ridiculous air. So many of the people, events and places which identify us, like Stonehenge, the Roman and Norman invasions, Shakespeare's plays, John Milton and John Donne, the Armada, the Union of the Parliaments, numerous wars, and the music of Thomas Tallis and Henry Purcell, all fall within its ambit. Yet there is one binding theme: during those millennia, for all practical purposes, everybody lived in an economy which was solar powered: by shafts of sunlight indeed. Not until the nineteenth century was this 'organic' economy truly converted to a fossil fuel base although coal was a minor fuel in earlier centuries. So the human-environment relations were mediated by an economy and a technology which harnessed the sun as biomass, as wind and as falling water. We can add the tides, too, since a number of tidal mills are recorded as early as the eleventh century.

But it seems unreal to allocate all those centuries to a single unit without acknowledging that the environmental relations of, for example, the eighteenth century AD were different in many respects from those of the eighteenth century BC, not the least because the population had grown considerably in the interval and because Great Britain had become incorporated into a world system at the latter date. So for the practical purpose of making divisions in an otherwise intractable stretch of time, we can think of three sub-eras of the agricultural phase. They are:

- The period between the introduction of agricultural practices into Great Britain with the adoption of a Neolithic culture (*c.*3500 BC) and the cultural change to Christianity which began in Ireland in AD 432 and in southern England in AD 597. The millennia before the Roman occupation are usually labelled **prehistoric** and then there is the early documentation by the Angles, Saxons and Frisians. These literary sources can be added to archaeology, palaeoecology and **epigraphy**. An end-date of AD 500 is arbitrary but helpful.

- The time between about AD 500 and AD 1700 which included the imposition of new laws of land-holding brought with the Norman Conquest of 1066 and the subsequent ways of holding land in common and under **feudal** rule, both of which gradually broke down. During this time there were discernible climatic fluctuations and the eruption of the epidemics known as the Black Death and the Plague. Throughout, the predominant economy was agricultural but there was perceptible intensification of resource use.

- The century of major **enclosures** and attempts at **improvement** of many lands by means of more intensive use, including the consolidation of large-scale reclamation of fenland and heathland, ending in AD 1800. The importance of recognisable industry rises during the period from 1700, with the discovery of the usefulness of the stationary steam engine when used to power a pump, and the way in which bringing manufacturing together in one building enabled machinery to be used efficiently. A national sense of optimism thought of Great Britain as a New Jerusalem, though scarcely with a unified vision.

Many differences between these periods can be identified but they are all united in the terms of this book by the kinds of energy to which they had access, and indeed in environmental terms they may have had more common factors than divergences. In some places, careful work has determined the main phases of environmental change. For example, at Shapwick in Somerset, the later Roman period with a villa economy, the tenth century when the land was reorganised by the landlord into a village with open fields, and the process of enclosure between the fifteenth and eighteenth centuries, are key periods.[1]

The stone-bone-wood technology of the hunter-gatherers was supplemented in about 1800 BC by the first metal, in this case bronze. Iron came later, at *c.*700 BC and that metal has remained central in its harder form as steel. Compared with today, the capability of society in getting resources was relatively limited and the subsequent environmental impact equally so. Autochthonous resources, however, could always be supplemented by trade if the right conditions existed. Simple technologies meant also that the buffering between the conditions of the environment and society was narrow: the effect of climatic change, to select our first topic, was always likely to be difficult to evade.

ATMOSPHERIC CHANGES: THE CLIMATE

The hunter-gatherer people had the best of the post-glacial climate. This warmest and driest period seems to have ended in Britain at around 3500–3000 BC and then there was a more pronounced deterioration in the 1300–1000 BC years. Yearly means in Great Britain fell by about 2°C to a level from which there has never been a full recovery; there was unsteadiness after 2000 BC, but the fall seems to have been complete by 1000 BC. Fluctuations in rainfall were more significant than those in temperature. High altitude cultivation on Dartmoor and the Pennines seems to have been abandoned after 1000 BC; a much wetter episode after 900 BC has been detected.[2] There were also swift and unpredictable changes, as with the eruption of the Icelandic volcano Hekla in 1850 BC. In Caithness, this was responsible for the deposition of over 1 million shards of **tephra** per cm^3 of peat.[3] At the same time, the deposits record a sharp fall in the pollen of the Scots pine (*Pinus sylvestris*). This may have been caused by acid deposition from the volcanic explosion or the eruption may have caused sufficient climatic cooling to stop the growth of pine, which may have been near its northern climatic limit in Caithness. If pine trees were so affected, then an agricultural economy based on crops would have suffered considerably. A pastoral regime would have been better buffered against loss since animals that could have been pastured over a wider area might make up for the lower growth rates in grass and herbs. About 150 BC the climate seems to have entered a milder period, though North Sea storms were especially frequent in the 120–114 BC interlude. The Roman conquest heralded a phase of climate similar to that of today (perhaps about 1°C warmer in the late third and the fourth centuries), which ended with the onset of colder winters and cooler summers around AD 400. Though parts of the third and fourth centuries AD brought warmth and dryness, it was colder and wetter by the 500s, with little improvement before 650–700.[4] Stormy periods were especially important in the Outer Isles of the Hebrides where they helped supply the carbonate sands that were periodically cultivated (as early as the Bronze Age on Benbecula and North Uist) and thus came to acquire the smoothed outline and distinctive grassland vegetation that is named *machair* in Gaelic.[5]

CULTIVATED LAND AND GRASSLANDS

Agriculture is the great novelty when introduced into environmental and cultural systems. It was not the first instance of environmental manipulation but was destined to become the material foundation of human societies until the nineteenth century. Cultivation was above all the main reason for converting land from wilder systems (such as woodlands, fens, heaths and moors, and salt-marshes) to more intensive uses. In turn it called forth more differentiation and eventually fragmentation of land cover as the land surface was used for settlements, communications and for leisure. At most times, the agricultural systems could produce a surplus which allowed some people to develop other skills, crafts and lifeways.

Availability

Radiocarbon dating has suggested that agriculture was present in Ireland before it came to Great Britain (the evidence comes from cereal pollen in peats and muds), with the first traces in Ireland at about 5750 BP (3800 bc) and in Britain at 5300 BP (3350 bc).[6] Archaeologists are now likely to suggest that an indigenous population adopted agriculture rather than being overwhelmed by immigrant cultures. This espousal often came in three phases: an availability stage when the existence of agriculture was known through contact with existing farmers but not widely adopted; a substitution stage when agriculture was becoming part of the ecology of a frontier zone; and a consolidation stage when the maturation of the social, economic and ecological structures and processes associated with the new economy took place.

The mainstays of the new economy were wheat and barley, sheep and goats, all of which had wild ancestors in the hill lands east of the Bosphorus, an early example of Britain's incorporation within wider cultural-environmental networks. The native wild cattle and pigs of temperate Europe were added to the suite of domesticated beasts and, rather later in prehistory, rye and oats.[7] Most of the evidence suggests that the penetration of agriculture (the substitution phase) into Great Britain was rapid and that by 2500 BC it was the dominant way of life in most lowland areas. It was also present on uplands which today we consider to be outside the range of crop production: where wheat and barley would not grow, then early farmers relied on a form of pastoral economy. Nevertheless, cereal agriculture in Scotland expanded in the early Bronze Age and intensified in southern Scotland after 500 BC. After being introduced into Shetland in 2750 BC, agriculture was implicated in the final phase of woodland destruction in 1150–950 BC.[8]

Consolidation

Between the Neolithic and, say, the fourth century AD, agriculture was the main cause of culturally driven environmental change. There were several waves of immigration and conquest from the east which brought new technologies and cultural systems. At the same time fundamental ecological relationships remained very similar, for an adequate amount of nutrition in terms of calories and protein had to be produced by the communities from mostly local sources. The majority of people had to get their basic needs within sight of where they lived but there was also trade, so that high-status individuals and groups could be supplied with food, stone, wood and metals over quite large distances.

To begin with, the dominant environmental system was woodland. Agriculturalists identified land which it was worth clearing, removed the trees and then either kept it open as permanent fields or abandoned it after a while and let it return to scrub and tree cover. As population increased or decreased, so the amount of wooded terrain might shrink or expand. One key factor was the type of soil beneath the woodland. It is likely that most rock and subsoil types which were below the tree-line carried a physiognomically similar type of forest in which more subtle differences indicated agricultural potential. The presence of lime trees (*Tilia* spp), for example, would have suggested that deep loamy soils with good levels of mineral nutrients lay beneath: examples from the Bronze Age have been suggested for the Welsh border. It seems certain that the 'lighter soils' over gravels, chalklands and limestones beloved of earlier generations of settlement historians would only have emerged as such after the forests had been removed: perceptible differences in the early Neolithic vegetation cover would not have been stark.

The tools available for environmental manipulation included axes (first of polished stone, then of bronze and then of the harder iron), digging-sticks, the wooden plough or ard (which became iron-shod in late pre-Roman times), mattocks and spades. Newer tools only came with the Romans. The main outcome was a set of fields in a wooded matrix, of which traces often remain in today's landscape. A variety of field patterns has been detected, in which there are often large blocks of land defined by continuous boundaries, with individual fields as subdivisions. The best examples come from southern English chalklands and from uplands such as Dartmoor.[9] Around Stonehenge, research has shown that the early phases of that monument were built within an area of grassland but that the landscape was not totally open. Nearby Coneybury Henge was, however, constructed in a zone of recent and localized woodland clearing. By the Later Bronze Age, the region was basically an open landscape with very few trees. The dominant system was short grazed grassland with a significant proportion of arable land. Some ploughed areas were bare enough to yield blown soil. Abandonment of arable land has been detected on the Marlborough Downs dated to the mid- and later Bronze Age. Even some barrows became overgrown. Cattle and sheep were the dominant animals here, with a gender mix that suggests dairy products as the most important. Feasting is inferred from some bone deposits, with pigs giving way to cattle as the main barbeque items when woodland declined.[10]

In upland England, the layout of field systems and the presence of banks on surrounding watersheds suggests that by the Bronze Age there may have been 'estates' in which a community had access to low-lying agricultural land backed by grasslands and perhaps relict wood or scrub, so that all the available resources were parcelled out in areas which foreshadowed the boundaries of medieval parishes. The land available might have included some hunting lands for wild boar or deer. Iron Age farming, though, is generally regarded as having been mixed cereal and animal husbandry in all environmental situations.[11] The field size was of the order of 0.16 to 0.23 ha, about one day's work; some were edged with wattle fences, others probably with hedges.[12] By Anglian times it seems certain that shifting cultivation had disappeared from the lowlands though versions might have remained at high altitudes in the north and west. A lowland version of the estate is proposed by Cunliffe (Fig. 3.1) for the chalkland in the Iron Age, with resources culled from water-side meadows up to scrub on the hilltops.[13] At Flag Fen in Cambridgeshire, the excavated remains suggest a system of fields and enclosures set up for intensive raising of large numbers of sheep.

Farming in the uplands of north and west Great Britain is subject to more constraints than its lowland versions. Soils have more acid substrates and are often poorly drained and the growing season is shorter. Where a plough was used, then an iron pan might form at the base of the soil turned over by the ploughshare, inhibiting drainage and encouraging the formation of peat. Adaptations included the introduction in the first millennium BC of hardy cereals like rye (*Secale cereale*) and oats (*Avena*) and in the uplands we find infield-outfield systems and the predominance of pastoral farming. These systems rely on a permanently cultivated infield on the best location, together with outfields that are abandoned when their fertility declines; they revert to grassland or scrub or even become covered in peat. Their soils can be improved by digging them into a ridge-and-furrow pattern to concentrate nutrients and enhance drainage. The environmental significance of pastoralist economies (mostly cattle at this time) was in the ability of animals to change the vegetation by grazing selectively. Any ploughing would have allowed the invasion of the very pervasive knot-grass (*Agropyron repens*). Economies which depend on animals have also a great impetus to kill their predators: in the case of Great Britain, the wolf must have started its decline (it eventually became extinct in the eighteenth century) almost as soon as cattle-keeping began. One overall effect of clearing, cultivation and settlement was faster runoff, so there is evidence of falling water-tables on the Dorset chalklands as early as the fourth century AD. Climate does not seem to be

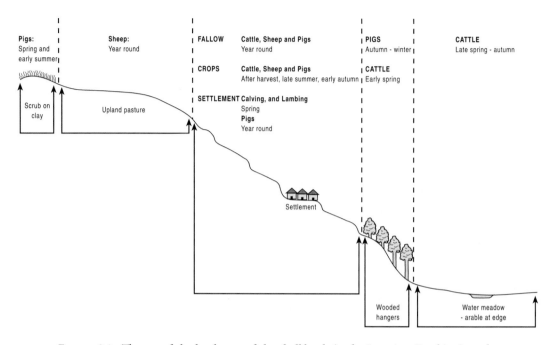

FIGURE 3.1 *The use of the landscape of the chalklands in the Iron Age. By this time, dense woodland was confined to steep slopes on valley sides ('hangers') and to scrub on the watersheds. Both areas were important fodder sources for pigs, as were most other land uses, at various seasons. The whole system has the potential to lose a lot of soil and thus the return of nutrients via animal dung is essential. We can imagine that the valleys would be subject to more flooding than in the more forested Neolithic period as at Avebury (Figure 3.2). The vertical range is about 100 m of height.*
Barry Cunliffe, 'Man and landscape in Britain 6000 BC–AD 400', in S. R. J. Woodell (ed.), The English Landscape: Past, Present and Future, *Oxford University Press, 1985, p. 61, Figure 2.9.*

FIGURE 3.2 (opposite)   *The environment of Neolithic Avebury. Top to bottom: initial Neolithic times; early–middle Neolithic; later Neolithic. The earliest stage of settlement takes place in a largely wooded landscape, with small clearings and isolated cereal plots. The middle stage is one of consolidation with more grazed areas, though not an immense increase in arable land. The last stage shows the landscape in which the ceremonial monument was built, which is still a working landscape producing cereals and animals (especially pigs) but where hazel scrub and bracken had replaced much of the forest.*
C. Malone, Avebury, *London: Batsford/English Heritage, 1989, pp. 31, 34 and 35, Figures 16, 19 and 20.*

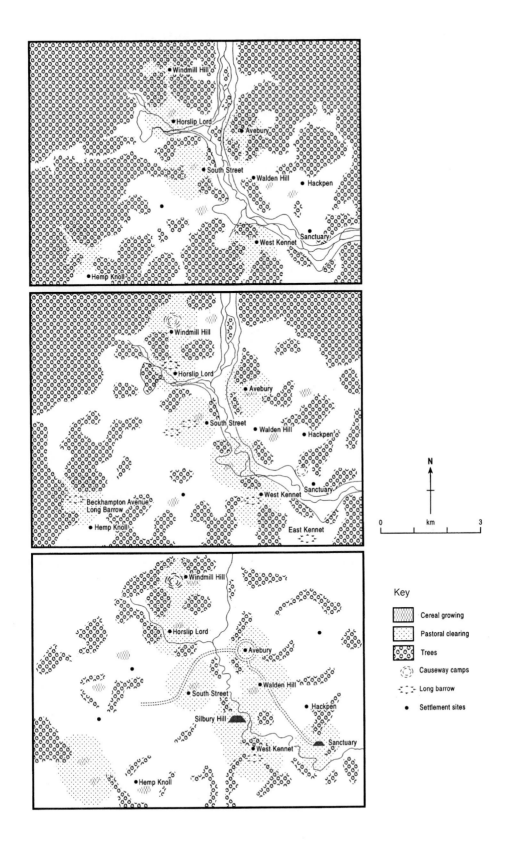

Key

▨	Cereal growing
⬚	Pastoral clearing
⬡	Trees
⟲	Causeway camps
⌇	Long barrow
●	Settlement sites

Windmill Hill
Horslip Lord
Avebury
South Street
Walden Hill
Hackpen
West Kennet
Sanctuary
Hemp Knoll

Windmill Hill
Horslip Lord
Avebury
South Street
Walden Hill
Hackpen
West Kennet
Sanctuary
Beckhampton Avenue
Long Barrow
Hemp Knoll
East Kennet

N

0   km   3

Windmill Hill
Horslip Lord
Avebury
South Street
Walden Hill
Silbury Hill
Hackpen
West Kennet
Sanctuary
Hemp Knoll

the cause, which points to the possibility of the poor recharge of aquifers in an era of relatively intensive agriculture.[14]

Agriculture was, then, responsible for hewing a humanised landscape out of a largely forested matrix. As Figure 3.2 shows, the major contribution to the new system was provided by the woodlands, together with the margins of open ground on upland heaths and moors. With early agriculture the main problem was the maintenance of the nutrient status of the soils: in common language, their fertility. As well as the losses from cropping, the baring of soils for planting and at harvest-time allows rainfall to remove nutrients in solution. Soil loss itself takes away minerals if erosion occurs. So, either new nutrients must be added or new ground must be broken. In terrain made finite, by the sea, by mountains, by climate and above all by the presence of other human communities, opportunities for the latter shrink and the process of intensification of yield becomes critical. Hence there is evidence from the Bronze Age onwards of manuring from stalled cattle, of tipping the settlement midden-heap onto the fields, of paring off top-horizons of heath soils to transfer them to the fields, of spreading seaweed and of folding cattle which had been grazed in the woods and pastures onto the arable enclosures. Provided that nature's nutrient cycles could be mimicked, then the farms could produce enough protein and energy to allow the maintenance and growth of the human population and its differentiation into those who tilled the land and those whose occupations were either to serve or to rule them.

RELATIONS WITH THE WOODLANDS

The forests were land banks in the sense that they could be cleared for tillage. Whenever this early period is reconstructed, the word 'clearance' is used in the same breath as woodlands. Yet there was sufficient of it that, even after 4000 years of attack, it must still have been a daily element in many people's lives in the Roman period, even though there may have been regions at that time from which most woodland had disappeared.[15] Certainly there are many place names of both Anglian and Norse origin which suggest the presence of woodland: -ley and -hurst denote clearings, as does the Norse -thwaite. At the same time as being land banks, woodlands were an asset of underwood and timber, fed grazing animals, and were certainly a fount of legends and stories.[16]

Without doubt, the clearance of woodland was the greatest environmental change of the period under discussion. It happened in all regions and through every climatic change, on all types of soils and at most altitudes. The major tool was the axe: in the Neolithic, axes were made of polished stone, but thereafter metal blades were possible, with iron being able to keep an edge much more readily than bronze. It seems likely too that people would know about the potential of ring-barking for killing trees. The largest stumps were left to rot in situ and the rest of the woody and leaf material piled up and burned, contributing minerals to the soil. Once opened, a clearing could be kept that way by killing any regenerating saplings. Herding grazing animals into a clearing would ensure that any young trees were eaten and would manure the opening as well; we imagine that clearings in woodlands were eventually converted into the small irregular fields sometimes visible today in, for example, Devon or west Wales.[17]

The conversion of natural forested systems to a humanised but nevertheless wooded set of systems may take more than one form. One common pathway is through the use of the forest for pasturage of cattle. Any juicy saplings are eaten and thus regeneration is halted. Further, people may ring-bark some trees to encourage them to sprout below the girdled area within the reach of the cattles' mouths. In this way, the forest gradually becomes more open and the type of vegetation usually called 'wood-pasture' is formed; here the cattle feed partly upon the grass of open glades and partly upon leafy material and twigs. The branches of many species of tree might be cut and then fed to animals: there is evidence for this from Neolithic Switzerland and from Yorkshire in recent times. Pigs would also have been allowed to forage in the woods and one casualty would have been the regeneration of trees like oak and beech, whose acorns and mast were so palatable. A second major category of woodland manipulation was coppicing,[18] when a tree (such as hazel or ash) is cut to a stump which then sends up a large number of regenerating shoots. These provide food for cattle and also a set of poles of varying size, vital to a rural economy. As early as 2200 bc, a trackway across a bog in the Somerset Levels was constructed using hurdles of coppiced hazel, and the practice is probably much older.

Clearance or conversion of woodland had noticeable effects on a number of systems. Notably, the number of mature trees diminished and that of younger trees and those of successional stages, like birch, increased. Clearings supported more bracken, grasses and, on acid soils, heather. The use of trees such as elm for fodder may have weakened them so that epidemics of disease could sweep through (as happened in the third millennium bc). The opening of the forest canopy encouraged ground flora dependent upon light: bluebells (*Campanula rotundifolia*) must have flourished as never before. More light and more grasses changed the insect and molluscan faunas: one lime-tree beetle (*Rhysodes sulcatus*) was common in the Mesolithic though now known from only two locations in Great Britain. Round c.1350 BC the Bronze Age saw the final demise of the wild cattle (*Bos primigenius*) in Great Britain.[19] Likewise, beaver, bear and wild boar populations diminished sharply. The creation of more edge may, we can surmise, have increased the fox populations. It may have augmented roe deer populations as well and in any event the post-Roman decades show an increase in deer bones at some sites, suggesting that in the post-imperial breakdown wild food had assumed an increased importance. Soils opened to the rain but deprived of nutrient input from trees became more acid. There is a natural tendency for temperate-zone soils to acidify during the course of an interglacial and this time it was accelerated by human manipulation of the tree-clad areas.

How much woodland was left? Palaeoecological evidence does not translate directly into maps but it looks as if in southern Britain there was extensive clearance in the Bronze Age whereas in the uplands this development was more pronounced in the Iron Age (Plate 3.1). In Wales and most of England, the 'Celtic' peoples almost certainly lived in an environment of managed woodland, for their needs were great. A large round house would have needed two hundred 50–90-year-old oaks, a year's supply of

PLATE 3.1   *At Reeth in Swaledale (North Yorkshire), visible strip lynchets indicate that, on that hillside, cultivation took place over an extended period during prehistoric times. There are implications for forest clearance and woodland maintenance, for soil loss and structural change, and for runoff rates, among other impacts.*
Photograph: Vicki Innes.

fencing stakes took perhaps three hundred ash trees, and 1 ha/yr of hazel for wattles.[20] Even in a Neolithic economy there would have to be a managed woodlot of 6 km² per 30 people, which confirms the kinds of reconstruction that accentuates a largely occupied and managed landscape. The Iron Age farm at Butser Hill in Hampshire, with exact reconstructions of houses, implements, crops and animal species, tells us the amount of timber needed for construction, the quantity of thatch for a roof and the amount of time needed to replace a roof. The need for coppiced wood, for example, led Reynolds to suggest that, 'The need for timber . . . suggests a complex and settled service industry'.[21] Yet in the North Tyne region of Northumbria, it seems as if there was very little forest removal until the Roman period and that some of the woodland near Hadrian's Wall was in fact removed for military purposes, mainly for construction timber, fuel and charcoal but perhaps also to deprive attackers of cover. Recent work in Fife and Aberdeenshire shows a regeneration of woodland during the Roman occupation which is interpreted as military interference with agriculture.[22]

Martin Jones suggests that by the Iron Age, 50 per cent of the forest present at the end of the Mesolithic had gone and that only 15 per cent remained by the time of Domesday Book in 1088.[23] On the eve of the Roman conquest, certainly, lowland Britain was a mosaic of woodland, pasture and arable, and wetlands. On the higher ground, moor and heath were usual. One effect of all this was to differentiate between Highland Britain and Lowland Britain: when both had been covered in forest there was less difference between them. Now that the matrix was being differentiated, this was one of the alterations that emerged. Comparison of Roman-period pollen diagrams suggests that in Wales and western England, most sites were more tree-clad in the Roman period than they are today. Northern England tells much the same story except for the zone near Hadrian's Wall, but central southern England seems to be far less wooded and more intensively agricultural.[24]

THE CREATION OF NEW LANDFORMS

The loss of soil from cleared and cultivated areas (especially those where autumn sowing was practised) resulted in the creation of valley-side and valley-floor landforms of a depositional type. Two types of material are principally found: colluvium, which is poorly sorted sediment often found on slopes and in minor valleys, and alluvium, which is sorted material laid down by running water. The source of both might be the kind of field systems described above for the Bronze and Iron Ages.[25] (Measurements made recently on sandy soils in Bedfordshire showed a yield of 17.7 tonnes per hectare per year of sediment to runoff under cultivation, 2.4 tonnes per hectare per year under grass, and none under woodland.[26] Smaller fields and ard ploughs probably mean the first figure would be less in earlier times.[27]) One result is the spreading of colluvium down slopes and into small valleys: colluvial sequences of 2 m of sediment are found spanning the period from the Neolithic to the Iron Age. A similar depth of alluvium has been detected in the valley floor of the Severn–Avon system in Warwickshire, starting at c.650 bc. At the base of these deposits a mass of drifted tree trunks and debris suggests floods following ploughing.[28] Soil loss in those times has also been held to account for the exposure of some areas of limestone pavement: the stripping of the soil revealed the bare limestone scraped by glaciers which had weathered into the characteristic pattern of **clints and grykes**.

As well as these human-induced changes, the climate controlled certain landforming processes. Even if human activity loosened sediments, then there was a significant redistribution of eroded sediment during relatively short periods of abrupt climatic change when there were major shifts in flood frequency and magnitude. Examples of alluviation in river valleys occur during the periods 9600–8400 BP in lowland Britain and 4800–4200 BP all over the country; a short episode in 3800–3300 BP is found mostly in southern England.[29]

Changes in channel form and in alluviation in river valleys are common during the Holocene and may well affect our perceptions of human occupation of the land since they bury archaeological sites.[30]

HEATHS AND MOORLANDS

Lowland heaths are a common feature of the twentieth-century landscapes of parts of Great Britain. Among other places they can be seen around the London area, on rocky outcrops in the Welsh borderlands, and at sea-level in the Scottish islands. They belong to a class of environments which are found only along the western fringe of Europe and hence must be related to climate in some way. Historical studies generally find that some of them are natural in origin: in the Faeroes, on St Kilda, parts of Orkney and Shetland, the Outer Hebrides and Caithness this appears to be the case. But in South Uist, heather-dominated plant communities existing from *c*.9500 BP (7550 bc) have been maintained, even if not created, by fire. Elsewhere, lowland heaths have been formed, by human activity, out of former woodlands in many different periods, probably with an intervening phase of agriculture. Since they are usually on acid sandy or gravelly soils, their fertility was perhaps soon exhausted and they were allowed to become grazing or fuel-providing areas. Certainly, some Bronze Age burial mounds are built partly of heathy turves and overlie **podsolic** soils.

On the moorlands of higher altitudes (above 300 m OD), the main environmental change during the period covered by this section was the growth of blanket peat. This has been described in the material on the Mesolithic and the same processes seem to have occurred whenever tree or scrub cover was removed (either by deteriorating climate or human activity) and especially if fire was present. Thus we can have the inception of this type of peat any time during the period of *c*.8000 bc to Roman times and probably later as well. The term 'blanket' thus becomes more and more appropriate as discrete areas of peat coalesce into great swathes across the moors. Where there was deforestation combined with limestone soils, species of arctic-alpine plants survived, as in Upper Teesdale. So some of today's rare plants only survive because of human activity at an early period.[31]

One English moorland exhibits probably the largest area of preserved prehistoric landscape in Europe. On Dartmoor, we can see stone boundary walls (**reaves**) associated with enclosures and the footings of dwellings, an example of which is seen in plan on Figure 3.3. These delimit areas of smaller enclosures which appear to be mainly pastoral in purpose, with only small areas of crop-growing. One of the systems covered an area of 900 ha with rectangular dimensions of 3 x 3 km; some cross river valleys. The reave systems date from the short period of 3300–3100 BP (1350–1150 bc) and were subsequently abandoned in Iron Age and Roman times to become moorland; some were taken into later, Medieval, reclamations back to cultivation.

INDUSTRIES

As early as the Neolithic, one scree on the Langdale Pikes in the Lake District was so popular as a source of material for stone tools that it is often referred to now as an 'axe factory'. While this perhaps stretches the usual definition of 'factory', it reminds us of the value of mineral resources. This is emphasised at the Neolithic flint mines at

FIGURE 3.3 (opposite) *'Reave' systems on south-east Dartmoor. These are the visible remains of linear systems of land division dating from the Bronze Age and portray a land which has been much altered by human presence and regulated into distinct land use/environmental systems. A. Fleming,* The Dartmoor Reaves, *London: Batsford, 1988, p. 51, Figure 28.*

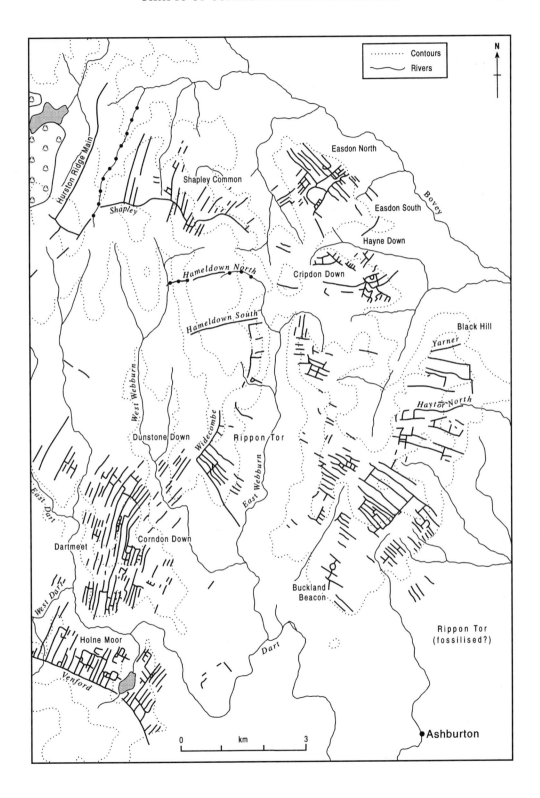

Grimes Graves in Norfolk, where there were some 300 galleried shafts, each of which produced 7–9 tonnes of flint. This material left the mining area already flaked, so was presumably a high quality product which entered trade. Wherever iron was smelted from ore or worked by a smith, then charcoal would have been needed. Roman ironworks in the Weald which produced 550 tonnes per year for 12 years would have needed 57,000 ha of coppice to fuel them sustainably. In fact in the last 2000 years BC, Great Britain produced not only iron but lead (often added to copper to make bronze), copper itself, tin, gold and silver. From stone came querns, rubbers, whetstones and building materials, and from clays the raw material for pots. This time saw the first major clearance event along parts of Hadrian's Wall. The need for wood for palisades, for fuel, and for forging iron, and set in the context of the military need for a northerly free-visibility zone together with the civilian impetus to grow crops to feed the garrisons, would have produced these changes. One interpretation of some remains suggests that the wall was whitewashed, in which case charcoal for slaking can be added to the environmental manipulations of the time.[32] There is only one Wall of that magnitude, but sites for more mundane production were widespread features of the landscape.

SACRED SPACE

For centuries, archaeologists and antiquarians have interpreted certain types of monuments (Plate 3.2) as having had a ritual or ceremonial function: the great henge monuments of south-central England such as Avebury and Stonehenge are the most obvious examples, but there are others. The stone circles of Dartmoor and Callanish, the many stone rows scattered through Great Britain and the temple of Aquae Sulis at Bath all point to cultures in which the non-material was an element. In our present terms, though, we are interested in environmental systems rather than indi-vidual monuments. Most of the latter were presumably set up within the living space of the people to whom they belonged and indeed were probably part of the belonging: in Mesolithic Scandinavia the creation of a cemetery seems to indicate the formation of a group territory. However, concentrations of 'ritual' structures as around Avebury and Stonehenge (Wiltshire) must surely mean that those locations had a special signif-icance: outsiders would likely approach them with lowered voices. The late-medieval French romance *Perceforest* describes round temples in a flat space surrounded by a ditch, on a low hill in a clearing. Round about was a forest of great oaks. Darrah has argued that this recorded an oral tradition that went back to the Bronze Age where it certainly sounds like the environmentally correct setting for many round barrows.[33] In the Fenland, it seems to have been common to deposit valuable objects (usually broken first) and bodies in east-flowing rivers on trade routes; in the west, Lindow Man was deposited in a small bog pool amongst cultivated land and managed woodland. The raven (*Corvus corax*) had a special place in Celtic iconography and was sacrificed along with domestic animals at Danebury.[34] The animistic religion of the Iron Age resulted in the consecration of natural sites such as lakes, springs and trees, as well as open-air enclosures and buildings.[35] The whole island of Anglesey was regarded as

PLATE 3.2 *As well as the great henge and circle monuments, there are many stone circles and rows in Britain which are usually dated to the Bronze Age and which are regarded as having had a ritual origin, probably of a religious kind. This small example in the Kielder Forest of Northumberland became known as the Three Kings. They seem in 1983 to have a polyandrous relationship with the Queen.*

holy by many of the British in the immediately post-Roman centuries. At this scale, the Stonehenge region's many monuments can be interpreted as displaying a socially constructed cognition of a solar cosmology: 'a complex sacred geography'.[36] This reminds us that human-environment relationships are not always purely functional and material. A wide zone of sacred space is proposed by Tilley for the region of Cranborne Chase (Dorset) in Neolithic times.[37] Marking the alignment and positioning of burial barrows, he suggests that they signified a transition to a sacred space beyond them. Cranborne Chase was not therefore permanently settled but periodically visited by a number of surrounding social groups to perform rituals, bury their dead and rebury the bones of their forebears. Hambledon Hill, north of Blandford, was a focal point of a 'death island' to the west and its surrounding terrain removed from the normal economic processes of the time: a kind of nature reserve for the ancestors.

THE SOCIAL CONTEXT

Writing about largely pre-literate times brings a temptation to stick rigidly to the palaeoecological evidence's immediate findings about the probable environmental constraints and the human impacts on natural systems. But given certain assumptions and some imagination we can proceed a little further towards understanding the ways in which the peoples of the time constructed their models of nature. We can be sure too that the number of people present was central to the ways in which they responded to, and affected, their surroundings. Yet any study is, obviously, limited by the lack of written evidence: for Roman times not a single word of literature remains which was written by an inhabitant of Great Britain or a visitor to it.

Population

Estimates of population are best regarded as bands of probability and those given by Peter Fowler (Fig. 3.4) are reasonable enough for our use. A late Neolithic population totalling perhaps 14,000 rises to tens of thousand in the Bronze Age and first hits the million mark in the Iron Age; the total population during the Roman occupation was 1–2 million, rising thereafter to between 2 and 4 million but falling somewhat to a Domesday estimate of 1.5 million. Major growth periods (as deduced from the number of settlements detected from the air and by field-walking) appear to have been 1500–1300 bc and 800–600 bc but population decline is also possible: growth is not inexorable. For an image of the population, think of 14,000 people lined up on a field: they would fit within the running track at a modern stadium. One million people would fit, seated, into the football grounds of the Premier Division clubs.

Great Britain in a wider system

Trade between groups within these islands is well attested. In the Neolithic, the Langdale tool-making materials are found well away from their geological matrices. The bluestones at Stonehenge come from the Preseli mountains of south-west Wales. But wider still, Neolithic contexts at Gwithian in Cornwall and Horridge in Devon contain German (including 'Bavarian') implements. By the time of the Iron Age, south-east England was part of a continental trading network, which was facilitated from the late second century BC onwards by the use of coinage in that region. Elsewhere barter was still probably the norm. Strabo's list of British exports (grain, cattle, gold, silver, iron, hides, hunting dogs . . . slaves) reads rather like the litany of Babylonian trade from the libretto of WILLIAM WALTON's oratorio *Belshazzar's Feast*. It omits only 'and the souls of men', together with some environmentally unlikely products such as ivory. The point of this perhaps is to emphasise that where the gaining and sharing of ideas (including those about environment no doubt) was concerned, Great Britain was not necessarily isolated by its insularity. Every place with which Britain traded, though, had a similarly solar-powered economy.

Sensibilities

These are difficult to gauge, for what survives has been so winnowed by time and filtered through so many newer perspectives that it is very difficult to reconstruct. Further, what was true of the Neolithic is unlikely to be valid for the post-Roman Angles and Saxons. The experience of landscape in the Neolithic may well have been mediated less through fields of wheat and herds of cattle than through architectural monuments which restricted space in both literal and metaphorical ways. Tilley[38]

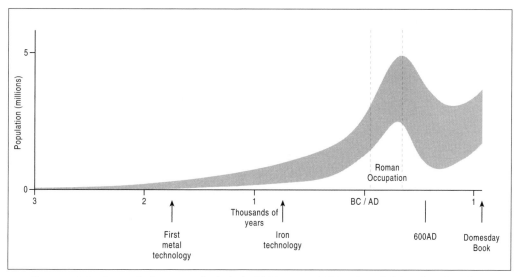

FIGURE 3.4 *A population curve for Britain 3000 BC–AD 1100 acknowledging that there is a degree of uncertainty about the numbers for that period. The general level of growth is well established, as is the decline after the Roman occupation.*
P. J. Fowler, The Farming of Prehistoric Britain, *Cambridge: Cambridge University Press, 1983, p. 34, Figure 7.*

argues that it was the monument which served to coalesce the natural and the cultural as a framework for understanding the world; the eating of meat from domesticated animals might have had a mythic significance that outweighed the nutritional one. None of this obviates the fact that animals have to be fed on something and that the vegetation of the pastures may be changed as a result.

We think too that the peoples of 'Celtic' culture assigned different values to colours: red was the colour of death and destruction, green was the hue of sophisticated divine beings (like members of the Green Party today, perhaps) and white animals were supernatural, especially if they had red ears. Among trees, the ash always had a special significance, as did groves, wells and rivers. But how many of these later sensibilities derived from, say, Neolithic times we shall probably never know. We might say, though, that the world on the eve of Christianisation was full of spirit: the arduous tasks of getting a living were enframed by the probably equally demanding responsibilities of aligning people's lives to the demands of the denizens of another world entirely. There may be survivals in our culture which give some clues about the environmental sensibilities of those millennia. Some medieval churches were built within or next to prehistoric monuments; dead chickens or cats were walled up in buildings from the Neolithic to the nineteenth century. A few folk ceremonies contain the seeds of very ancient practices, though most are much more recent than local Tourist Offices care to admit.[39] The New Year custom of rolling burning barrels downhill (to represent the sun) in Allendale in Northumberland could be early Bronze Age but is in fact developed out of a Methodist watch night service in 1858.[40] The attested dressing up of a man as a stag, with the head and skin placed over the dancing body was anathematised by the Archbishop of Canterbury in 700; revivals are still seen in the presence of Hobby Horses and the Horn Dance at Abbot's Bromley in Staffordshire. This is taken up into Ted Hughes's long poem *Gaudete,* set in the 1970s, where a local priest conducts orgiastic gatherings of the local women[41] at which the Revd Mr Lumb,

> bobs under stag antlers, the russet bristly pelt of a
> red stag flapping at his naked back

If such ritual connections to the animal world are not the stuff of the tourist and educational industries, then reconstructions are often acceptable. The open-air museum concept lends itself readily to this period, especially if penurious students can be paid to dress up appropriately.

Overview

The evidence suggests that a reasonably stable human-environment system had developed by the time of the Iron Age. A human population subsisted off cereals and some meat and produced surpluses for exchange and for its own non-productive citizens. The Roman economy had to be more productive since many of its surpluses were mulcted off to satisfy the imperial core. New technology was imported: asymmetric ploughshares and the coulter brought up more nutrients to the rooting zone of crop plants; the scythe and balanced sickle reduced energy input into harvesting; water mills harnessed the sun's energy in a different way. The Chalk downlands of southern England may have turned from cereal production to wool for export as another consequence.[42]

CHAPTER FOUR

Closed and open systems,
AD 550 to AD 1700

This chapter starts with the major cultural change brought about by the introduction of Christianity into Great Britain between AD 500 and 600.* It eventually became a social and economic given since it provided not only local frameworks such as the parish but was also a source of values, some of which have had environmental connotations.[1] During this extended phase there was a growth in population from perhaps 1.0 million to 3.75 million in 600 (but maybe as low as 0.4 million) to 5.3 million in 1695, the date of the first serious contemporary estimate. Only in *c*.1700 did the population of Scotland reach 1 million.[2] The agricultural base also underwent change in some of its environmental relationships in the form of the presence of **open fields** (which were not, however, ubiquitous) and their subsequent enclosure and parcelling-out. At times there were agricultural surpluses which made it possible for other environmental systems (such as those concerned with pleasure) to be developed. Manufacturing industry, too, began to grow; it was often founded upon biological products such as wool, and was the basis for the growth of some towns. International exchange ensured that Great Britain was not isolated in terms of either materials or ideas: witness the introduction of the potato in the one instance and enhanced notions of the virtues of work and thrift with the Puritans in the other. So while there are no great transformations like that of the Neolithic or the Industrial Revolution, there was nevertheless an increase in social and economic complexity which meant, among other things, the further differentiation of the land cover of Great Britain into a more separated set of environmental systems than had previously been the case.

ATMOSPHERIC CHANGES: THE CLIMATE

The next major documented interval is a period of dry climate called the Medieval Warm Epoch. Its limits seem to have been 1000–1250, with the best time between 1100 and 1200. Researchers vary in their estimation of these limits and the extent to which this epoch is actually much more complex in its sequence of weather periods:[3] the decades 1230–50, 1290–1300 and 1370–1400 were all warmer than average, for

---

*From here on, all dates AD are quoted without any prefixing letters.

69

example. At best, the summer temperatures were 1°C warmer than in 800–1000 and rainfall was 10 per cent less, yet in 1205 the Thames was frozen over from 8 January until 1 April. On some English moorlands, the moss *Rhacomitrium lanuginosum* came to dominate, signifying a prolonged period of drier climate.[4] After about 1250 the climate worsened and the early fourteenth century contained periods of poor years like 1315–21. In 1315, for example, there was torrential rain and a poor harvest; 1361 was also bad, so that famines followed the flooding, the liverfluke and the rinderpest.[5] The end of the fourteenth century saw an increasing frequency of storms and shortly after 1500 the long stretch (until 1850) of the Little Ice Age begins, with its depression of temperatures.[6] In the fifteenth century there were years like 1436 when there was continuous frost between 7 December and 22 February and an ensuing fuel shortage. Increased storminess was also a feature: a Venetian galley on its way from Lisbon to Bruges in the winter of 1431–2 was forced to circumnavigate the British Isles and ended up in Norway. In the 1590s cool summers led to poor harvests with subsequent food riots in both England and Scotland. The 1690s also saw seven bad harvests in a row in Scotland. In England, the year of note was 1684 when the winter was so severe that the frozen Thames became an auxiliary market place and carriageway, details of which were noted by JOHN EVELYN in his diary.[7] Any period might be interrupted: the years 535–6 seem to have been bleak in many countries (the evidence is especially good from Ireland) and a large-scale volcanic eruption in Iceland is the most likely cause.

We easily forget how important the weather was in a pre-industrial society: if there was famine, then there were no international relief convoys. The prosperity that came from trade was dependent on the weather, too. Port towns such as Sandwich in Kent may have employed the night watch to call out the wind direction as a service to merchants and mariners.[8]

LANDFORM CHANGE

Sea-level was not constant. It rose rather suddenly in the late Roman period but receded during the Saxon years, starting to rise again in *c*.1250. In low-lying areas such as the Fens, rising sea-levels meant marine transgressions, laying down bands of clays. Such incursions were mostly stopped when effective sea walls were built. Coastal erosion during periods of storminess and rising sea-level has robbed the coast of the East Riding of Yorkshire of a strip of land about 60 km long and 0.5 km wide since Domesday; in Suffolk the whole town of Dunwich was lost by the seventeenth century. Intensive gathering of marram grass (*Ammophila arenaria*) allowed the Culbin Sands of the Moray Firth to overwhelm settlements in 1694 and led to legislation by the Scottish Parliament.[9] In contrast, many harbours silted up as sea-level fell or as silt loads increased from the increase in bare soils or from the initiation of mining in their watersheds. Coastal accretion is therefore recorded in this period: Spurn Point (east Yorkshire), Dungeness (Kent) and Orford Beach in Suffolk all reached more or less their present form by the sixteenth century.

AGRICULTURE AND GRASSLANDS

Here we come to one of the key passages in the whole history. This is the creation of the fields, farms and grazings of Great Britain in the years between the Saxon settlement and the breakdown of feudally controlled social groupings during the sixteenth century. We have to examine the cropland and grazing systems which produced food and fibre and see how these changed the pre-existing environmental systems (few of which, we must remind ourselves, were 'natural' in the sense of being unchanged by humans) and in turn produced new alignments and combinations of cultural and 'natural' elements.

The spatial relationships of a food- and fibre-producing system which is solar powered are to a large extent regionally self-sufficient. That is not to deny the existence of a great deal of trade, but most communities depended on locally produced food for their basic nutrition. Most research suggests that the development of markets meant that by 1400 economics was more important than ecology in determining the use of many natural resources. One consequence was the differentiation of distinct agricultural regions, with diverse environmental consequences.[10] Nevertheless, basic stresses in the **food system** appear from time to time in the shape of episodes of famine and of 'land hunger', when it seems that Malthusian limits have been reached. The three main types of response will structure the next few pages:

- The extension of the agricultural area by conversion of land areas into cropland or grazing land. Woodland especially, but also heath and moor, salt-marsh, fen and bog, received this treatment. The addition might be intended as permanent, but temporary extensions (analogous to shifting agriculture in the tropics) are also found.

- The intensification of production from an already existing productive area. Enclosure of common fields can be seen as an instance of this process. Other ways include the improvement of nutrient status of the fields by fertilisers of numerous kinds, the use of new crops and tillage systems, and the recognition of the virtues of extensive animal husbandry in some places.

- The cultivated area may in fact contract, with an extension of grasslands instead of arable land, or indeed a return to less intensively used systems of a wilder or more 'natural' kind. Behind such movements there may be economic changes (like an increase in the demand for wool) or the actual reduction of the number of people, as after the Black Death. Luxury crops and industrial plants such as dyes or herbs may be added to the repertoire at such times.

Technology

Most scholars agree that the Middle Ages were not a time of enormous innovation in agricultural technology. This is true if the standard of comparison is the earlier introduction of iron or the later transition to tractors but there were nevertheless some improvements like the introduction of the mould-board plough. It turned over the spit and so brought up nutrients from a greater depth in the subsoil than had previously happened, and some weeds were stifled by being buried under the turned clods. An important development in energy relations was the padded horse-collar which allowed the use of the horse for traction without choking the animal; this was introduced sometime between the tenth and twelfth centuries. Yet the horse only slowly replaced oxen in ploughing. Maybe 25–30 per cent of manorial land was ploughed by horses at the end of the fourteenth century though their major use was for hauling loads in carts, that is, not as pack-horses. Disadvantages centred on the amount of oats needed to feed a horse compared with an ox, which was higher by a factor of 6–20 times.

Another labour-saving machine was the mill. Tide mills (Plate 4.1) are mentioned in Domesday Book and both wind- and water-mills attract mention in the late twelfth century, including the more efficient overshot waterwheel. Early windmills were mostly **post-mills** and the provision of the central 'post' demanded a timber piece 12 m long by 600 mm thick and weighing 3–4 tonnes, so that capital was needed to buy it.

Extensions of the cultivated area

Increases in population, or desires to try out new economies, often result in attempts to extend the cultivated area. In a solar-powered economy there may well be environmental constraints, such as a climatic limit (altitudinal or latitudinal) beyond which a crop will not ripen, problems with soil drainage, or soils which easily become exhausted. Means of tackling some of these could be highly effective to the point where apparently poor soils were no barrier to a healthy rural economy.

Not all such changes were permanent. In the uplands of Great Britain, for example, one common agricultural system involved the permanent cultivation of a field or fields near to the main dwelling. This 'infield' received the most attention to its nutrient status but its production was supplemented by one or more 'outfields' beyond the limits of the permanently cultivated zone.

Such a system in Scotland in about 1500 is characterised by an infield (which comprised about one quarter of the total cultivated area possessed by a community) which was kept in continuous cultivation: there was no fallow period. Its fertility was maintained by manuring it with the accumulated products of the winter-housing of cattle as well as the dung from animals pastured on the stubble after harvest, together with discarded thatch, seaweed, shell-sand and peat. Livestock were prominent in the farming systems not for export but for manure.[11] A hardy form of barley (*Hordeum*) called bere and oats were the staple crops. Wheat might be grown in the more fertile parts of the lowlands of Scotland, along with intercalations of peas or beans. The outfield was more extensive, cropped only with oats, and manured largely by beasts folded in at night during the summer months. The outfield might also be discontinuous, avoiding patches of especially rocky or boggy ground. It was nevertheless a key element in cereal and then by 1800, potato, production.[12] Cultivation by spade or a foot-plough was often found in these patches. Where an animal-drawn plough was used, ridge-and-furrow built up, with ridges as high as one metre which can often still

PLATE 4.1   *The harnessing of energy from sources other than biomass was most evident in the form of mills. This is an example of one of the less common types: the tide mill at Woodbridge in Suffolk, as repaired and presented in 1996. In the foreground is part of the tidal basin which stores the water.*

be seen. Both infield and outfield were enclosed by a **head-dyke** of stone or turf. In the Outer Hebrides it is often only 30 m ASL but it goes as high as 365 m in the eastern Borders. On small islands, the head-dyke might be a lower limit which separated the cultivation zone from the common grazing of the foreshore. As further south, upland grazing was a valuable resource used in common by a number of tenants; in Scotland it was called 'commonty' and was carefully regulated even though the area might be very large: Mey in Caithness had over 5665 ha.[13] Agriculture was found even in very remote places: on St Kilda, islanders grew barley and oats, cabbages, and from the later eighteenth century, potatoes, carrots and turnips. (In 1896 the minister grew strawberries, which sounds strangely hedonistic.) Fish were not very popular but seabirds and their eggs were both eaten and exported. The fulmar (*Fulmarus glacialis*) bred only on St Kilda until 1878 and was valued for its food, feathers and the oil which it spat at intruders: this was sold as a machine oil.[14]

Environmental change in uplands might often respond to market forces. In Wales, for example, the production of cattle which were driven to the Midlands and London intensified by two or three times between 1400 and 1650. The nutrition of the animals required both the encroachment on open hill land (Fig. 4.1) and the conversion of hill, woodland and meadow to pasture, meadow and arable land respectively.[15] Elsewhere, population increase seems to have been the force that drove farmers as high as 305 m onto the shoulders of Bodmin Moor in the twelfth and thirteenth

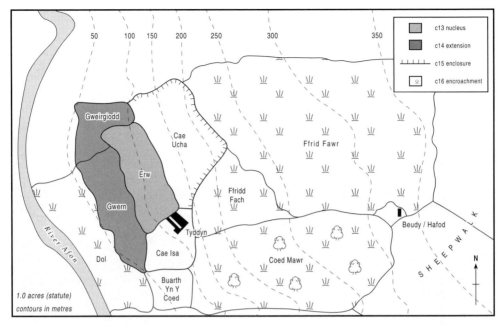

FIGURE 4.1   *A sixteenth-century farm in upland Wales. Most of the woodland has gone, with just an open wood-pasture area remaining of the 'Big Wood'. The rest of the farm shows a progressive intensification of land use, with the ffridd areas critical in supporting the cattle that would be driven to the English Midlands and London. By the 1640s the intensity of grazing by cattle was perhaps three times its level of 300 years before.*
C. Thomas, 'A cultural-ecological model of agrarian colonisation in upland Wales', Landscape History 14, 1992, p. 46, Figure 3.

centuries. Deliberately planted hamlets at 250 m ASL were created in northern Bilsdale (North York Moors) in 1250–1300: the new settlers were to reclaim (**assart**) the moorland waste and manage the woodlands for coppice. Most of the hamlets except the one now called Urra were abandoned in the first third of the fourteenth century. (Interestingly, the whole process was repeated 350 years later.)[16] Lower in Ryedale, the monasteries were especially influential in reclaiming both waste land (that is, moorland) and *wasted* land, the product of the depredations of King William's army in the winter of 1069–70: some wasted vills were replaced by monastic **granges**.[17] These examples may have inspired thirteenth-century landlords such as the de Laceys to build 28 vaccaries on the Lancashire uplands of Rossendale and Pendle as cattle ranches producing cows and oxen, an experiment that lasted for about 25 years but eventually failed since the rate of successful reproduction was not high and there were losses from disease and predation by wolves.[18]

Withdrawal from such heights in the fourteenth century encouraged a shift to sheep as important elements of an upland economy. They required less labour for instance, which was important in the period after the Black Death. Demographic recovery meant that by the end of the sixteenth century, land hunger was a problem for the first time for 200 years and there was a renewed interest in marginal lands which brought moorland into cultivation once again on Dartmoor and the North York Moors. The environmental effects of taking in moorland are considerable: the turning-over of podsolic soils opens the leached, sandy horizons to wind and rain, so that the loss of minerals and silt fractions is accelerated, even though walls and banks may trap some of it. More people are brought into the presence of the upland peats which form a major source of fuel, increasing the intensity of their use. The peaty tops of the soils may be pared off to be burned or worked into the sandy horizon to form a more friable substrate for crops.

An example of temporary use which was sometimes converted into permanent settlement and cultivation is the shieling. This was originally the summer out-station of a farm where the cattle were taken to graze during the warmer months. The animals were folded at night near to a simple building and fertilised an enclosure (still often visible as a green patch) which might become the nucleus of a hamlet if land hunger drove settlers there permanently, for a crop of oats might be possible.[19] The *shieling* is most associated with Scotland but the place-names *shield* and *sett* (from the Norwegian *saeter*) in northern England are a reminder of the practice there, with presumably a Norse origin or at the very least a take-over of an existing custom.

The main source of permanent assarts was the woodland. Since deciduous forests accumulate a brown-earth type of soil in which the nutrients are relatively evenly dispersed through the top metre or so of the profile, they are good for cultivation once the tree roots have been grubbed out. In Hertfordshire, one manor had cut so many trees by 1363 that it had nowhere (no **pannage**) to depasture its pigs. The fields created by cutting woods were added to the already existing set of fields dating back to late Roman or later prehistoric times. The twelfth and thirteenth centuries were particularly important stretches of time for felling trees: in Witney (Oxon) the first 50 years of the thirteenth century saw the clearing of 400 ha; Laughton in Sussex affixed 395 ha to its cultivated area between 1216 and 1325.[20]

Obviously, the ecology of the fields was different from the woodland that they replaced. The soil was now turned over by the plough not the pig and the wide turning-curve of the ox-team created the S-shaped pattern of ridge and furrow that is a relic of medieval cultivation often still visible now. The practice of gradually throwing up a ridge helped the drainage on wet soils but there would always have been wet patches in an epoch that preceded under-drainage with clay pipes. Many assarts were enclosed and privately owned from the start and were early recruits to the process of privatisation of common fields which ended with enclosure by Parliamentary Act or by agreement. In this they joined with areas of Great Britain which never had open fields at all. A remnant of an open-field system persists at Laxton in Nottinghamshire, though only 196 ha were in common fields in 1988, compared with 579 ha in 1635: some 66 per cent has been fully enclosed.[21]

The boundaries of a field are thus an essential feature of its ecology and environmental relations. In the south-west of England where open fields were less common, the stone wall surmounted by trees and bushes was dominant; in open-fields and closes, there were hedges on the outer perimeters with grassy baulks separating holdings within open fields. Temporary divisions would also be made with hurdles woven from hazel or willow and these would fold cattle or sheep upon a furlong that was due to be manured. Each type of boundary and its adjacent strip of land had a characteristic fauna and flora: long-established hedges, for example, accrete species, so that the number of trees and shrubs in a 30-metre length is allegedly equal to the age of the hedge in centuries.[22] Some hedges are the 'ghosts' of woodlands and keep the species of that habitat. Fields too have characteristic species: weeds of crops are an obvious example, few of which have survived modern herbicides though an occasional swath of poppies reminds us that colourful plants may have been less attractive to the medieval eye. Certainly, it is now uncommon to see the weedy plants which were so troublesome to the pre-industrial farmer, like the corncockle (*Agrostemma githago*), cornflower (*Knautia arvensis*), hairy vetch (*Vicia hirsuta*) and, worst of all, *Anthemis cotula*, the stinking mayweed.[23] Today's readers of Shakespeare's *Coriolanus* need an explanatory footnote to explain that, 'the Cockle of rebellion, insolence, sedition' was indeed 'plough'd for, sow'd and scattered' and not sieved out of inter-tidal mudflats.[24] Many of these weeds became commoner as ploughing became deeper since their seeds could tolerate long periods of burial.[25]

Birds feeding off ploughland, which nevertheless could nest undisturbed, would have been common: the lapwing (*Vanellus vanellus*), a devourer of wireworms, is a likely instance; its nesting habits suggest that fallow fields would have been ideal. As the cultivated area expanded but left patches of woodland, the rook (*Corvus frugilegus*) populations must have grown, as would those of the native grey partridge (*Perdix perdix*) and the skylark (*Alauda arvensis*). There was a widespread lack of generosity towards wild birds and small mammals. In 1566 Parliament authorised churchwardens to pay bounties for the corpses of foxes, polecats, weasels, otters, hedgehogs, jays, ravens and, among the total of 15 taxonomic groups thus identified, kingfishers. Acts for the control of rooks in 1533 in England and 1424 in Scotland probably mark the progress of enclosed arable land, and the monks of Durham apparently had a rook control problem as early as 1348.[26]

Whenever there was 'land hunger', lords looked keenly at wetlands. Peasants were less inclined to do so because bogs and fens were likely to yield them some food or even a living as a wildfowler or thatcher. From Roman times onwards, however, most of the major wetlands were the subject of piecemeal reclamation at the margins or at the edges of the more solid islands that interrupted, for example, the Fens of East Anglia or the Somerset Levels. Although always intended to be permanent, drainage of these lands carried the risk of reversion. If there was a slight rise in sea-level or an increase in storm surges, then flooding might reclaim the land, as happened in the medieval Fenland. A poorly maintained sea wall might fail at an especially high tide. It might be helped to deteriorate by those who were reluctant to build it in the first place: gangs of displaced fen-men destroyed sluices and blocked drains. Above all, the drainage of peatlands near to sea-level allows them to shrink in volume, so that continuous pumping became mandatory thereafter.

None of these disadvantages deterred attempts to win for agriculture parts of the Fens, Somerset Levels, Pevensey and Romney Marsh during the medieval period, with the late twelfth and the thirteenth centuries being periods of intense activity. The targets were mostly wetlands underlain by silts, and the peaty areas had to wait 200 years unless they were very shallow, as at Wrangle (Lincolnshire) where the common was a meadow enclosed from the fen by 1200. But a good market for produce and the knowledge of how to construct walls of packed clay, stiffened with timber and hurdles, and faced with turf or stone, all made it possible to turn winter into summer by improving the pasture. The newly won lands were not usually divided into common fields. In this way, major landlords like the Archbishop of Canterbury and other ecclesiastics dried out 9300 ha in Romney Marsh by the end of the thirteenth century. In Lincolnshire some 25,890 ha (259 km²) were won not only by the great abbeys of the Holland district but the peasants as well, displaying the qualities of independence and enterprise so characteristic of those brought up as 'yellow-bellies'.

The Fens of East Anglia were the biggest challenge of all, stretching roughly from Cambridge to Wainfleet, with the former only marginally more important in medieval times. The pre-reclamation era was definitely not one of an unproductive waste, even though fens were reported to be covered with water in winter so that they resembled the sea. As Darby showed,[27] the resources were widely used. Many parishes had elongated shapes so as to include upland, fen edge and fen within their territories; the fen yielded summer grazing, sedges and reeds, birds, eggs and peat. The extraction of peat left 'deepes' of open water (and see the account of the Norfolk Broads p. 114) and was used for domestic fuel and as a source of heat used in salt extraction from sea water. The waterways and open meres were famous for their fisheries, especially those of eels. The taking of birds was the basis of a lucrative trade with London, Cambridge and other southern cities: all kinds of ducks, swans and waders were sent to the tables of the rich and of the yuppies of the day, including even the bittern (*Botaurus stellaris*). Doubtless the High Table of Porterhouse College (a fictional creation by Tom Sharpe) was thus furnished in the early sixteenth century. All this was aided by the use of guns in wildfowling from the mid-sixteenth century.

The users of the Fenland were aware that the region underwent inundation both from sea and land. There were sea-floods in 1607 and 1613, for example, with the latter being the worst since 1236. The salt of 1613 was followed in 1614 by a freshwater flood from heavy snowfall in January and February. Such floods increased the demand for full drainage from those who had access to the benefits of marginal drying-out and in the first quarter of the seventeenth century numerous inquiries were set up, plans made and arguments conducted to remove the flood hazard once and for all by reclaiming the whole area for agricultural use. The engineers, as always, said it would be possible. Those against included those who argued that it could not be done in practical terms, those who said it should not be done since God had clearly placed the Fens there for his purposes, those who were for it provided it was done somewhere else, those who pointed out that there would be a class division between beneficiaries and losers, and those who sharply noted that there would be an influx of foreigners into the strategic east coast ports. All arguments that have a certain familiar ring.

The advantages of draining fenlands could be seen close at hand in the Netherlands and eventually it was a Dutch engineer, Cornelius Vermuyden, who created the grand

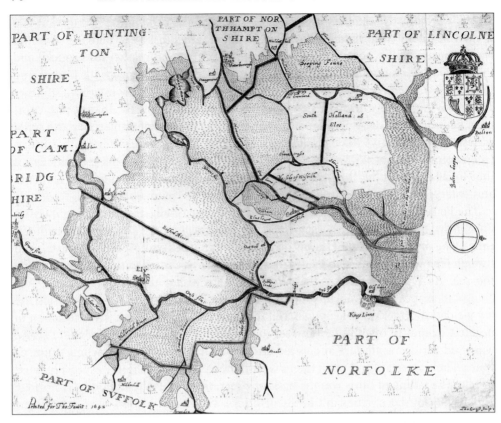

FIGURE 4.2   *The Vermuyden map of the Fens of 1642. As well as the original drainage, the new cuts are shown, as are the great tracts of marshy ground which were the habitat of so many wildfowl. Whittlesey Mere (near the Holme Fen, see Plate 7.2) is also shown. North is to the right of the map.*
C. Vermuyden (1642), 'A discourse touching the draining of the great Fennes . . .' T. Fawcett: London. BM Catalogue number E. 143.(14.).

plans and saw them to completion in the southern Fenland by 1653; others were responsible for most of the northern Fenland at about the same time. Vermuyden's plans hinged around a series of straightened rivers with cuts to shorten their courses and thus have a greater fall (Fig. 4.2). Towards the seaward end, sluices were key features of the plans: they would keep back the high tides but allow the rivers to discharge at low water. To enable the land to cope with excess water at times of very high tides or with exceptionally heavy run-off, areas of 'washland' were designated to be allowed to flood: that between the straightened Old Bedford River of 1637 and the parallel New Bedford of 1651 is the largest, surviving to this day.

The effects were numerous. First of all, there was land which was available for growing crops and for fattening stock. There were still some fisheries and wild birds were still taken with decoys running off small ponds left for the purpose: a remnant of the old economy, just as remnants of unreclaimed fen still persist (with engineered high water tables) as nature reserves at Wood Walton and Wicken in Cambridgeshire.

Wicken in particular was heavily managed in the nineteenth century and so it little resembles the original fen. However, it and other fen remnants retain some of the original flora of the acid peats, like sweet-gale (*Myrica gale*) and alder buckthorn (*Frangula alnus*).[28] The spoonbill (*Platalea leucorodia*) had gone by 1700 and is now a rare visitor to the east coast. Few of the seventeenth-century engineers foresaw the main impact of their manipulations, which was (and is) the shrinkage of the peat. The fall in level was rapid: perhaps 2 feet (0.6 m) in the first 20 years. In the mid-twentieth century large stretches of the southern levels were about 5 feet (1.5 m) below mean sea level; high tides in the Wash may reach 17 feet (5.2 m) above mean sea level. In the seventeenth century, the first response was a horse-powered pump, to be followed in the eighteenth century by an explosion in the use of the windmill; thereafter steam pumps, diesel and then electric power have been used. Indeed the drainage of the fens is a microcosm of the periods of access to energy which structure this book.

One other source of new land was the salt-marsh. In quiet inlets like the Wash and the Solway Firth, the tides bring in silt which accretes round the roots of the salt-tolerant flora and slowly builds up so that a wall can be constructed parallel to the old shoreline which will exclude the sea. The land thus enclosed needs a few years to be leached of its salt content but thereafter is a stone-free, flat field. The Wash is the best-documented region of such intakes: recent research suggests that the 'Roman Bank' around the shore is in fact Saxon in date and that there have been a series of later inclosures with the medieval period being especially important, but carrying on until the 1950s north of Boston, for instance.[29] To keep out the sea, walls have to be kept in good order. Then, a graduated set of hydrological controls can be introduced which move the marsh away from brackish water and salt-marsh plants to one of interconnected ditches and even underdrainage. At the latter stage, the wildlife diversity is low and eutrophication of the ditches is common.[30] The salt-marsh grass (*Puccinellia maritima*) has been replaced with cultivated fescue and the swirl of the dunlins along the tide is displaced in favour of the eddy of gulls behind the plough.

Intensifying production 700–1700

If demand for food exceeds supply and 'new' land is difficult to acquire then another response is to produce more crop from the existing area: the process of intensification. This requires a greater rate of throughput of crop and hinges on two features. The first, obviously enough, is that there should be more crop than non-crop plants and animals ('weeds and pests'). The second is that the nutrient levels of the soils should be maintained, so that the removal of the crop and exposure of the soil does not presage a long-term decline in the nutrition available to plants. Nitrogen is especially important, though it is not the only necessary element. The nutrient status of a soil was also tied in with its soil structure. If this broke down then nutrients were lost to the drainage; hence the input of organic material was important since it maintained the crumb structure that physically kept the soil in place except in exceptionally heavy storms.

There were several sources of nutrients that might be used to keep up or even improve the fertility of soils. Deep ploughing brings up nutrients from the sub-soil and a mouldboard also buries weeds for a while, though it does not damage their seeds. It was customary to allow a field to lie **fallow** for one year in three (there were regional variations) so that soil nutrients might accumulate. Any grasses or open-ground plants were simply ploughed in as 'green manure'. Ploughed-in stubble was also a fertiliser, but the demand for straw for many uses meant that most of the cereal stem was removed from the field, though that portion used as animal bedding might eventually return.[31] Intensification meant, eventually, that forage crops were introduced. These formed the food of folded animals which then contributed their dung to the fields: examples are the turnip (*Brassica rapa*) (from 1580) and clover (*Trifolium*) (from 1650), though the latter was only slowly adopted before 1700. The field might also be a focus for nutrients imported from other ecosystems: pigs, cattle or sheep pastured in woods, on heaths, moors or downland for instance would be folded onto the arable at night, so that the nutrients of trees, shrubs, grasses, salt-marsh or fen plants could be transferred to cropland. In East Anglia, sheep dung was carted from coastal marshes to inland arable.[32] Given such attention, it is surprising to hear that in the fourteenth century less than 30 per cent of arable was manured by animals and that only 15 per cent of fields received manures from off the farm.[33] Some additions to the nitrogen budget could come from **legumes** as elements of a rotation or they might be grown intercropped with cereals on the wetter parts of ridge and furrow. The ability of peas and beans to fix atmospheric nitrogen directly was certainly valued long before science illuminated the reasons for their capacity.[34]

Organic materials for manuring were brought from all kinds of sources: seaweed, unusable grain, straw and stubble, dead leaves, pigeon dung, deer droppings, crushed shells and the ash from burned grass have all been recorded. In western Scotland, seaweed was second in importance to animal dung.[35] Inorganic materials are dominated by alkalis such as lime, marl (recorded first in 1095) and chalk. Towns might be sources also: butchery wastes could be sold to fertilise fields. The use of human excreta (politely referred to since the nineteenth century as 'night soil') was low compared with Asia but the officers of the Bishop were in the market for it in the Winchester of the fourteenth century. Whether the episcopal product itself was so used is not, alas, recorded.

The ecology of intensified production is only one strand of the story: the economics of demand are also important, as are the control systems which keep the ecology stable. On St Kilda even in the nineteenth century every possible local source of input to fields was used but the yield, in that environment, was still very poor; fortunately total self-sufficiency was not required otherwise the remote island would have been abandoned well before 1930. Market-based demands also increased overall production by breaking down local and regional self-sufficiency in favour of specialisation. So cities developed fringes of market gardens and penumbrae of fruit farms and vine-yards; fen 'slodgers' sent thousands of wild birds to London, and industrial crops such as rapeseed (*Brassica napus*) and woad (*Isatis tinctorum*) were grown. In general, intensification came at periods of Malthusian pressure on food resources. Once intensification had increased production then more luxury crops and more industrial materials (such as herbs and dyes) could be grown.

Scholarship on pre-industrial agriculture in Great Britain emphasises the progress of enclosure. Much agricultural land in England and Wales had been enclosed since at least Saxon times and so by the Middle Ages it was a only a great swathe of the middle of England that held most of the open fields (Fig. 4.3). Even here assarted land

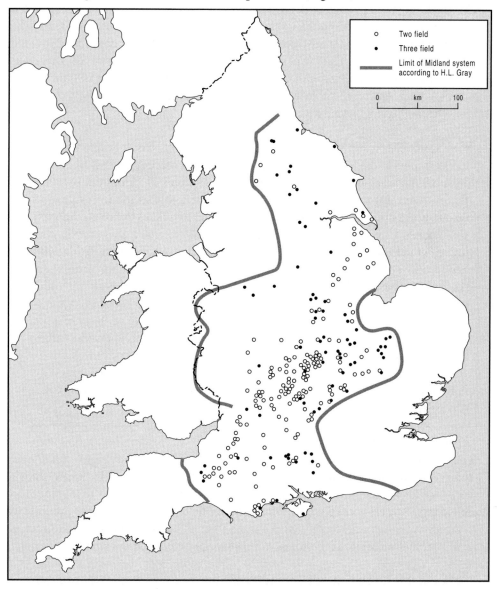

FIGURE 4.3 *Field systems in England to the mid-fourteenth century. These represent the major areas of open fields: the areas where the impact of agriculture on wild life would be strong. The woodland had undergone maximal retreat and field boundaries were less likely to be hedged or banked. Ploughing and fallow systems, however, would have encouraged rooks and lapwings.*
R. A. Donkin, 'Changes in the Early Middle Ages', in H. C. Darby (ed.), A New Historical Geography of England before 1600, *Cambridge: Cambridge University Press, 1976, vol. I, p. 82, Figure 23.*

from wood, heath and moor would very likely be assigned to an individual or corporate body rather than added to common fields. The ecology of enclosed land focuses on more individual control of soil fertility and on the ancillary yield of fuel, timber and edible fruits like hazel nuts and blackberries. Enclosure also meant more isolated farms which might be planted with sheltering trees like the elm.[36] The hedge expanded the populations, we must suppose, of the hedge sparrow (*Prunella modularis*) and the hedgehog (*Erinaceus europaeus*), and small woodlots (*copse* = coppice) provided habitat for foxes and badgers (*Meles meles*). So in the centuries before 1700, there was a basic division into a landscape that had 'always' had hedges and that middle swathe where they were more recent (often dating from the high years of enclosure here, 1415–1600) and had accumulated fewer species.

Concern over hedgerows in recent years has led to a perception that all of lowland Britain was enclosed with hedges at a late date. The essential diversity that resulted from a field pattern which might be as old as the Iron Age or as recent as the nineteenth century was rather ignored, just as the contemporary ecology has been ignored by farm managers told to make large profits. So, indeed, in some parts of the country the pattern of irregular fields with hedges or walls is older than there are records to show; other areas went through a phase of large open fields which were then enclosed; yet other regions have a dominance of land enclosed from heath and moor at a time of high prices for agricultural produce. Policy can either reflect these historic divisions and try to frame regional policies for, for example, spending public money on restoring hedges, or it can assume that hedges (for example) are a good thing in today's environments and that they should be encouraged no matter that they are not part of the 'traditional' scene. Clarity of purpose is, however, to be encouraged.

The environmental impact of these centuries is unmistakable as Figure 4.4 shows. The depth of the human imprint on the land is considerable even from prehistoric times onwards: we need not be surprised to be told by archaeologists that some Iron Age field layouts were incorporated into medieval field systems. As a clear example of such depth, we can quote Hooke's work on the Vale of Evesham in Anglo-Saxon times (Fig. 4.5). Her evaluation of the human impress can be quoted *in extenso*:

The natural landscape of hill and valley, stream and marsh was still of fundamental importance and large areas of waste and wilderness survived, both as dense woodland and undrained marsh. At the same time it is evident that the landscape revealed in the West Midland charters was a very long way indeed from any primeval countryside. Woodland was obviously controlled and managed, cropland often already inherited from a more distant past, and much of the countryside already portrayed the influence of man's activity over a long period

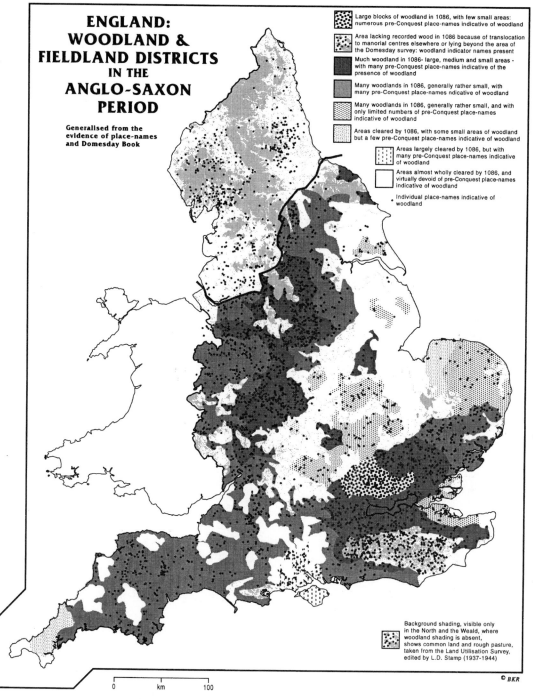

ENGLAND: WOODLAND & FIELDLAND DISTRICTS IN THE ANGLO-SAXON PERIOD

Generalised from the evidence of place-names and Domesday Book

Large blocks of woodland in 1086, with few small areas: numerous pre-Conquest place-names indicative of woodland

Area lacking recorded wood in 1086 because of translocation to manorial centres elsewhere or lying beyond the area of the Domesday survey: woodland indicator names present

Much woodland in 1086- large, medium and small areas - with many pre-Conquest place-names indicative of the presence of woodland

Many woodlands in 1086, generally rather small, with many pre-Conquest place-names indicative of woodland

Many woodlands in 1086, generally rather small, and with only limited numbers of pre-Conquest place-names indicative of woodland

Areas cleared by 1086, with some small areas of woodland but a few pre-Conquest place-names indicative of woodland

Areas largely cleared by 1086, but with many pre-Conquest place-names indicative of woodland

Areas almost wholly cleared by 1086, and virtually devoid of pre-Conquest place-names indicative of woodland

Individual place-names indicative of woodland

Background shading, visible only in the North and the Weald, where woodland shading is absent, shows common land and rough pasture, taken from the Land Utilisation Survey, edited by L.D. Stamp (1937-1944)

0   km   100

© BKR

FIGURE 4.4 *A map courtesy of Professor B. K. Roberts showing the chief types of regional wood-land distribution in the Anglo-Saxon period of England. Some of these areas were 'fixed' by their designation as Royal Forests and others, in the north especially, became even more wooded after the harrying by King William's armies.*

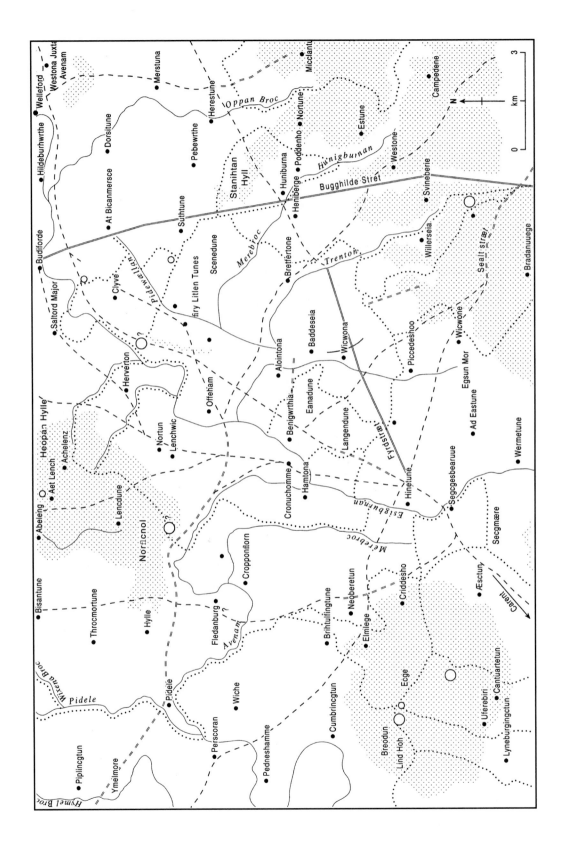

of time. Large areas of waste and woodland might await eventual colonisation but already settlement had become so dense that few communities can have been more than a few miles from a neighbour. Trackways and routeways penetrated every corner of the kingdom and even in the most remote recesses the deer were not safe from controlled hunting by man.[37]

A reconstruction of the land use in Hanbury (Worcestershire) in the late thirteenth century, emphasises this point even further.[38] The map (Fig. 4.6) shows that the mid-Holocene wooded matrix has been differentiated into a mosaic of arable, pasture and meadow and that the woodland shows largely as a series of relic patches, one of which was emparked and all of which were no doubt intensively utilised and managed. Individual trees or small copses too were managed, for example, for the production of withies and osiers for basketwork. So in the depths of rural England, the year 1300 saw a more or less completely human-dominated environment. More marginal settlements than Hanbury were likely to be abandoned after the Black Death and their lands reclaimed by scrub followed by woodland.

Contraction of tillage

Another route to intensification is to produce more animals from land formerly in cereals or pulses. Land for sheep farming might be transformed from arable by enclosure, be taken in from heath, down or moor, or be reclaimed fen and marsh. Sheep losses in pre-industrial times were high by today's standards with perhaps 30 per cent loss each year from sheep scab, wolves and foxes. In spite of this, flocks were sometimes large: as early as 1209 the Bishop of Winchester had 15,000 and doubled that number by 1260; the Cistercian monasteries were important in the wool trade (see below under moors and heaths) but never overshadowed other producers. One ecological result was the creation of downland on Chalk substrates. The close nibbling by the sheep and the shallow calcareous soils meant that, especially on slopes, a rich herb flora developed of plants unable to tolerate acid soils (**calcicoles**) or being shaded by more luxuriant growth, but capable of withstanding being eaten from time to time. Conservation of relics of this flora today often involves mimicking old ways of herding sheep. Species such as the bee orchid (*Ophrys apifera*), which is one of several orchids found on chalk soils, salad burnet (*Sanguisorba minor*) and dwarf thistle (*Cirsium acaule*), are often the objects of such management. The great bustard (*Otis tarda*) or gant (= goose, as in *gander*) was a common bird of the great sheepwalks of Salisbury Plain, though shot to national extinction by 1832.

Meadows and other permanent grasslands were an important part of the feed

FIGURE 4.5 (opposite) *A section of the Vale of Evesham during Anglo-Saxon times combining various sources of evidence derived from charters. Though none (apart from the shaded land over 200 ft ASL) are directly environmental, the impression given by place-names, trackways and roads is of a well-occupied, even 'busy' landscape in which the human impress was strong.*
D. Hooke, Anglo-Saxon Landscapes of the West Midlands: The Charter Evidence, *Oxford: British Archaeological Reports British Series 95, 1981, p. 345, Figure 4.4.*

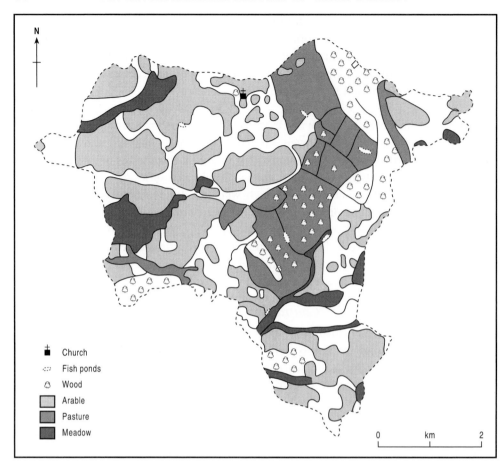

FIGURE 4.6 *Land use in Hanbury (Worcestershire) in c.1300. The predominance of arable land is noticeable, with its mirror image the small amount of woodland. Some wood-pasture is also present and together with the grazing and meadow represents a major source of nutrient return to the arable fields, as do 69 marl pits noted in field surveys. Fish ponds allow a diversification of animal protein source.*
C. Dyer, Settlement and Society in a Woodland Landscape, *Leicester: Leicester University Press, 1991, p. 47, Figure 12.*

resources of farm animals. Most common meadows had a high value and were strictly controlled. The hay crop was the first priority so that no grazing was allowed after about Candlemas (early February) until the last mowing had taken place. So, depending on local custom and conditions, grazing might start on Lammas Day (1 August) or Michaelmas (29 September). Ecologically, this regime favours plants which flower and set seed before the hay is cut in June or which reproduce vegetatively. The cowslip (*Primula veris*) and the green-winged orchid (*Orchis morio*) flower before the rest of the vegetation overtops them; other species have tall flowering stems so as to disperse their pollen: the meadowsweet (*Filipendula ulmaria*) and the meadow foxtail (*Alopecurus pratensis*) are instances. Flower-rich meadows which have survived modern agriculture usually have a management regime which imitates the pre-industrial pattern if they are to have conservation value, as in the City of Oxford meadows, for example, or at Cricklade (Wiltshire), where the North Meadow retains the ancient pattern of grazing rights in the August–February period but is laid up for hay the rest of the year. One result is the presence of 80 per cent of the British population of the fritillary (*Fritillaria meleagris*).[39] Lowland meadow might be overgrazed and one response was the development after *c.*1500 of the water-meadow. A system of small channels allowed controlled flooding of the meadow between mowing-time and the beginning of the following May. The river's silt was dropped in the still water so that the nutrients thus imported added to the lushness of the meadow: hay yields might rise four-fold. The general practice of channelling water and using it to produce a flush of nutrients is probably much more widespread than has been generally recognised.[40] A similar practice was followed on the upper margins of estuaries where the water was not too salt: this was called warping and was especially characteristic of the Humber lowlands.

Overview

By 1700, crop yields had increased by about one-third upon their 1400 level. This meant among other things that there were fewer famines and so population growth was set to become constant. That trend was reinforced by the disappearance of the plague in 1649 in Scotland and in 1679 in England and Wales. But there had been breakdowns: times of land hunger and of famine punctuated the millennium we have just considered. It is difficult to assert simply whether or not these catastrophes were basically ecological and Malthusian in nature (that is, that there were too many people for the carrying capacity of the land at the contemporary level of technology and knowledge) or whether the socio-political system demanded too much of the producers, both in nature and in society. Greedy manorial lords, the increasing number of owners of large enclosed estates, and the taxation demands of the state to support high-spending monarchs or to prosecute wars (sometimes the same thing) were also inextricable parts of the equation. Yet there was always sufficient contrast of terrain to support a diversity of species. Brian Vesey-Fitzgerald held that the period from the accession of the Tudors to 1750 was a Golden Age of British wildlife.[41]

HEATHS, MOORS AND MOUNTAINS

These environments are characterised by their overall treelessness, though patches of woodland remain, like Wistman's Wood on Dartmoor or the Coille na Glas-leitire by Loch Maree. If left to nature, the lowland heaths would soon become woodland since it was directly out of forest that they were formed. Most moorlands were also formed from a matrix of woodland, though at high altitudes the growth of peat may have been mostly a response to climate. Leave them alone and woods might return but only very slowly above about 400 m and perhaps not at all on deep peats. Many mountainous areas were forested but so much soil has gone that recolonisation would be a patchy affair. Reclamation for cultivation has never been very difficult for heaths, technically feasible but often uneconomic for the edges of moorlands, and impossible away from the valleys for true mountains like the Lake District and Grampians. All can be used for grazing, hunting and mining.

As an example of the relatively intensive use of some moorlands, we shall look at the occupation of some moorland areas of northern England by the Cistercian monasteries. From mother houses in France, Great Britain was colonised from 1128 onwards, starting in southern England. But an edict of the Order in 1134 laid down that houses were to be

In civitatibus, castellis, villis, nulla nostra construenda sunt coenobia sed in locis a conversatione hominum semiotis	Far from cities, castles, towns and away from the noise of men . . .

This was interpreted to mean that wild and waste places were especially suitable and so the order accepted invitations to settle where for instance there was a great deal of moorland, where the land was still 'waste' after the harrying of the North, and where the existing population could be moved out. Eventually, 13 out of 75 Cistercian houses in England and Wales were established in northern England. Their first sites were often selected more for their spiritual qualities than for practical attributes and so sometimes had to be moved. So might the local people: evictions ceased only in the early fourteenth century.[42] Isolation might also result from proximity to a Royal Forest, for one-third of Cistercian houses were in such protected zones in 1250.

An early impact on the land might have been the building of the abbey: Vale Royal in Cheshire absorbed 40,000 cartloads of stone in the years 1278–81. Thereafter, the basic unit of environmental management was the grange, which was staffed by lay brothers, the *conversi*. A grange had typically 120–60 ha of arable land but was responsible for a total of about 405 ha. Its holdings were delimited by an earthen bank and ditch, hedges or a wall. Granges often added to their area by assarting from woodland or moorland, which was no casual act. A Cistercian house in Germany in the twelfth century has left an account of making an assart in forest. The abbot accompanied the workers when they started to fell the trees. He first planted a cross in the ground to be cleared and then sprinkled holy water around before making the first, symbolic, cutting of some shrubs. Then a group of men cut down the trees (they were the *incisores*), a second took out the trunks (*exstirpatores*), and a third burned up the roots, boughs and undergrowth: these workers were the *incensores* in a reference to the smoke given off by their fire.[43]

In keeping with the upland environment, however, pastoralism was the main use away from the cultivated land of the valleys. Cattle were very important in the north of England, as they were in the Fenland houses further south. Cattle-keeping centres (vaccaries) were established by 1140 at Byland in north Yorkshire and by the end of the twelfth century in most northern dales (Plate 4.2). Nidderdale's economy was almost 100 per cent devoted to cattle but it was more common for the granges to keep sheep as well (in beccaries); typically cattle comprised 10–25 per cent of the number of sheep. But the ratios were not uniform: at the Dissolution, Fountains Abbey (Yorkshire Dales) had 1326 sheep, 536 oxen, 738 cows and 1000 head of other horned

PLATE 4.2   *The influence of the monasteries is symbolised by the size and grandeur of their mother houses (as here in 1986) even when ruined. This Cistercian example at Rievaulx in North Yorkshire had huge estates in both upland and lowland areas and would have been responsible for environmental manipulation on a large scale, from the management of the meadows and watercourses in the foreground to the woodland of the surrounding area that supplied its iron works.*

cattle. In general, the smaller houses (especially those in Wales) were dominated by cattle and the larger units had more sheep, as we might expect. The sheep were the abbeys' means of contributing to the agricultural specialisations of the time and of earning money in the export market: their prices were competitive since they often had exemption from road and port tolls and the *conversi* were not paid. These granges were in later years often leased out rather than run by the monks and *conversi* themselves.

Other environmentally linked economic activities included the smelting of iron, the production of salt (Byland Abbey had a salt works on the Tees estuary at Coatham) and the production of lead: even some of the granges' outbuildings had lead roofs. Though not necessarily the innovators, the Cistercian houses acted as early adopters and diffusers of technologies such as the ribbed vault, the water-driven hammer forge, the fulling mill, windmills and the use of the horse for ploughing. Excavations at the former grange of Laskill five miles from the parent abbey of Rievaulx (North York Moors) suggests that an early blast furnace had been developed alongside the less productive bloomeries run by the abbey. Taken together, the environmental impact of all these enterprises was considerable for much woodland was felled, much converted to sustainable-yield coppice rotation, and the ratio of sheep to cattle changed the mix of plant species on moorlands. There was as well the lasting effect of the granges which passed into private hands after the Dissolution: these were often higher and more remote than ordinary settlements of the same period and so maintained a frontier of relatively intensive land use.

Outside the realms of the monastic houses, animals were often used to convert some plants which are inedible by humans to food and skins. The introduction of the rabbit (*Oryctolagus caniculus*) in the twelfth century was one such move, the first record being a document from the Scilly Isles in 1176. This animal cropped herbs and grasses even closer than sheep and produced good meat and saleable skins. The twelfth-century rabbit was much more delicate in its constitution than today's specimens: it needed supplemental winter feed (hay, furze and even oats were often provided) and was often unable to dig its own burrow. So where colonies ('warrens') were established they might be on sandy soil so that the warrener could the more easily dig burrows or erect mounds of loose soil ('pillow-mounds') for housing the rabbits.[44] The rabbit might be introduced onto common land by the manorial lord without infringing the grazing rights of the commoners but, since it was an animal subject to hunting laws, the lord might amass all the benefits. In the Breckland of Norfolk, the Bishop of Ely set up warrens in the twelfth century big enough to be surrounded by 16 km of bank and ditch; like many warrens, it was enlarged after the Black Death when there was a glut of unused arable land. The warrener had to be paid extra for killing foxes and polecats and eventually had a flint-built house as a base against poachers.[45] In some places, warrens were a good use of small patches of grass or heath: high parts of the Failand ridge (west of Bristol) bear names such as Conygar, Conygar Hill and The Warren. A triangular warren lodge at Rushton (Northampton-shire) is said to make reference to the Trinity and thence to the theological meanings of the rabbit.[46] In Leicestershire a warren dating from the 1280s has now been designated as an Ancient Monument.

The climatic amelioration of the late fourteenth century was good for the rabbit and by 1695, GREGORY KING estimated that there were a million of them in the British Isles. But they were mostly not the pest that they are now reckoned to be, although even in the thirteenth century the Bishop of Winchester's rabbits were accused of devouring wheat year after year. It appears that they learned how to dig their own burrows and exploded into their familiar ecological niche and cultural role in the late eighteenth century. Medieval and early modern warrens, however, must have had very short-cropped grassland with some open soil that blew in strong winds. Any rare plant that was inedible might well have owed its survival to the lack of competition thus produced but no examples are given by ecologists.

The process of reclamation of moorland for arable crop production is subject to stringent economic controls: nobody wants the effort unless the profits are high. Most uplands show a set of old field systems, ruined buildings and fallen head-dykes well above the present limit of cultivation, and historical research soon shows that the limit of intaking fields has gone up and down like the tides. Such land can be called 'recurrently marginal' (Fig. 4.7) and it has been shown clearly for the Lammermuir

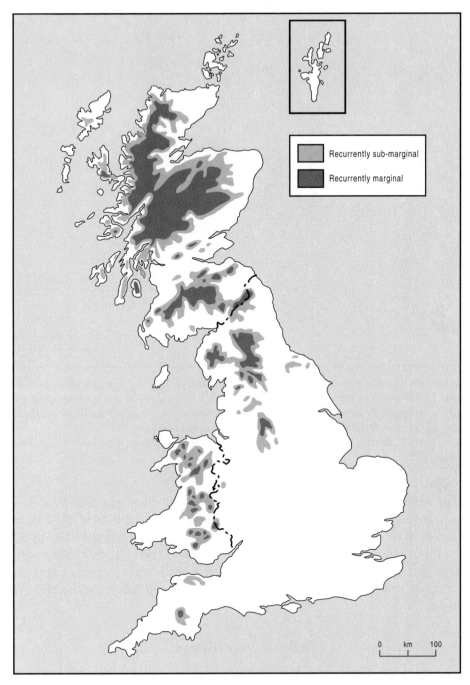

FIGURE 4.7   *Land designated as 'recurrently sub-marginal' is land which since the Conquest is unlikely ever to have been cultivated. 'Recurrently marginal' land may under favourable climatic and/or economic conditions become cultivated but it is likely to relapse when such conditions deteriorate.*

M. L. Parry, Climatic Change, Agriculture and Settlement, *Folkestone: Dawson, 1978, p. 104, Figure 27.*

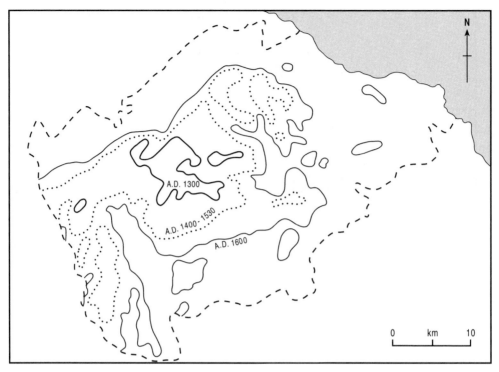

FIGURE 4.8   *Cooling between* AD *1300 and 1600 produces a lowering of temperature as mea-sured by day-degree accumulations and rainfall totals. Thus in 1300 all but the summits of the hills could be used for cereal cropping in the Lammermuirs of south-east Scotland. Late medieval cooling progressively extended the areas where cereals would not ripen.*
M. L. Parry, Climatic Change, Agriculture and Settlement, *Folkestone: Dawson, 1978, p. 101, Figure 25.*

Hills of south-east Scotland that between 1300 and 1600 there was *c*.160 m of down-ward movement of the upper limit of arable cultivation. In 1150–1250, the Medieval Warm Epoch set the limit to cultivation at *c*.450 m OD; by 1300 it had fallen by 50 m and in the next three hundred years by another 75–90 m (Fig. 4.8). By 1600, therefore, the core of uncultivable upland had doubled in size which is not surprising since there was by then a ten-fold increase in the probability of crop failure on the high-lying farmland.[47]

WOODS AND FORESTS

The British Isles are now the least forested parts of Europe. The hewing-out of numerous land uses from the forest matrix began in prehistoric times, as we have seen; between then and the present there has been a continuous interchange between felling, replanting and management, leaving us with a mixture of woods which are patently new and woodlands which are at least in part remnants of much older tree-clad areas.

The medieval woods

In Scotland, the processes begun in prehistoric times seem to have continued, with the far north of the mainland and the Isles losing their tree cover in the face of peat growth. Harris had good pine forests in the second millennium BC but timber soon became scarce thereafter. In the Pictish times of AD 500, animals carved on stones included wolves, boar, deer, eagles and salmon; the Romans sent home a Caledonian brown bear for the circus. But by the end of the fifteenth century there was a serious timber shortage and the lowlands were seriously denuded by 1600; the beaver and the wild boar disappeared after c.1550 and the wolf by the end of the seventeenth century.[48] With the loss of the pine woodlands (Plate 4.3), populations of birds like the crested tit (*Parus cristatus*), crossbill (*Loxia curvirostra*) and the turkey-sized caper-callie (*Tetrao urogallus*) fell. As early as 1180, woodland near Dumfries with grazing for pigs and cattle was replaced by moorland with sheep.

South of today's border, where the ecological hold of the woods was less tenuous, it seems as if the Saxons entered some landscapes in which woods were plentiful, and other regions where they were scarce. The Cotswolds, for example, seem to have been densely wooded in the valleys which dissect the escarpment and along the scarp face itself. The high country was, however, largely open pasture for sheep.[49] Such terrain

PLATE 4.3 *The Coille na Glas-leitire in 1966. A relict pinewood on the slopes above Loch Maree in the Highlands. The fragmentation of the pine forests is long-standing and remnants such as this regenerate only poorly due to the numbers of deer. Controlled fire might also encourage seedlings.*

might be more wooded if it formed a political boundary, as with the kingdom of the Hwicca in Anglo-Saxon times.[50] Domesday Book records 7800 woods for those areas of England which it encompasses and the average tree cover seems to have been *c*.15 per cent. The less wooded zones probably had as little as 3–4 per cent cover. Some of the wood was recorded as *silua minuta*, that is, coppice; other records are of *silua pastilis*, or pasture wood. The earliest Arthurian tale in Welsh, *Culhwch and Olwen*, contains a kind of chronicle of woodland demise and regrowth: the ancient Owl of Cwm Cawlwyd says, 'When first I came hither, the great valley you see was a wooded glen, a race of men came thereto and it was laid waste. And the second wood grew therein, and this wood is the third.'[51] The fourteenth-century Welsh poet Dafydd ap Gwilym probably summed up one view of woods[52] when the 'I' and his love went:

> Together to plant birches – task of joy –
> Together weave fair plumage of the trees
> Together talk of love with my slim girl,
> Together gaze on solitary fields.

By 1200, all the woods in England were owned. Title meant that the woods were often demarcated by banks and ditches and were intensively managed; Rackham suggests that the last virgin (that is, unmanaged) woodland was in the Forest of Dean in *c*.1150. These trees were felled to make timbers for the Dominican friary at Gloucester and were huge, being about 68.5 cm in diameter and 15 m in useable length.[53]

Thus by early modern times, all the **wildwood** had gone; the remaining woods included, of course, woodlands that were the direct, but managed, descendants of it, usually now called 'primary' if the site has been continuously wooded since the time of wildwood. If trees have grown over unwooded ground then it is called 'secondary'. A parallel classification is also used in Great Britain: woods on sites continuously wooded since 1600 are called 'ancient'; those wooded since 1600 qualify only for the term 'recent'.[54] Thus we can have ancient secondary woodlands, which are probably more common than is thought. Since most of the woods were owned by a manor, they were usually bounded by a bank or a bank and ditch. Trees were also a feature of hedgerows and indeed also of fields, as they were of orchards, gardens and some urban streets. As long as there was oak, there were probably pied flycatchers (*Muscicapa hypoleuca*) (associated with oak-tree caterpillars) and the redstart flourished on the caterpillars found in edge habitats in fragmented woodlands.

The uses of woodland

The first mention must be a reminder of woods as land banks and indeed as potential sources of cash for monarchs and monks alike, but this topic has already received discussion. Here, we are concerned with the management for woods for organic produce: wood itself but also animal products and the pursuit of pleasure.

For most medieval and early modern communities, the small trees and shrubs were key elements in their resource flows. Firewood and fencing were essential products. Both came mostly from pole-sized lengths which were produced by coppicing. Many species of tree respond well to being cut at their base, for they produce abundant and fast-growing shoots which could be harvested in rotation. This would likely be between five and seven years when applied for example to ash, oak, hazel, maple (*Acer campestre*), elm, lime and birch. During the fifteenth century, the cycle seems to have got longer and more regular.[55] The long, flexible poles taken from coppice were especially useful for making hurdles that controlled cattle and sheep in open fields, and for wattle to be daubed with clay as a filler panel for timber-framed buildings. There seems to have been a widespread transition from wood-pasture to coppice in the later Middle Ages as the demand for iron grew, for example, for nails and edge tools. Some trees were usually allowed to grow to greater size ('standards') to provide timber, over half of which went into buildings. Here, oak was sought above all, with trees from 25 to 70 years of age providing the major sources.[56] A farmhouse might need about 80 oak trees and the main roofs of Norwich Cathedral (fifteenth century) consumed 680 oaks around 36 cm in diameter at the base. The Octagon lantern of Ely Cathedral appears to have been adapted for smaller trees than in the original design: sixteen struts 12 m long by 35 cm square were apparently impossible to obtain. Deficiencies in other kinds of timber may be inferred from the import from Norway of pine spars for the scaffolding of this cathedral during its construction. The great impact of shipbuilding upon woodlands came from 1780 onwards and was directed mainly at parks and hedgerows: it was not a significant manipulator before then.[57] All trees were worked green immediately after felling and nothing was wasted: the chippings, the twigs and the leaves all found a use in pre-industrial economies.

Common grazing in woods and wood-pastures was widely practised until the eighteenth century. Cattle, sheep and pigs were the usual beneficiaries and their impact on the woods was regulated by a manorial court, as with other resources held in common. Grazing and browsing do little to improve the reproduction of trees; pigs eat a lot of acorns and beech mast though they also loosen the soil and provide a seedbed in a minor act of compensation. Woods regularly grazed and used for swine-herding ('pannage') develop an open character, with extended grassy glades and swards. An open area once created is likely to be kept that way since the grass is differentially attractive to cattle and the manure in turn allows a lush carpet of grasses and sedges to develop. Grazing and browsing is inimical to the growth of young trees but favours spiny shrubs like holly. These in turn may protect saplings and allow some regeneration of oak and beech, for instance.[58] Only about 14 pasture woodlands are now present in Britain: they are mostly in lowland England (for example, Windsor Forest and Park, New Forest, Wyre Forest in Hereford and Worcester) with one each in Scotland and Wales.[59] Conflict between grazing benefits and wood-production interests are nevertheless endemic; fencing of parts of the wood after felling was sometimes practised, to allow the establishment of saplings. Queen Elizabeth's lease of Hayley Wood (Cambridgeshire) in 1584 specified that John Spurlinge should

> after each felling . . . shall well and sufficiently enclose and encoppice it with . . . fences . . . and shall guard it and protect it from the bite Trampling and damage of animals . . . which may be able to hurt the shoots and regrowth . . .[60]

One way of reducing strife was to provide cattle with browse by pollarding trees. They were cut 2.5–3.0 m above the ground and allowed to grow again; the consequent branches could then be cut and fed to the beasts. If allowed to grow further, large poles could be produced. The active pollarding of oaks has been carried out since the Middle Ages at Hatch Park in Kent and old pollarded trees are now an excellent habitat for invertebrates. At Dunham Massey Park in Cheshire, over 180 species of wood-dwelling beetles have been recorded in pollarded wood-pasture.[61]

The woods were also a source of food for everybody. Mushrooms, blackberries and hazel nuts could be gathered in season to add both nutrition and variety. If discreet enough, poaching might bring in some welcome fresh meat. None of these gathering activities would have exerted enough ecological pressure to alter the ecology of the woodlands significantly, compared with underwood management and grazing.

One ecological consequence of all these uses can be singled out: they affected the ground flora. This is most easily seen in the diminution of the quantity of bluebells and wood anemone in lowland woods if the canopy is allowed to grow over. Coppice-with-standards, however, provides a wonderful carpet in the spring. Another facet of woodland history is the fact that most ancient woodlands have some plants which are confined to those areas. In Lincolnshire, *Carex pallescens* is strictly confined to them, and both *Milium effusum* and *Oxalis acetosella* have very strong associations. The small-leaved lime (*Tilia cordata*) is also confined to primary woods and very old hedges. The presence of such species can hence be used today as an indicator of ancient and primary woodlands.

The management of small deciduous woodlands today can be part of a diversified rural land use in which the biodiversity can be raised by adherence to some of the older management systems. Coppice, for example, creates attractive ground floras in which the bluebell is often dominant, and provides cover and food for a variety of birds. Combining coppice with some high forest and large old trees on internal and external boundaries allows an association of protection and production. A more enlightened use of rural subsidy would improve the ecology and we may note that woodland management can be labour-intensive.

Impacts on the woodland systems

All these uses of wood (along with its role as an industrial fuel) suggest one important story: that it was vital not to diminish the productivity of woodlands. In effect, almost all interests are best served by a renewable resource rather than one which is becoming depleted. Nevertheless, the area of woodland shrank steadily as the area of cultivated land and grassland rose: the land-bank function became the most important as population grew. Accurate data are impossible to obtain: perhaps 15 per cent of England was under woods at the time of Domesday but between 1086 and 1350 this had fallen to 10 per cent. Big losses in the eighteenth century brought a nadir of around 4 per cent in 1900. A steady diminution in Wales down to the Black Death has been postulated and it is known that Edward I's armies were accompanied by woodmen who cut down the woods on either side of the main roads, to discourage ambush. The loss of wooded area in this period was fastest, though, in agricultural areas rather than in places with nascent industry. In Scotland, the usefulness of wood was emphasised by the development in the Highlands of a birch-based economy, with that tree providing many important aids to rural life. Indeed, in some glens neither harvesting of pine for ship's masts nor cattle breeding nor even commercial sheep farming caused significant deforestation in the seventeenth and eighteenth centuries. The events of 1745, however, resulted in the ravaging of entire glens and their economy by troops and fire.[62] The dynamics of land use change have meant that woodlands sometimes reveal their history in their shape. Wood-pasture held in common, for example, is typically straggling in outline, with concave boundaries which funnel out where they are crossed by roads; there is normally no boundary bank. Sinuous or even zig-zag outlines are common with ancient woodland and there will be the remnants even today of a bank with an external ditch. Much wood-pasture was in time converted to farmland or to single-species woodland aimed at timber production.

It is inevitable that such a loss of one type of habitat should bring with it a loss of species. In the case of Great Britain, we can note especially the extermination of the wolf, the wild boar and the beaver. The demise of the beaver was probably most tied up with the loss of extensive woodlands: in the late twelfth century the chronicler Giraldus Cambrensis said that the beaver was gone from Ireland and England but remained on the Teifi in Wales and one unnamed river in Scotland. The wild boar was the next most affected, with the last free-living group in the Forest of Dean c.1260. After that some herds were kept in a semi-wild state and later records are probably of escapes from parkland herds; one such breakout occurred in 1540 in Savernake Forest (Wiltshire). The wolf is not so specifically an animal of woodlands but is a clear target for bounty hunters. In Surrey in 1212, the large sum of 5s. was paid in bounty for one wolf; the last attestable record of a wolf in Great Britain is of a bounty payment in Scotland in 1621, of the remarkably generous sum of £6. 13s. 0d. The wolf seems to have gone from Wales in the reign of James I and the last English record is from the monks at Whitby in 1395–6, when they paid the tanner for dressing 13 wolf pelts.[63]

No records exist of the effect on bird species of the shrinking area and changing types of woodland. Using existing unmanaged woodland as a datum, it has been suggested that with the increasing amount of edge, there would have been more birds but fewer species. Central European deciduous forests of a 'primeval' character may have as many as 107 species of forest and forest edge birds and today's British woodlands no more than 75. Some changes in habit are assumed: that swifts (Apus apus), for example, began to forsake high trees for buildings, and the long-tailed tit (Aegithalos caudatus) learned to nest in bushes rather than in the canopy. Fewer old trees would have reduced the nesting sites available for the tawny owl (Strix aluco) and woodpeckers. In the seventeenth century, some bird species acquired an identity as pests: raven, kite and buzzard were seen as predators on domestic poultry and on lambs rather than as scavengers. The development of orchards encouraged bullfinches and jays, which ate the buds of fruit trees. All attracted bounty payments, as had the wolf.[64]

Towards the seventeenth century, the question of woodland shortages began to attract the attention of Parliament, though both Henry III and Edward I had issued royal edicts against certain forms of timber sales in 1258–9 and 1290 respectively. The first timber preservation Act came in 1543 which demanded that 12 young trees be left on every acre (0.4 ha) whenever woods were cut down. Under Elizabeth I this was extended to the protection of ship timber within 14 miles (22 km) of navigable waterways. A succession of Acts flowed from Parliament in the seventeenth century and John Evelyn's book *Sylva* of 1664 was symbolic of the disquiet felt over timber supplies (as distinct from underwood or fuel) for an island nation. By the later sixteenth century the idea of plantations had taken root from a continental origin and an oak plantation was made in Windsor Great Park in 1580. Its purpose was to produce timber and soon it became common for conifers to be used. In the late sixteenth century, too, the sycamore (*Acer pseudoplatanus*) was introduced: at first it was a favourite for shelter round isolated farms but subsequently spread into many woodlands as well.

The woodland as a cultural phenomenon

The human ecologist FRANK FRASER DARLING once said in a public lecture that woods were useful for urination when young, for copulation when older and for contemplation when older still. Two at least of these themes are echoed in overtly cultural material relating to woods and forests between 700 and 1700: as Glacken puts it in talking about Europe as a whole in the Middle Ages, forests are involved in the need for change and the need for stability.[65] This is echoed in simultaneous delight in the woods and the perception of them as places of threat. In the early Middle Ages, the woods were places of mystery and of testing. The later Middle Ages saw the control and parcelling-out of woodland resources, which no doubt excited the later romantic and Golden Age perception of their hey-day: the Robin Hood stories may well come out of the forest as a space for a counter-history to the then prevailing practices of ownership and lordship.

Another binary opposition is seen in the popular medieval image of The Green Man, seen carved in numerous churches. The male face both devours vegetation (usually oak tree leaves and branches) and disgorges it, and so can be seen as either a fertility figure or a harbinger of death and decay or indeed both. His greatest flourishing was probably during the twelfth-century rise of high culture but continued to attract attention into the seventeenth century, when Andrew Marvell (in *Upon Appleton House*) turns into a Green Man, when 'The Oake leaves me embroyder all,/Between them Caterpillars crawl' (v. 74, l. 4). This fondness for the oak spreads from the medieval carving of leaves on roof bosses and misericords into the romantic stories of Charles II hiding in one, of the people waving oak branches at the Restoration, and the naming of many public houses as The Royal Oak.

No writers recorded the demise of the Wildwood in explicit terms, though in the late fourteenth century alliterative poem *Sir Gawain and the Green Knight*, we have in Gawain's journey what sounds like an encounter with primary woodland:

Bi a mounte on the morne merily he rydes	Merrily in the morning by a mountain he rode
Into a forest ful dep, that ferly was wylde,	Into a wondrously wild wood in a valley,
Highe hillds on uche a halve and holtwoodes under	With high hills on each side overpeering a forest
Of hore okes ful hoge, a hundred togeder.	Of huge heavy oaks, a hundred together.
The hasel and the hawthorne were harled al samen,	The hazel and the hawthorn were intertwined,
With roghe raged mosse rayled aywhere	And all was overgrown with hoar-frosted moss[66]

This description is paralleled in about 1370 by another from Geoffrey Chaucer's *Book of the Duchess*,[67] where the height of the unbranched trees and their closeness suggests once again that this is a fragment of very ancient woodland indeed:

Where there were so many grenė greves	Where there were so many green groves
Of thikkė trees, so ful of leves.	Of thick trees, so full of leaves
And every tree stood by hymselve	Every tree stood by himself
Fro other wel ten fete or twelve	Ten or twelve feet from the next
So gretė trees, so huge of stengthe,	Such great trees, hugely strong
Of fourty, fify, fedmė[1] length	Forty or fifty fathoms high
Clean withoutė bowgh or stikke	Without a bough or branch
With croppės brode and each as thikke;	With a spreading and thick canopy
They were nat an ynch asonder,	There was not an inch between them
That it was shadewe overal under	That was not in total shadow underneath
And many an herte[2] and many an hynde[3]	And many a hart and many a hind
Was both before me and behynde	Was before me and behind
Of founės,[4] sowrės,[5] bukkės,[6] does,[7]	Fallow deer of various kinds
Was ful the woode, and many roes,[8]	And roe deer, filled the wood

1 Fathom (= six feet)	5 Fallow deer in the fourth year
2 Stag	6 Male fallow
3 Female red deer over 3 years	7 Female fallow deer
4 Fallow deer in the first year	8 Roe deer

The forest is sometimes linked to the Gothic architecture of cathedrals and abbeys. The pillars, especially where they end in fan vaulting, are compared to the tall trunks of forest trees branching out into a canopy of interlaced foliage. Where the style was imported from more wooded lands, then it may be so. In the case of the major native development of Gothic, the Perpendicular style, then its unfolding after 1350 could only have looked back to such forests, for little actual woodland would by then have been anything like such a vault. This type of roofing is sometimes linked onwards to the polyphonic music of the Renaissance in England and Scotland, but to suggest a continuous thread of inspiration from lime and oak woods through to John Taverner at Tattershall in Lincolnshire or Thomas Tallis at the Chapel Royal is altogether too fanciful.

FRESH WATER AND FRESH-WATER WETLANDS

Here we look at the streams and rivers, the lakes and ponds, and the lowland peat bogs which are constituents of the environment, essentially small in area, though not necessarily in importance.

The river has an important place in any environmental history, for it is not only a source of water and a route of trade but a fountain of metaphor as well, central to forming environmental attitudes and sensibilities.

> Along the shoare of silver streaming Themmes
> Whose rutty Bancke, the which his River hemmes
> Was paynted all with variable flowers

wrote Edmund Spenser (?1552–99) in his *Prothalmion* (11. 11–13), setting the ideal against which the Thames has been judged to this day. But most of the human activities in this period, though not directed against the flowers, have tended to diminish them in the cause of the increase of rutty banks. Most of the rivers' uses have required some diversion of the early channel. Fishing was often intensified by the use of fish weirs and fish-garths which trapped or corralled the animals. There were 40 such weirs on the Shropshire part of the Severn before 1700. Weirs could not often span the whole river, so that an island might be constructed to divide the weir from the navigable part of the stream. A mill, too, might require diversions in the form of a weir to hold up a good head of water or else a stream might have to be led from higher up to fall onto the wheel; such a channel when parallel to the main river is called a leat. Other mills might need the construction of reservoirs on tributary, non-navigable, streams in order to pond back enough water in a reservoir to power a forge or a set of ore-crushing hammers. In the Sussex Weald there are numerous 'hammer ponds' of medieval date.

Rivers carried much of the internal trade of Great Britain before canals and surfaced roads. Some 1100 km of navigable rivers existed in the period 1600–60, increasing to 1500 km by 1700. Thus there were incentives to improve river navigation which meant at first simply dredging, though that failed for instance to keep Wainfleet (Lincolnshire) in the premier division of English ports after *c.*1600. An integrated system of sluices, weirs and locks became important: on 40 km of the Wey in Surrey in 1651–3 were constructed 10 locks, 4 weirs and 11 km of straight new channel (a 'cut') as well as 12 new bridges. This followed hard upon the first locks in Britain, built on a 4-mile canal between Exeter and the river Exe at Topsham. The locks were of the mitre-gate pound type, already known in Italy for some time. Locks are interesting in the sense that they divide the river into sections: a kind of fragmentation.

One effect of lock-building must have been to slow the rivers down. Hence, silt loads acquired upstream were more difficult to flush out and the need for dredging increased. The fauna, however, might change in favour of species of slower and deeper water: the trout giving way to the tench, we might imagine. Slower and more regulated rivers would be better habitats for birds nesting in marginal vegetation, so perhaps moorhen populations were beneficiaries; mute swans almost certainly so. This assumes a clear river but we know that many towns had ordinances requiring the butchers and tanners, for example, to dispose of their wastes at particular spots on the river bank: the tanners were usually downstream of everybody else. The householders of Berwick-upon-Tweed in 1249 were required to dump their household wastes in the river and not in the streets: happily, tidal scour would be an aid to civic pride.

Where a body of still water was required, then ponds might be built. The fish pond is the best example, often stocked with eel (*Anguilla*), pike (*Esox*), bream (*Abramis*), perch (*Perca*) and tench (*Tinca*) from the native fauna, and the carp (*Cyprinus carpio*), which was introduced before 1496. Medieval Catholicism was a great spur to fish production and the pond's heyday was that of the great monasteries, declining somewhat after the Dissolution. Ponds were flat-bottomed and only about 1 m deep. Ornamental ponds gained the goldfish (*Carassius auratus*) by 1691. A pond's function might be combined with that of a moat. Originally a defensive device, it became a status symbol for country houses between 1150 and 1325 and was often fed by a diverted stream. They are so common in parts of Arden (Warwickshire) that they point to a commercial activity.[68]

Waterlogged environments may accrue peat, and many such wetlands accumulated during post-glacial time, often developing from an open stretch of water which became smaller as the peaty margins grew inwards. A few even acquired a floating mat of bog-moss of the genus *Sphagnum* above several metres of rather murky water. Before 1700, only a few of these with open water had been converted to crops, but many peaty hollows between drumlins or other glacial deposits were drained in order to add to pastures. Drainage at the margins also characterised the early uses of some of the larger lowland mosses, such as Whixall Moss in Shropshire and those of Lancashire, the southern Lakes, the Solway lowlands and Ayrshire. These were **raised bogs** in their natural state and were a source of peat fuel and wildfowl. The areas largely stripped of peat might have some of the remaining organic matter worked into the sandy soil beneath and, with added manure, the new soil could be worked.

Few streams seen today are anything like wild. Apart from upland rocky streams (and not all of those, if there was mining in the vicinity), any watercourse is likely to have been diverted or had its flow or morphology altered since Roman times or even before.[69] Looking at the attempts at flood control, at various scales of drainage and of water management generally, it seems as if medieval and early modern Britain was a wetter place underfoot no matter what variations in rainfall there were. The wellington boot, as we now know it, had to await the coming of vulcanised rubber technology.

RURAL SETTLEMENT

The environmental linkages of villages, small towns and hamlets converge on the supply of basic needs. The local surroundings would also provide some grazing for, for example, pigs, but other animals such as cattle and sheep might be subject to forms of transhumance: up hills or down into fens are both possible. People's livelihoods might also be tied locally: the service of the manor required a man's presence unless it had been commuted to a cash payment. Service in a Royal Forest or a Chase needed actual attendance and might indeed determine the employment mix of a village, with wardens, verderers, foresters, rangers, woodwards and agisters.[70] But environmental determinism must be avoided, for there were areas of rural specialisation and trade between them, all helped by the money economy introduced in the late Iron Age.

Environmental connections might not, however, be shown in the morphology of rural settlement, for a straggle of buildings along a road could have essentially the same relationships as a planned village around a green. The latter is likely to exhibit the power of men rather than of nature. But the siting of a settlement within a parish may also show an environmental awareness, as in those parishes in Surrey that straddle Chalk upland and clay vale, or those of Norfolk and Lincolnshire that stretch from dry land through fen edge to the summer-only pastures of then unreclaimed water-land. Rural settlements are not necessarily permanent. After the Black Death and the subsequent move to sheep, many villages were abandoned and some remain thus, as humps and hollows in grassland, as reversions to a more 'natural' state.

URBAN SETTLEMENT

In 1600, about 20 per cent of the population of England was urbanised; the proportion was lower in Wales and Scotland and by 1700 the proportion of males in the primary occupations of farming, fishing and forestry was about 40–45 per cent. Of the urban population, the greatest single concentration was always London. It had about 20,000 inhabitants* in 1200 and 40,000 by 1340, growing to perhaps a quarter-million by 1600 after suffering a decline to about 37,000 in the second half of the fourteenth century. In 1700 it held 60 per cent of the country's entire city-people. Other cities were always smaller but still gaining in size and population through the years from 700 to 1700. In the forefront of the second rank in 1600 were Norwich

*About the size of the crowd at a mid-week First Division league football match in the UK.

(15,000), York (12,000), Bristol (12,000), Newcastle upon Tyne (10,000) and Exeter (9000). A typical medieval market town, though, had only about 60 households.

A city of 250,000 (such as London in 1600) is the site of much material input and storage. In outline, a pre-industrial city provides a market for near-at-hand produce such as fruit and vegetables, meat that can be brought on the hoof from some distance, and a nucleation for timber, fuelwood and stone. The demand for fuel for domestic heating and cooking as well as the civic versions of these and for the industries of the city, such as brewing, meant that coal was a constituent of the fuel mix by the thirteenth century in London. Yet all these cities were small enough to walk out of into the countryside and to get most of their water from wells within the gates, though that water was easily fouled by sewage. The morphology reflected the site to some extent, especially if there were strategic considerations but once launched economically, disadvantages of terrain were less important than the maintenance of economic and perhaps political power. Ports, for example, were subject to all kinds of construction works and dredging on banks and estuaries to keep them open even when coastal silting and bigger boats combined to threaten them with obsolescence.

The city was not necessarily a major industrial site, as was the case after the Industrial Revolution, but any material input brings the production of wastes. Rubbish in medieval towns was subject to laws and ordinances from at least the twelfth century, with London taking the lead.[71] Most bye-laws pertained to public places rather than private property and related to the disposal of human and animal wastes. Sewage went into private cesspits but they soon undermined other property and contaminated water supply by seepage; nevertheless this was held to be an adequate disposal system until the nineteenth century. Animal refuse was another major problem: there was dung and straw from stables, especially those of the inns, the corpses of dead semi-feral pigs and dogs, and the refuse from the butchers. The meat purveyors slaughtered their wares near to their stalls so that blood and guts fell into the street and so had to be made to take unsaleable material to special places by the riverside to chop it up into the water; the fishmongers had the same problems. Dyeing and tanning produced noisome wastes as well and the latter were usually forced into the furthest downstream position of all. Households produced wastes other than excreta: rushes and straw from floors and food preparation wastes were thrown out. It was usually forbidden by a bye-law to pitch this into the street or into a market space but no doubt amnesia often took over and these materials mingled with the sludge of the unpaved streets. Surface drainage was normally inadequate and so mud was a constant element of town life. Within the city, one environmental impact was often the disappearance of small streams into culverts as properties came right to their edges and so had to be protected against flooding: few of the 22 rivers of London are visible within the central area today.

Overall, the town provided a good habitat for scavengers such as pigs and the red kite (*Milvus milvus*), for parasites such as fleas, and above all for the black rat (*Rattus rattus*). This rat was probably a Roman introduction which seems then to have virtually disappeared until the ninth century; it probably suffered from the collapse and desertion of Roman cities in Great Britain.[72] The common shrew (*Sorex araneus*) often appears in deposits from medieval towns along with rats and mice. Attempts to clean up were made by many civic authorities from time to time, for putrefaction was associated with disease. Streets might be dressed with layers of flint or limestone to keep them dry until covered with mud and muck once again and the process had to be repeated: archaeologists sometimes find a kind of club sandwich of such layers. Then as now, a royal visit called for some cosmetic action: Winchester in 1330 and York in 1332 had to spend money on doing up the streets.

By 1700 a few urban centres were beginning to specialise as spa towns, synonymous with cleanliness. The great development of the environmental resource of mineral-laden water was yet to come but mineral springs in towns were becoming increasingly frequented during the late sixteenth century. The Reformation had closed many holy wells but their curative properties were put into a secular frame so that Bath and Buxton became well known for their waters from the 1570s. Something of a mania developed in the 1580s and many, like branch lines on a railway, had a brief life only.[73] But Hotwells in Bristol remains, in name at least, as a reminder of the rush to find suitable waters for drinking and bathing.

Cities became big enough to develop an urban climate. Under calm conditions, a late medieval city could develop a **heat island** of 4°C and the higher temperatures allowed town vineyards to persist into the Little Ice Age.[74] The warmth came from burning fuel and this in turn affected the atmosphere. Smoke reduced visibility and produced fogs and these increased in step with the use of coal in cities. Damage to furnishings was noted in London in 1512 and the corrosion of buildings by 1661. Cities that used coal would have a sulphur dioxide concentration which was 20–30 times that of the surrounding rural areas. Table 4.1 shows the inexorable rise of such pollution in London. Disamenity, poorer health and damaged plants were inevitable concomitants. In his Diary of 1661, John Evelyn wrote that at Court in White-Hall, 'Men could hardly discern one another for the Clowd . . .' and that the smoke from coal caused '*Catharrs, Phthisicks, Coughs and Consumptions* [to] rage more in this one City, than in the whole Earth besides.'[75] The onset of complaint about air pollution in London in the later sixteenth century followed a period of quiescence that had lasted from the late fourteenth century, when the more familiar and less toxic wood fuel was sufficient for the urban and national population of those years.[76]

TABLE 4.1 Air pollution in pre-industrial London

Date	City width (km)	Population (thousands)	10^3 tonnes of wood fuel	10^3 t of coal	Smoke in $\mu g \, m^3$	SO_2 in $\mu g \, m^3$
1300	2.0	40	40.0	0.6	17.0	1.5
1500	2.5	70	70.0	10.0	25.0	17.0
1650	3.0	300	0.0	200.0	43.0	256.0

Source: adapted from P. Brimblecombe, 'Early urban climate and atmosphere', in A. R. Hall and H. K. Kenward (eds), *Environmental Archaeology in the Urban Context*, CBA Research Report 43, London: CBA, 1982, pp. 10–25.

INDUSTRY

There was a great number of dispersed sites where minerals were extracted and refined, where wool was processed and salterns built, to quote a few examples. In many cases the materials were too bulky to be easily transported except by water and so the sources of the materials was also the site of production and of environmental impact. Yet the pull of the markets (especially the cities) became stronger with time, so that London was one of the world's leading industrial cities by 1600. But to some extent, industry in Great Britain was controlled from outside the nation, another familiar cry to us, though the Flemings and the Italians are not now the targets of the nationalists.

Organic products

Before the days of cotton (not significant until after *c*.1770) textile processing was dominated by wool with its associated operations of fulling, spinning, weaving and dyeing. The environmental linkages of wool production are not obvious but, as was shown for moorlands, the close nibbling of sheep can produce vegetation change and the short swards of chalk downland were very much produced by large flocks of walking fleece. The processing of wool needed mills and controlled water supplies for dyeing and fulling. It also produces wastes but, except in concentrations such as around Stroud (Gloucestershire), the contaminants would soon have been dispersed. At one remove, the demand for alum as a mordant (a substance which fixes dye in cloth) and for dressing leather led to an extractive industry along the cliffs of North Yorkshire on either side of Whitby; the alum stone was smelted in brick furnaces fired by coal (mostly brought by boat into small 'holes' blasted into the rocks at the cliff-foot. Kelp ponds, water ponds and wood ash stores added to the developments, which gave the appearance of burning cliffs when the shales were being fired. Local incomes were supplemented by selling urine to the alum works: that of the poor was preferred since they drank only ale and small beer, not strong drink.[77]

The popularity of leather for all kinds of articles from clothing to buckets placed quite severe demands on the sustained production of oak-bark for tanning: the hides were soaked for months in pits containing a liquor of shredded oak-bark and water. It was noisome and contaminated streams considerably when released.

Inorganic products

The dominant theme of industry in this sector must be iron. Given its many uses, its ability to be fashioned and re-fashioned (at least 10 per cent of the stock was recycled material in the seventeenth century),[78] it was an essential material in the economy. Produced in quantity in Roman times, the medieval industry was located notably in the Weald, the Forest of Dean and Cleveland, moving by 1600 to concentrations in West Yorkshire, South Wales and the West Midlands. The environmental significance of the wastes from iron-working can be judged from a Roman example at Laxton (Northamptonshire) where a cluster of furnaces has slag tips extended for over 400 m around them and a 100-m wide valley is filled with iron slag.[79] In the Lake District, centuries of charcoal production along the shores of Coniston Water have left pitstead remnants every 100 m or so in some of today's woodlands and there is evidence that juniper and alder were encouraged in coppices which supplied the gunpowder industry. The workers lived in the woodlands and so plants of nitrogen-rich soils such as nettles (*Urtica dioica*) still form stands where the packhorses were once tethered.[80]

The early technology was that of the bloomery, where iron ore was heated with charcoal, the air being forced through by bellows. Water was used to power the bellows after 1350 and after 1490 to drive the hammers which beat the iron 'bloom'; the first recorded instance seems to have been the Bishop of Durham's bloomery at Bedburn in Weardale. Charcoal was used as a fuel, under a system of highly regulated woodland management practices.[81] In Britain generally after 1500, the blast furnace was brought in, which also needed water-power. Unquestionably, the focus of the economic historians has been on the fuel supply. Earlier generations of scholars tended to suggest that iron-working produced deforestation and hence the move to coal as the major fuel. Rackham, on the other hand, has adopted the contrary argument that it was largely agriculture which fragmented the medieval woods and that iron-masters were much in favour of the sustained production of charcoal, of which their chief source was the coppice.[82] Earlier data for a Roman military ironworks in the Weald allow an unusually accurate estimate of output, namely that of 550 tonnes per year of iron for 120 years. Such an installation could be supplied by 23,000 acres (93 km²) of coppice. The management of woods for this kind of output is suggested by documents like that of 1237 in the Forest of Dean which called on a landowner

> and to sustain those forges . . . with thorn, maple, hazel and other underwood; so that no oak, chestnut or beech be cut down . . . to be . . . fenced lest any deer or other beast be able to get in to browse these[83]

and the later statement by the polemicist ANDREW YARRANTON in 1677 who talked of the planting of new woods in Wyre Forest (Shropshire-Worcestershire), 'knowing by experience that the Copice woods are ready with Iron Masters at all times'. In spite of the management of coppice on renewable rotations, there were national regulations on the use of wood for iron-smelting in 1558, 1581 and 1585, with the last of these carrying restrictions on the building of new iron mills as well. Given the concern about the renewed use of coal in London in the sixteenth century and the use of coal in iron-smelting from *c.*1600, it seems as if either the wood supply was no longer sufficient or that coal was becoming much cheaper anyway. So while we associate coal with later times and above all with the nineteenth century, its earlier use became so common that charcoal was little used in the iron industry after 1800. Indeed, some coppices in the Midlands were either grubbed up or converted to high forest after 1760; others survived by finding alternative outlets such as ash-poles for hop production.[84]

There were many small bloomeries in medieval and early modern Scotland. Bog-iron and bands of iron-stone in Carboniferous strata were the raw materials from Kintyre to Sutherland: the east side of Loch Lomond has a concentration of them. At Letterewe and Red Smiddy on Loch Maree are the remains of the first blast furnace (1610) in Scotland, which used ores brought from Fife and from Cumberland by water. As in England, there was a perceived threat to woodland from iron-working and regulation was threatened in a statute of 1609. As late as the eighteenth century, though, 4500 ha of coppice oak was still fuelling a furnace at Bonawe on Loch Etive over a 20-year rotation period. Although the Highlands had only 8 per cent woodland cover in c.1750, the demands for fuel and for tanning bark helped conserve oak woodlands against grazing animals, since the trees were usually enclosed as well as coppiced.[85]

Other metals were mined and refined. Lead was noted in six places in Derbyshire as early as Domesday Book and it was mined in the northern Pennines and Mendips as well. Alston Moor in the Pennines was also a source of Royal silver from c.1170. Primitive drainage pumps might be powered by treadmills and the water shed to the runoff inevitably was contaminated with lead and other heavy metals, especially if refining was carried out on the same site. The bulk of the wood needed to smelt lead in Swaledale (North Yorkshire) in the seventeenth century came from lowland Cleveland and Durham since for various reasons coppice was not well developed there.[86] Remoteness is no guarantee of lack of impact: the mobilisation of lead, copper and zinc and their deposition in loch sediments has been detected on Islay in the Inner Hebrides from c.1367 onwards. This is a consequence of a history of lead mining: an eighteenth-century account records one thousand 'early' workings in the north-east of the island.[87]

Tin was regulated by Stannary Courts, of which there were four in Devon and five in Cornwall but mines in the former county were exhausted by 1400 and mining had gone underground in Cornwall by 1600. The earlier source had been river alluvium and the longitudinal 'streams' of gravelly waste can still be seen in the river valleys of Dartmoor, as can the remains of Lydford Castle where offenders against the Stannary laws were imprisoned. An example of economics outweighing environment can be seen in the case of copper: German miners were brought in to the Keswick area of the Lake District in 1560 to develop the industry. Its main market was London and when continental supplies became cheaper the indigenous supply was competed out by 1700. Stone was another commodity which needed water transport if its use was not to be merely local: abbeys built of Caen stone were not rejecting British sources on anything but the grounds of cost. Barnack stone from eastern Northamptonshire was widely used partly because of its proximity to water transport via the Welland; likewise, sea transport made it possible to take Portland stone to Westminster in 1349. Otherwise it was the cart and perhaps from 2 to 3 km distance downriver, as at Durham, to a rather soft sandstone ill-prepared, as it happened, for the ravages of the nineteenth century.

Several monastic establishments of north-east England were producers of salt. In this, they joined in a major industry, for the demand for salt was a major economic force. It was partly satisfied by imports in the thirteenth century and by the development of rock salt mines after 1670, but mostly it came from coastal salterns. The main method of production was to collect wind-dried, salt-encrusted coastal mud, extract from it a brine and then evaporate the water using peat as a fuel. (The strength of the brine was measured by the depth to which a hen's egg sank in the fluid.) The Wash coast of Lincolnshire was a centre of the practice from at least late Saxon times until the early fifteenth century (Fig. 4.9), and between Wrangle and Wainfleet the mounds of waste up to 3 m high created a new landscape element, locally identified as 'The Tofts'. A strip of land some 10 km long by 500 m wide is easily spotted on the ground or from the field patterns on the 1:25,000 OS map.

Recycling

A characteristic of pre-industrial societies, then as now, is to re-use materials rather than throw them away. In the Middle Ages and early modern period, much recirculation went on.[88] Clothes were a good example: there was a brisk market in used

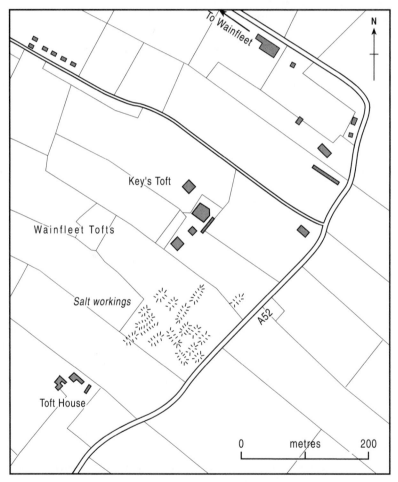

FIGURE 4.9 *Part of the 'tofts' along the Wash coast of Lincolnshire shows evidence of salt working and the field patterns allow the inference that this slightly raised land between the Fen and the sea was in fact largely created by the debris from the boiling of salt water and mud.*
F. Macavoy, 'The marine salt industry', Fenland Research 1, Part 1, 1984, pp. 37–9, Figure from p. 38.

garments and cloth. At the harder end of the market, buildings such as mills and monasteries readily found other uses when disused, if only as temporary quarries. Most metals were used over and over again: this was especially true of iron, lead and pewter. Over half the marketed items of pewter, for example, were made of recycled metal. Paper was made of rags, initially of linen and then of cotton and linen and a decent living could be made collecting them for re-use as paper, since wood-pulp was a later development. All production depended on such re-use, so that export of rags was forbidden in the late seventeenth century. Glass, too, was re-used and there are records of shipments of **cullet** from Yarmouth to Hull and Newcastle in the early eighteenth century. The incidence of recycling generally declined as production levels of almost everything soared in the eighteenth and nineteenth centuries, a situation which we have barely begun to change except in the case of metals.

General observations

The most obvious observation is that almost everything industrial started out earlier than most people think. We have accumulated a perception of the eighteenth century and onwards as the period of the Industrial Revolution and forget the many earlier developments. True, they and their environmental consequences were small scale compared with what came later but they developed a knowledge of the whereabouts of raw materials, of the uses and drawbacks of technology and of the institutional and financial needs of these economic activities. So while 'revolution' in hindsight is not a term to be dismissed, it came as an acceleration of an already turning wheel rather than a 0–60 in eight seconds affair.

PLEASURE AND THE ENVIRONMENT

There was no lack of response to nature by the literate and the artistic during this period and some monasteries have left structural evidence in the form of holiday-houses away from the main priory, as with Durham's recreation house at Finchale, a few miles downriver. The larger proportion of the population have, in this instance, left few direct observations. Since all hunting and gathering within the manor was by licence or enforceable custom, then perhaps poaching might count as a recreation. In fact, a great deal of poaching was carried out by the lesser gentry rather than the peasants.[89] For the well-off there were the pleasures of gardens, parks and the hunt.

Gardens

The castles, great houses and monasteries of Europe all possessed gardens but little is known of the medieval examples in Great Britain, except for those of the monasteries, which were mostly devoted to culinary and medicinal herbs. We hear of the later knot garden introduced in the early 1500s, where intricate patterns of low box (*Buxus sempervirens*) hedges were filled with herbaceous plants or, in later years, with coloured earths and brick dusts. (Were it not for its Italian origins, the knot garden might be seen as a representation of the enclosure of open fields proceeding at that time.) Native flowers were brought into cultivation: foxgloves, violets, primroses and daisies among many others. To this flora were added introductions from across the sea: damask and musk roses, the Madonna lily and the hollyhock were novelties in Elizabeth's time. Topiary was another Italian art which arrived (via France) in the sixteenth century. Yew (*Taxus baccata*) was the favourite shrub for it grew slowly but closely. It could be elaborated into a maze which might be either low-growing or allowed to be so high that getting lost was possible. Fountains became popular but never on the continental scale. In every design, however, small arbours, garden houses and summer houses were frequent: presumably they were often more private than the big house which at this time lacked the corridor so that virtually anybody could pass through almost any room at any time.[90]

From 1550, stately gardens under the windows of great houses dominated the iconography of such pleasures. Henry VIII was a trendsetter (as indeed in other matters) with his 2000 acres (810 ha) at Hampton Court containing two deer parks as well as formal gardens. The hunting demands of monarchs meant that Hyde Park and St James's Park, then on the edge of London, were stocked with deer in 1536 and 1532 respectively. After 1660, the influence of French ceremonial gardens became stronger, though nothing was attempted on the scale of André Le Nôtre's park at Versailles. Most of the late seventeenth-century formal gardens of Great Britain vanished under the re-formations of the eighteenth-century landscape gardeners, though a good idea of their nature can be gained at Melbourne Hall in Derbyshire and Moseley Old Hall just north of Wolverhampton. Happily, we have retained some of the botanic gardens established to nurture exotic plants from overseas.[91] That at Oxford was laid out to the same plan as the first European garden at Padua (a layout no longer retained at Oxford), being founded in 1621, opened 1632 and completed in 1690.

Parks

By way of demonstrating ownership, the lord of a manor might enclose an area with an oak pale or an earthen bank and devote it to keeping deer: red deer were an essential possession and deer leaps were erected in the boundaries which allowed wild deer to get in but not out. There might also be a herd of the smaller and highly decorative Fallow Deer (*Dama dama*) which had been introduced from the continent some time before the twelfth century. The first deer park is known from a will of 1045 which mentions a *deerhay* at Ongar in Essex.[92] At their height in the fourteenth century, there were about 800 parks in England and 30 in Wales. In the Weald, 46 in 1300 had become 68 by 1350; in 1498 Wilstrop in Yorkshire was depopulated to make a park and this resulted in a riot in which hedges, orchards and 100 walnut trees were burned down. Regrettably, it seems unlikely this was the origin of the adjective 'stroppy'. At the end of the sixteenth century, some 10 per cent of the Weald was emparked and the Chilterns had 69 parks. By that time, the hunting function was often secondary since the parks had acted as reservoirs of timber and sales of timber were then highly profitable.

Hunting in these parks was more restricted than in Forests and Chases. The deer were nearly always driven to the hunters and indeed might be corralled with fences and nets in order to be shot from a safe distance with bow and arrow; even ladies might participate. Small fenced holly (*Ilex*) woods were sometimes maintained to provide winter food for the deer. The parks were often the scene for falconry as well, though that sport was not confined to them. Trained birds of high status were mostly the stooping falcons which killed in the air, of which the peregrine (*Falco peregrinus*) was the favourite. Short-winged hawks like the goshawk (*Accipiter gentilis*) and the sparrow hawk (*Accipiter nisus*) were flown at ground targets such as rabbits or pheasants, which were both introduced species. Women might be allowed one of the smaller falcons such as the merlin (*Falco columbarius*). It seems probable that the populations of neither predator nor prey were much affected by this sport.

Forests and chases

Even before Domesday Book was compiled, the kings had hunted in personal game preserves.[93] A great extension of this privilege came with Norman rule, for William I, it was said, loved the deer as if he were their father, which seemed not to prevent him killing large numbers of them. The designation by decree of a tract of land as a Royal Forest meant that the king had control not only of all the hunting but also of tanneries and forges, and first call on any cattle and sheep depastured there, as well as owning any timber. The term, however, meant that there were deer and not necessarily trees. The Forest was nevertheless a resource-laden asset which could be sold ('disafforested') when money ran short: King John for example disafforested Dartmoor to its commoners for cash. He might also confer hunting rights on others and the land specified was called a Chase.

The high tide of afforestation was in the late twelfth century, when 70 Forests and Chases were declared, covering at least one-fifth of England. In 1327–8 there were 71 Forests, a diminution of one-third since 1250. Their distribution (Fig. 4.10) was uneven, for about one-third of Wiltshire was covered but none of East Anglia. The management of the Forest was intensive, for the deer were the king's beasts and so poaching or any other interference was subject to heavy penalties. A very similar set of laws and practices was transferred to Scotland by 1153.[94]

By the end of the sixteenth century, the Forests were in decline. The monarchs were less fanatical about hunting and the greater value of their area as timber reserves took precedence, for in 1608 it was calculated that there were 580,000 valuable timber specimens in the Royal Forests. The Civil War resulted in a great deal of damage and in 1651 there was a plan to vest all the Forests into the hands of trustees. The remnants are still to be seen in names on the map, in fragments such as Epping Forest, and in the best survival of all the New Forest, which once had 30 to 40 villages under forest law, and still has its verderers, its purlieus (land disafforested by order of Magna Carta after ill-received expansiveness by the Plantagenets) and its laundes (hence the modern word 'lawn') or grassy glades. In the whole country, however, so much land was never again placed under restriction by a national authority, not that the National Parks Commissioners after 1949 saw themselves as being quite the same as the Norman kings.

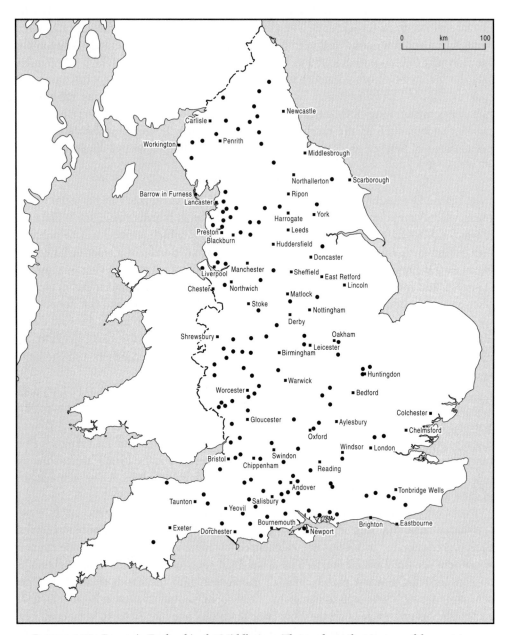

FIGURE 4.10 *Forests in England in the Middle Ages. These refer to the presence of deer not trees but their ubiquity is remarkable, with the exception of East Anglia and the far south-west. (The towns are named for location purposes only.)*
N. D. G. James, A History of English Forestry, *Oxford: Blackwell, 1981, maps 1 and 2, pp. 79–80.*

ENERGY

As in all pre-industrial societies, the economy was dominated by the sun. At a low level until perhaps the 1580s, there was also a source of energy from a non-renewable source, that of coal. This mineral came to great prominence in later centuries but it underlay some of the nascent pieces of industry which characterised the early modern period.

Coal

Used in Roman times, coal appears to have been rediscovered at the end of the twelfth century and its use expanded in the thirteenth century. Until 1500 nearly all consumption was local to the deposits except for the export trade from the Tyne, especially to London. Extraction was from shallow bell-pits or from adits into hillsides which could use gravity for drainage. Commonly, bell-pits were less than 30 m deep and worked by 5 men, who could produce 1200 to 1400 tons per year. Little machinery was used (though water was taken out by a horse-powered pump as early as 1486 in County Durham) but a great deal of timber was needed for shoring and for access ladders. By the fifteenth century, most of the major coalfields were in production but only that of north-east England had any significant non-local use. In 1700 the total production in Great Britain was *c.*2.5 million tons per year. In colliers of 75 tons burden, some 40,000 tons per year was exported from the Tyne; CELIA FIENNES rode that way in 1698:

> this country all about is full of this Coale the sulpher of it taints the aire and it smells strongly to strangers; upon a high hill 2 mile from Newcastle I could see all about the country which was full of coale pitts.[95]

Unless it could be moved by water, however, coal was too expensive where there was any wood at all; equally, its greater development awaited the steam-powered pump. An intermediate stage from the 1550–1600 period in the Tyne-Wear region used a lined air tunnel and chain-driven force pumps to allow mining often down to 120 ft and occasionally to 300–400 ft. One mine could then produce 25,000 tons per year.

Apart from the effects of the burning, this type of coal mining produced a pockmarked landscape (mostly now built over by later developments but occasionally causing a structure to slump into its remnants), and any drainage water would have been quite acid and also full of particulate matter, so that the quality of runoff water declined markedly. A secondary environmental linkage could be seen in the uplands of Scotland south of Edinburgh (for example, the Moorfoot Hills) in the 1620s. The acid soils could be cultivated if lime was worked into them and this could be produced by burning gypsum with coal. The limit of cultivation could then be pushed up to 180–220 m OD. Higher energy input in this case extended the agricultural area.

The development of coal output was one response to an energy crisis in the seventeenth century. The price of charcoal doubled in 1630–80 and a lot of energy was imported in the form of Swedish iron. But the expansion of coal output solved the problem by c.1680, aided by the popularity of woodland plantations, which had been rising since the Restoration. Coal was the major fuel in lime-burning, brick-making, glass manufacture, the production of salt, soap and alum, and in malting, brewing and sugar refining (though not in iron until the eighteenth century) and it reached its full transformative potential only in the nineteenth and twentieth centuries. Nevertheless an astute observer in c.1725 might well have seen the way things were going and, like Johann Sebastian Bach in Thuringia, bought a share in a coal mine.

Peat

Most areas of blanket peat and lowland peat bogs large and small were exploited for peat, which was a cheap fuel. Peat is mostly water, so that a laborious digging time was succeeded by a prolonged drying period, but there was no competition for its use as in the case of woodlands and thus manorial lords might be more ready to grant rights of turbary to commoners. The most spectacular result of peat-digging in Great Britain is without doubt the Norfolk Broads. These are simply 'holes left by a huge industry of peat digging'.[96] In the thirteenth century, Yarmouth and Norwich were large towns but in a region without much woodland. The deficit was made up with peat: in the 1270s South Walsham earned as much from peat sales as from farming and in the region generally the sales were of about 12 million tonnes per year. This all came from a series of pits about 3.5 m deep by 1.6 km or more across: Fritton Lake is 5 km by 600 m by 5 m deep. The pits were divided by linear baulks, and similar embankments separated the pits from the rivers. At the time, sea-level was about 4.5 m below its present level but rising and a surge in the North Sea in December 1287 reached nearly all the workings. Further years of rising sea-level killed off the industry. The result is a series of lakes with vertical sides and chains of small linear islets which are the remnants of the baulks; the lakes are now smaller because of subsequent silting and infill from marginal vegetation.[97] A smaller example is to be seen at Kilconquhar Loch in Fife: here medieval peat digging was also responsible for creating a large area of open water.[98]

SACRED SPACE

Since Classical times at least, European cultures contained the idea of space devoted to the gods or God. The Greek temple became the Christian church and indeed after the Renaissance looked back to that model. The sacred grove was almost certainly present in pagan Britain but discouraged as a concept thereafter. But medieval and early modern Britain participated in a religious culture: until the Reformation everybody was a Catholic in a Catholic part of Europe.

In cartographic terms, the actual setting aside of space, other than buildings and their immediate surroundings, is seen to greatest extent in the large land holdings of monasteries, of which the Cistercians are a highly visible example. To them, sacredness resided in solitude and silence, following the example of the Desert Fathers. Thus they were glad to receive gifts of wild space, as with the vills that were described as *wasta*, which were presumably mostly moor, woodland and scrub. Where their gifts were not quite silent enough, then people were moved out to make a kind of *cordon sanitaire* in which only the abbey and its buildings and the granges might persist. Islands too were popular as retreats: Bardsey was a monastery from 'Celtic' times through to Augustinian rule but was worth only £36 at the Dissolution; Caldey was rated at £5 in 1535. Both suggest solitude being valued over production. By contrast, Lindisfarne (Holy Island: Northumberland) housed an agricultural community both in its earliest phase and as an outpost of the priory of Durham.[99] It is interesting that many of the great abbeys, although mostly in ruins, still retain a certain numinousness to this day: except perhaps on a Sunday, visitors tend to lower their voices inside these shells.

The real popular attachment in pre-Reformation times was probably not to land-scapes but to relics, and pilgrimage to one of the major shrines was a serious undertaking for both pilgrims (though, as Chaucer so sharply depicts, not all that serious all the time) and for the receiving communities. Housing the visitors meant particular patterns of urban building and of stop-over hospices on the way and of food production for their needs: an embryonic tourist industry, lacking only the Boeing 737.

SOCIAL CONTEXT

Once writing became firmly established, then our access to the social realm is vastly increased. Whether it be legal systems or abstract theology, the chances of having some insight are much greater than before 700, albeit millions of documents have not survived as neither have most buildings and other landscape traces. We can, however, say some useful things about questions like the role of government, the actual popu-lation changes and the sensibilities of the time.

Population change

Since many references to demographic trends have been made, this opportunity will be created to bring some of them together.[100] The basic population data are shown in Figure 4.11, which demonstrates the course of total numbers from an 1100 popu-lation of 1.5 million through to 9.5 million in 1800. Interruptions to growth came with the Black Death of 1348–9, which exerted an intensifying effect on a decline already under way by 1315 and lasting until 1375. The Plague of the seventeenth century dropped the population from 5.2 million in 1656 to 4.8 million in the 1670s and not until 1721 was the 1656 total passed. (All figures are approximate until the first Census of 1801.) The growth rates varied, as always, with the age and incidence of marriage: in the 1556–61 period, some 6.3 per cent of the population never married, but in 1600–49 this became a startling 24.1 per cent. Occupations showed steadier trends, for in 1700 between 40 and 50 per cent of adult males were in agriculture, fishing and 'forestry', which became 36 per cent in 1810. Even as late as 1780, about half the population was on the land, though this conceals the statistic that, in the 1690s, 30 per cent of adult males were occupied in mining and manufacturing. In the century 1540–1640, the proportion of agricultural labourers who had some other form of employment (for example, in wool, hemp, linen, leather or mining) was as

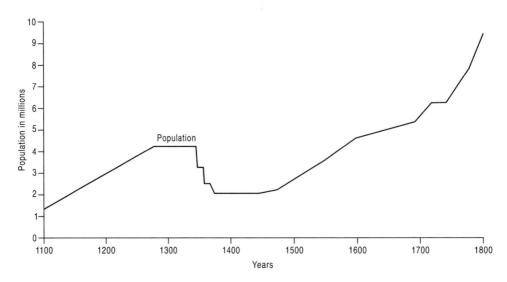

FIGURE 4.11 *The population of England 1100–1800.*
J. D. Chambers, *Population, Economy and Society in Pre-Industrial England, Oxford: Oxford University Press, 1972, p. 18, Figure 1.*

high as 60 per cent. At any one time, one-third of the 15–25 year-olds were servants so that, as feudalism broke down, there was a mobile element in the population. The relation between population growth and food supply never became totally Malthusian.

Disease

Data for population growth show the importance of infectious disease, and until the seventeenth century privileged groups in society did not live any longer than anybody else. The sources of everyday mortality are well known. No more than 70 to 75 per cent of births survived to adolescence, being killed by diarrhoea, dysentery, measles and other fevers. Influenza affected all age groups, as did tuberculosis, which caused 16 to 28 per cent of deaths in rural areas and 20 to 24 per cent in the towns. Third in incidence was smallpox, which arrived during the reign of James I (1603–25) and which normally killed 5 per cent of the population but in epidemics rose to 18 per cent. In wetland areas, malaria ('ague') came later in the early modern period to affect 47 to 71 per cent of the local population. Apart from the Malthusian pressures of the late thirteenth century, famine and bad harvests killed fewer people than on the continent though a European mortality peak in 1740–2 included the British Isles. The winter of 1739–40 seems to have been the coldest and longest in modern European history and 1741–2 was nearly as cold; their summers were drought-ridden. More people spent time indoors so that louse-borne typhus and relapsing fever both had a better chance of spread. If the annual death rate for 1735–44 is indexed to 100, then in England in 1741 it was 118 and in 1742, 124, recovering thereafter. There were fewer wars as well. War was not a great demographic factor within Britain but soldiers and sailors who went abroad often died there: only 1 in 10 returned from Cartagena in 1739. For the active upper-class male, death in battle or by execution was a real possibility: between 1330 and 1479, some 46 per cent of peers died violently.[101]

The most stressful disease of all was the plague. Its last appearance in Scotland was in 1649 and in England 30 years later. Episodes often came soon after each other: London was affected in 1603, 1625 and 1665, for example, with 95,000 out of 400,000 dying of it in the 1665 outbreak. Both the Black Death and Plague caused the withdrawal of farming from marginal lands or into practices that needed less labour, like sheep herding; the immediate environmental consequences can still be seen in Green Park in London where the hillocks and hollows reflect the plague pits in which thousands of corpses were buried.

Social control

This work is not a political history and the detail of institutions is not central to it. Nevertheless, the control of access to environmental resources and the disposal of wastes has necessarily been mentioned several times. The medieval manor was an all-embracing resource-allocator; the lord and his courts might decide on the apportionment of arable, the number of cattle grazed on the meadow, the prohibition of selling manure off the manor, pest control and many other matters. As feudalism broke down and was replaced with capitalism, many more decisions became private rather than corporate. The one institution that grew as the manorial system faded was that of the state. During early modern times especially, the notion of the state as a unitary body with interests in resources and environment grew. This is first seen in high profile during the later sixteenth century with the concerns about timber, particularly for the shipbuilding deemed necessary for the defence of the realm.

CONNECTIONS: AT HOME AND ABROAD

As we know from today's less developed nations, 'pre-industrial' does not necessarily mean 'only locally based subsistence'. One theme of this section has been the increasing openness of productive systems to trade and exchange. Certainly, agriculture in the early fourteenth century reflected broad physical controls and exhibited a pastoral west and an arable east. Between then and the fifteenth century, a number of regional economies developed with a distinctive type of husbandry: East Anglia was different from the Midlands, just as the Welsh borderlands were distinct from south-west England. By 1600, 'Regional differences ... were the outcome of more than eight centuries of human effort in reaching a compromise with nature', as Joan Thirsk has put it.[102] Within such a division there were a few areas of specialist production plus some environmentally linked non-agrarian regions such as the fens which produced birds and fish for the tables of several towns and cities. Yet for the seventeenth century, it is possible to produce schemes of farming regions for England and Wales which suggest that soils and topography still had as much influence as markets.

There was also a vigorous foreign trade. We know that Anglo-Saxon merchants traded with Rome, and that in the eleventh century there was trade with Lorraine in fish, pepper and wine. England was drawn into European trade systems in significant measure in the latter part of the twelfth century, with wool as a leading commodity. In the thirteenth century, some three-fifths of production was exported: 33,000 sacks (1 sack is approximately 165 kg) in 1273, much of it to Flanders and northern Italy. In the late fourteenth century and in the fifteenth century the emphasis changed to the export of cloth (only 19,000 sacks of raw wool in 1390).

Almost as important as these flows of goods, there was also the exchange of ideas. In moving towards a discussion of the sensibilities of the age, then perhaps the primary import of significance was the modification of religious sensibilities caused by the Reformation. The long-term consequences in joining Great Britain largely to Protestant Europe will eventually be apparent but for the present we should note that the dissolution of the monasteries released a great number of estates into private hands and with them their wealth, to appear in developments with environmental significance in the seventeenth and eighteenth centuries. But for all these years, the environmental event of greatest moment must have been the Black Death, when the decline in population not only allowed the natural world to take back some of its lost ground but also altered the land cover patterns of great areas.

SENSIBILITIES AND ATTITUDES TO ENVIRONMENT

There is a great span to the environmental attitudes of medieval and early modern times. When forests were plentiful and inhabited by wolves and perhaps bears, then it is easy to see that the woods as providers of resources might have another side as a source of terror. When the country became more 'tamed' then a more relaxed set of views might emerge. Environmentally, the most famous statement comes from Emile Mâle who asserted that the people of the Middle Ages gazed at every blade of grass with reverence. Elaborating this statement, Glacken suggests that the predominant thinking was concerned with a continuous Creation, which acted as a canopy to everything.[103] Under that baldachin, **realism** prevailed: in art and life alike things were as they were and could be changed where advantageous. Exegesis, Glacken's second category, allowed an examination of Creation that turned nature into something which might be either or all of mystic, sacred, symbolic or secular. The Renaissance altered this to some extent. If we follow FRANCIS BACON (1561–1626), then we attain control over nature by cultivating the arts and sciences so as to mitigate the effects of the Fall. But experiments should be like the Book of Genesis and create light rather than profit, that is, derive from a time before the Fall. The predominant view in early modern times, argues Sir Keith Thomas, was taken over from the Greek **Stoics** who thought that the Earth had been designed for humans alone. In a 'breathtakingly anthropocentric spirit', the Tudors and Stewarts were sure that domestic animals were there to be labour and wild animals to be hunted.[104] The animals were to minister to humans: to be useful, aesthetic or moral. Even weeds and poisons had a purpose, for they exercised ingenuity; eating a lobster provided exercise in itself. The discriminating criterion was that of reason and therefore the possession of a soul, something which animals certainly lacked, as quite possibly did women, according to more than one theologian in the late sixteenth century. Human domination was hence to be complete even if it was dressed up as stewardship. There were some counter-currents. The idea and language of stewardship made it more difficult to indulge in slaughter for sport, and local laws were enacted in Chester against bear-baiting in 1596 and elsewhere against, for example, cock-pits in 1654, though alas this practice was still alive in County Durham in 1994. The objections brought by the Puritans were often sustained by Dissenters and Quakers after the Restoration. The hunting of deer, however, persisted because of its aristocratic and clerical connections.

In many ways, agricultural production was the site of much advocacy about the relations of society and its surroundings. From c.1700, certainly, there was a feeling that the natural world had been stabilised and would not experience dramatic changes like the Flood any more. Early geologists confirmed this about a century later. So the world was there to be developed and there flowed a series of texts on gardens, orchards, drainage, soil fertility and the reclamation of wastes. Soil was a treasury in both spiritual and material terms. The Flood, for instance, had dissolved the soil and thus it was the job of the husbandman to restore its virtues by his own labour. On the other hand, soil might be a repository of chemical treasure whose profits were accessible by the application of reason accompanied by individual enterprise and little disturbance of the social order.[105] The second of these qualities can be seen as contributing to the fragmentation of the non-human world as well, when common resources were assigned to single owners, sometimes under the guise of reclamation from a clearly post-Edenic state.

So there is a great chasm of thinking between *Beowulf* in the eighth century, with no mention of agriculture or even the bee-keeping needed for the production of the mead consumed in such large bumpers but only of a nature where there was:

> rocky broken ground . . . meagre tracks . . . uncertain ways, and beetling crags . . . [and] a dismal grove of mountain trees overhanging a grey rock.[106]

This rather contrasts with the view of John Milton (1608–74) in *L'Allegro*:

> Streit mine eye hath caught new pleasures
> Whilst the lantskip round it measures,
> Russet Lawns and Fallows Gray,
> Where the nibling flocks do stray,
> Mountains on whose barren brest
> The labouring clouds do often rest (11. 69–74)

Wherein we see one old attitude and one new. The old is that mountains were barren, a view which would be changed in the eighteenth century; the new is one of the first uses in English of the word that became 'landscape', which has turned into a major channel for feedback about human-environmental relations.

CHAPTER FIVE

Building Jerusalem:
the eighteenth century

WILLIAM BLAKE (1757–1827) contributed one of the most resonant poems to the English language when in a mere 16 lines he juxtaposed a desirable pastoral past and an unwanted industrial present. 'Mountains green' and 'pleasant pastures' are contrasted with 'clouded hills' and, most famously of all, 'dark Satanic Mills'. No wonder that often since then, people seeking inspiration in the past have often launched with added fervour into the last lines, to the tune added by Sir Hubert Parry[1] in 1916.

> Till we have built Jerusalem
> In England's green and pleasant land.

All this encapsulates the idea of a turning point, which in many ways is a suitable image for the eighteenth century. Though the full development of an industrial economy with its inevitable environmental accompaniments was still some time away, the signs were all there. So while in 1700 the population in England and Wales was probably about 5 million, by 1801 it was 9.2 million: from 1750 the population increased yearly for 226 years. In 1740 the population produced from 10 to 11 lbs (4.5–5.0 kg) of wrought iron per head but by 1790 it was twice that figure. The growth had many causes. Diet was better: by 1750 wheaten bread was an almost universal symbol of a better-fed population which was able to resist some diseases (differentially by social class for the first time), and after 1760 it was common to be inoculated against smallpox. Agriculture and empire began to generate significant surpluses which allowed savings and generated capital for investment. Britain in the nineteenth century soon became the most urbanised country in the world, the one with the fastest growing population and the first primarily industrial nation. The eighteenth century was the junction from which that road was taken with the substitution of inanimate for animate sources of power and the much increased use of mineral rather than organic substances as resources.[2]

THE ATMOSPHERE

From the eighteenth century, we can designate this compartment of the environment as both influencing and being influenced by human affairs. As the use of coal grew the loading of particulate matter and compounds in the air increased. Whereas in 1650 the use of coal in London had been 200.010^3 t/yr, in 1800 it had gone up seven times (see page 151) and the consequent levels of sulphur in smoke and as sulphur dioxide had risen by 2.8 and 1.9 times respectively. Since the Little Ice Age was not over until $c.1815$, the need for warmth in houses clearly added to the increasing demands for fuel. After the disastrous years of the 1680s and 1690s, the climate improved until $c.1740$, after which lower winter temperatures were recorded in the 1750–1800 period. The year 1782, for instance, was a particularly bad year in Scotland, with the temperature falling 12 per cent below normal: no oats were harvested until after Christmas. For eight months in 1783–4, the Icelandic volcano Laki discharged acid gases and particulate material, some of which fell out over Britain, a fact noted by many at the time.[3] These included the diarist and natural historian Gilbert White, of Selborne in Hampshire, who talked of the summer of 1783 as an 'amazing and portentious one, and full of horrible phenomena'. Mists and fogs, damage to vegetation and crops, intense heat and ferocious thunderstorms are all recorded, with most of the events being concentrated in the summer of 1783, when the volcanic eruptions were at their height.

WOODS AND FORESTS

Much of the discussion of the previous period applies to the eighteenth century. The underwood was used as before and coppicing was still a key management practice. One new element came into prominence: a plantation on the estate of a landowner, using a fast-growing and sometimes imported tree to grow a timber crop. In spite of such new plantings (some of which were conversions of existing oak woods), the total amount of woodland in Great Britain fell to a new low, of perhaps 2 million acres (810,000 ha) in England and Wales. Though precise estimates are misleading, most of Highland Scotland was poorly wooded by 1750.

Intensified demands

Today's bikers might be surprised to learn that they had antecedents in a rapid demand for leather clothing in the 1780s. This had the environmental spin-off, as it were, of creating an intensified demand for the bark of oak trees which was needed for tanning leather. Oak was coppiced for this purpose and the last such tannery in Wales closed only in the 1950s. As with most coppice products, a sustained yield was desired and this was usually possible unless the wood were grubbed up for agriculture or converted to a more profitable timber yield on a large estate which could afford to wait for the income. Coppices that could produce pit timber for the growing coal industry were also favoured: they needed a long (25–35 years) rotation period. Such poles were taken by sea from north to south Wales in the period before the railways. This century saw the last great flourishing of coppice before its major product – charcoal – was replaced by coal.

The high profile use of timber (as distinct from underwood) in the eighteenth century is for shipbuilding, though Rackham suggests that in that century more oak was consumed in total by the tanneries than the shipyards.[4] A 74-gun ship would consume about 50 acres (20 ha) of timber, representing about 2000 large (over 2 tons each) trees, about half of which would end up as waste. Areas like the Midlands which still had considerable timber resources largely because of poor transport connections began to contribute to the building of gunships and East Indiamen. Travellers in that region in the early nineteenth century were able to remark on the depletions which had occurred in the previous 40 years.

The demands were met reasonably well, it seems, for the price of oak stayed roughly constant in real terms in the second half of the eighteenth century, thereafter rising most steeply after the battle of Trafalgar. Nevertheless, Commissioners of Royal Forests were appointed to ensure the supply for the Royal Navy and they issued 17 reports between 1787 and 1793.[5] Their Fifth Report of 1798 is especially interesting in its attention to the New Forest (Hampshire) which makes it clear that the proper function of that area was no longer to be a reserve of deer but a store of timber for shipbuilding. It can be inferred from their statements that there was good regeneration of trees even in unenclosed woodland as a result of planting beech and oak on unshaded land cleared of its former tree cover with plenty of holly as a protecting layer.[6] Some of this generation of trees still persist. A preference for denying access by grazing animals to young trees was shown by the Commissioners' desire to remove deer altogether from the New Forest (there were then about 6000 so it was never achieved) and their taking of powers in 1808 to enclose woodlands on a rolling basis to keep out the deer along with cattle, pigs and ponies as well.

In Wales, the eighteenth century saw a shortage of naval timber throughout the Principality. In this case the stated reason was usually said to be the management of the woods for charcoal production to feed iron furnaces, and oak-bark for tanning.[7] Thus the large trees that furnished the straight trunks and branches for 'great timber' were less common; smaller trees might yield the curved pieces and smaller planks which constituted 'compass timber'. Masts had to be sought in Scandinavia or even Russia.

The growth of plantations

After 1825, many species of trees from North America were brought in to enhance the pace of timber production, but their immediate precursors were the plantations of continental European species made during the eighteenth century with the aim of restoring timber as a profitable crop from large estates. There was much conversion of coppice to planted high forest with the height of the transition in 1820–50. The conifers were most popular: larch (*Larix decidua*), Norway spruce (*Picea abies*), silver fir (*Abies alba*) and the Scots pine all had their enthusiasts. The larch was especially favoured in cold and wet places so that, for example, the Duke of Atholl on his estates at Dunkeld and Blair planted 14 million of them in the years up to 1830.[8] (Larch also had the advantage that it was said to be unpalatable to sheep.) On an estate at Holkham in Norfolk, two million trees were planted between 1781 and 1801 but only one in six were oaks, the rest being conifers. In Wales, the period 1750–1825 was a major planting era, with landowners buying thousands of trees from London nurseries. As well as the popular conifers, sycamore was much used for shelter and as a source of dairy utensils. All these plantings took place in the context of 'pleasure, profit and patriotism', in which parks with planted trees, timber sales and the defence of the realm were the pride of landowners, who took great advantage of such conditions to alter the ecology of their estates as at Welbeck, Thoresby and Clumber in the Dukeries.[9]

AGRICULTURE AND GRASSLANDS

A pervasive slogan of the post-1945 period, still to be seen in the streets, has been 'Keep Britain Tidy'. The cultural mood which gave rise to it may well have begun in the eighteenth century, where a new and pervasive quality of neatness, allied to opulence and orderliness, became a feature of the British scene. This was most obvious in the countryside but the towns were often straightened up by Georgian planners as well. The cultural shift occurred during a period when not everything was stable: there were poor harvests in the 1750s and 1760s and complete crop failures in the 1794–1800 years, so that the perception of impending famine was often present even if not actually realised. The new nation formed by the Union of the Parliaments in 1707 was, further, at war from 1701 to 1714 and from 1793 to 1815. The later conflict, against Napoleon, caused interruptions in the supply of imported foods and so had consequences for agriculture. Some of the imports came from an empire which by the 1760s included large portions of North America, India, the Caribbean and West Africa.

An improved husbandry

In many ways, the 1700s saw the consolidation of many earlier changes rather than radical innovations. Ley farming, or convertible husbandry, became almost universal in enclosed land though it had not been altogether absent from common field systems. For a while grassland was usually part of a rotation and hence replaced fallow with a grazing period and much manure was cast onto the soil. Demands on cropland might be heavy, for the price of wheat tripled in the 20 years after 1780. In the pursuit of higher agricultural productivity, new crops were important. The virtues of roots had been known since the sixteenth century but turnips became much more widespread after 1724 (earlier in some regions like East Anglia),[10] and well into the century were followed by swedes. The virtues of legumes meant that clover was often sown, usually mixed with grasses such as perennial rye: there was a whole book on British grasses in 1790. Other crops were encouraged: oil seeds such as flax and rape, utility crops like hemp and woad, and even liquorice.

To maintain the fertility of enclosed land which was not allowed a fallow period took more than, as it were, a ley-off. Many fertilisers were used, of which animal dung is most frequently mentioned. Most organic substances from town and country were used (soot and sheeps' trotters, for instance; animal bones were ground to make fertiliser after 1794) with the apparent exception of night soil from major cities. Horse manure was another matter: Middlesex alone received 150,000 cartloads per year from London. The nascent soil science of the seventeenth century had recognised different types of soil and so the usefulness of marl, lime and chalk on acid but well-drained soils was well understood. The popularity of burnt lime meant that the remains of many local kilns are present in rural areas today. To make lime effective on wetter soils required underdrainage and the possibilities of mole drains were being realised after about 1725. Soils that became drier and more alkaline must have changed their flora, though the transitions seem to have been unrecorded by the naturalists of the time.

Tidied patterns

In the eighteenth century, there were about 4.5–4.8 million ha (11–12 million acres) of arable land, with about 1.5 ha of grassland for every arable hectare. Enclosure in this century was by Act of Parliament rather than by feudal dictat (though the result was much the same for fauna, flora and the tenant farmer) and great swathes of land in, for example, the Midlands were now enclosed which had possessed open fields as late as 1700. Between 1760 and the mid-nineteenth century there were some 2800 Acts of Enclosure in England and Wales, of which half were passed by 1800. The resulting landscape was regular, with large rectangular fields showing their origins in mapping and the surveyors' chains. The various Parliamentary and private Commissioners for enclosure must have reorganised about one-third of the English landscape (Plate 5.1).

The cultivated fields were dominated by cereals, of which wheat was most important as a bread grain, yielding to barley, oats and even rye only in areas of poor climate and soils; the grassland supported sheep and dairy herds of cattle. Regional differences were consolidated: Leicestershire was three-quarters grass, Norfolk two-thirds arable. In spite of wheat prices, a great deal of the better enclosed land, both old and new enclosures, was out to grass in a kind of 'greening' of the countryside. To keep up arable production, the poorer land was converted to cropland with its fertility maintained by a flow of lime, dung and other organic materials so that many cultivated soils could be said to be virtually artificial by the end of the eighteenth century. A momentous landscape change was the use of quickset (that is, alive, as in 'the quick and the dead') hawthorn (*Crataegus monogyna*) hedges as field boundaries, usually with trees set in them. In 1787 the agricultural writer and propagandist WILLIAM MARSHALL said of Norfolk that 'the eye seems ever on the verge of a forest, which is, as it were by enchantment, continually changing into inclosures and hedgerows'.[11] Though it is an error to assume that all hedges are eighteenth-century features of our landscapes (since many are older), this was certainly a stage when the regions in which they were familiar was doubled.[12] Where stone was easily available from local pits or had been dragged out of the soils, then drystone walls might replace the hedge, as in the Cotswolds and in the Isle of Purbeck in Dorset. Another landscape element from earlier times which was improved in the eighteenth century was the water-meadow, which ever more complicated systems of channels brought either water and silt to river valleys or simply distributed water from a higher area to the grassy levels below. The influx of water would ameliorate winter temperatures, and the silt enrich

PLATE 5.1   *A typical landscape of the English–Welsh Marches with the rectangular patterns produced by Parliamentary enclosure or by local consent. Such landscapes encouraged the birds of the hedgerow and of small copses. Here, near Longtown in Herefordshire, animal husbandry has resisted field amalgamation, though in 1997 the lighter strips of set-aside can be seen in a few places.*

the soil,[13] providing a good nutrition at lambing time; the Golden Valley of Herefordshire was especially noted for its system of 'floating meadows'.

Floating or 'warping'[14] was not without consequence, however. In the preparation phases, after the removal of surface brush vegetation and if the soil was not properly broken to admit water or lowered excessively in the break up of its lower layers, permanent depressions would be left in the land post-irrigation.[15] Water was frequently lost to rabbit burrows or in land too steep to be successfully irrigated.[16] During the flooding stage, achieved by opening hatches to natural and artificial watercourses, an excess of flow could entrain too much sediment from the natural channel as well as deprive downstream landowners. Too low a current might allow ponding and stagnation, poisoning the sward beneath and leaving a surface scum post-flooding.[17] Towards the end of the eighteenth century, the practice spread from its southern and western origins, accompanied by widespread field-underdrainage and drainage of wetlands, until common fens and marshes had largely disappeared.

The demise of the common field and the growth of the quickset hedge meant changes for the wildlife. Most hedges contained some timber trees, though the eighteenth century may have been a time of the shrinkage of their number, as well as of the custom of pollarding of, for example, willows. The quickset hedge is not of itself

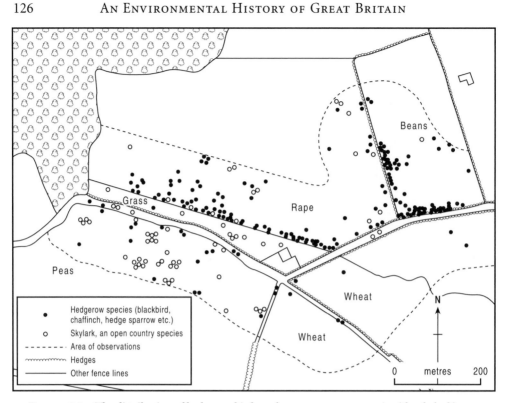

FIGURE 5.1   *The distribution of hedgerow birds and an open country species (the skylark) on a farm in Huntingdonshire in 1966. The attraction of boundaries of at least one other kind besides hedges seems to be shown.*
E. Pollard, M. D. Hooper and N. W. Moore, Hedges, Collins New Naturalist, London: Collins, 1974, p. 120, Figure 34.

a habitat with specific plants found nowhere else, though as noted earlier it might act as a refuge for relict woodland species. The hedges of the eighteenth century, however, were not rich in plant species though they have often accumulated biodiversity since then.[18] A number of bird species seem to benefit from the feeding and nesting potential of hedges (Fig. 5.1) and we presume that their populations increased: the thrush family, including the blackbird (*Turdus merula*) and robin (*Erithacus rubecula*), the wren (*Troglodytes troglodytes*), the dunnock or hedge sparrow (*Prunella modularis*), the chaffinch (*Fringilla coelebs*), yellowhammer (*Emberiza citrinella*) and also warblers like the whitethroat (*Sylvia communis*) may have found even new quicksets a positive addition to their living space (Table 5.1). The expansion of the rabbit from its cosseted warren to being an aggressive pest may be connected with the expansion of hedges at this time. Other mammals to benefit were the mole (*Talpa europea*), shrews, field mouse (*Apodemus sylvaticus*), bank vole (*Clethrionomys glareolus*) and, from *c.*1760, the brown rat (*Rattus norvegicus*) as it spread out from towns. Though the hedge would seem a natural baulk against soil erosion by wind, the areas of England most vulnerable to soil blow in the Fens have never had many hedges until quite recently, when low willow windbreaks have been tried. The eighteenth century saw some quite strong responses from wildlife. Some species were persecuted and remained only in the wilder parts of the north and west: buzzard, kite and raven are examples. The great bustard was hunted and subject to habitat change, but wheatear and stone curlew simply saw their preferred food and nesting sources disappear. The great gainer from enclosure and mixed farming was the house sparrow and many parishes formed sparrow clubs to try to exterminate them, carrying on from Vestry bounties as early as the turn of the eighteenth century. A parallel scourge was the wood pigeon, a relatively rare species until the seventeenth century, but encouraged by deep plantations for nesting and clover leys, turnips and young corn off which to feed.[19]

TABLE 5.1 Common bird species of hedges

Part of hedge	Nest only	Nest and feed	Feed only
Upper branches of hedgerow trees	Carrion crow Rook	Wood pigeon Greenfinch	Blue tit Chaffinch
Trunk and holes	Barn owl Little owl Stock dove Jackdaw Great tit Blue tit Starling Tree sparrow	Wren	Treecreeper
Shrubs	Turtle dove Magpie	Wood pigeon Cuckoo Long-tailed tit Song thrush Blackbird Lesser whitethroat Wren Goldfinch Linnett Chaffinch Bullfinch House sparrow Hedge sparrow	Fieldfare Redwing Mistle thrush Robin Great tit Blue tit Whitethroat
Herbs and low brambles	Whitethroat Yellow-hammer Reed bunting	Goldfinch Greenfinch	
Ground	Skylark	Robin Corn bunting Pheasant Partridge Red-legged partridge	Hedge sparrow Blackbird Song thrush Wren

Source: extract from E. Pollard, M. D. Hooper and N. W. Moore, *Hedges*, London: Collins, 1974, p. 123, Table 13.

The net effects of all these changes was a strong growth in output from 1700 to 1730. After 1741, something like zero growth was the best that pertained, though some areas with easy access to cities (and especially to London) did better. The common element to all the century was an increase in labour productivity, though this should not be seen axiomatically as shedding a surplus of men to take part in the industrialisation of the economy.

Scotland after the Union

The Scottish context must begin with the reports of rapid population growth of agricultural 'peasants' especially in the second half of the century, reaching a maximum in 1841. Increases of 60 per cent were common and from 75 to 100 per cent in the Outer Hebrides. Malthusian clamps were avoided by the introduction of the potato after 1730 but by 1780 several regions seem to have reached their carrying capacity for food production. The population of the Highlands was never higher than in 1750–1850 and large herds of cattle and goats were kept. The pressure on wild habitats meant, for example, the loss of the capercaillie (extinct by 1770), the goshawk and the great spotted woodpecker (*Dendrocopus major*); the roe deer and red deer populations were reduced, especially since the pine forests were whittled down in the 1600–1800 period.[20]

The onset of 'improvement' in Scotland is generally taken to be the years immediately after 1745. Some large estates of Jacobite owners were confiscated by the crown and re-assigned; other larger landowners decided that it was the paths of Hanoverian England that led profitwards. So after *c.*1760 real changes could be observed in the agriculture of Scotland (Plate 5.2). The quality of farmland was improved by enclosure, by intensive cultivation in tenanted smallholdings, by stone walls and hedges, the planting of shelter belts and the construction of a good road system. The quantity of farmland also increased from an intaking of middle moorland, though there was a retreat from the highest settlements of medieval foundation. The result was a more or less total revolution in the appearance of the landscape in lowland Scotland and in parts of the Highlands as well. In a way not found in most of England, it was geometrised by all these changes. A further departure came in the development of new ways of making a living, in the hopes of relieving the pressure upon the cultivated lands. Forestry was one, as was fishing. There were also transient occupations like **kelp** burning to supply an alkaline ash to industries further south. This started in Orkney in 1722, spread to the mainland in the 1750s and peaked in the 1780s. The boom collapsed in 1815 when cheaper imported materials became available.[21] Like fishing, though, there had always been a difficulty with the kelp trade because it required male labour in summer, at the same time as agriculture.

The other great change in agricultural production in the Highlands was the coming of commercial animal farming. Between 1720 and 1814, the number of cattle exported from Scotland multiplied by four times to be followed by a sheep boom. This started in the southern Highlands in the 1760s but came to the North and West Highlands only by *c.*1790. Sheep are most profitable when kept on a large scale, in flocks of at least six hundred. Estates began to relocate their tenants to make room

PLATE 5.2   *Near Monymusk, Aberdeenshire, 1998. The regularised landscape of Scottish improvement, creating habitats similar to the Enclosure movement to the south: hedgerows, small spinneys and larger plantations on any marginal land.*
Photograph: David Simmons.

for the Blackface and Cheviot sheep in the upland rough grazings, and so the infamous clearances began, with first reports in the 1790s and extending northwards to Sutherland in the early nineteenth century. Some landowners relocated their tenants within the estates so as to clear the upper glens; others simply and brutally displaced them. Fire accompanied the clearances: 'muirburn' to improve the forage for the sheep and also the clearance of some woodland to extend the area suitable for sheep, though the area of woodland removed for them has subsequently been exaggerated. Most of the native woodland had probably disappeared before the coming of commercial sheep farming: they were unlikely to have been the agents of the woodland decline before 1760.[22]

One aftermath of the clearances and their feudal context was the development of coastal crofting, where larger farms were split into small holdings (0.5–1.0 ha was common, often as a strip leading back from a road, with the dwelling on the croft and not in a village: Tiree has some good examples) and there might be common grazing, for example, upon the *machair* of the Hebridean islands.[23] The day of the sheep was short-lived: as the foundation of an economy, it was over by the 1870s because of competition from refrigerated imports. The landowners then took to deer and grouse as their next crop. In the meantime, the close grazing of sheep must have altered the relative proportions of plants on the muirs since they nibble where cattle pull. The movement of productive emphasis from valleys to the uplands meant that there was less attention to drainage in the valleys. The appearance of more standing water, combined with possible shifts in climate, has been held responsible for the surge in populations of the Highland midge (*Culicoides impunctatus*; in Gaelic *mhbeanbh-chuileag*, the 'tiny fly') from the late eighteenth century onwards. Before 1750 it was rarely mentioned by either natural historians or travellers: if plagued by these biting insects, Dr Samuel Johnson was sure to have mentioned them during his journey through Scotland in 1753. There may, however, have been a temporary peak in the 1730s and 1740s, badly affecting (no doubt among others), Bonnie Prince Charlie on Uist in 1746, who scratched the midge-bites so that he appeared to be 'cover'd with ulsers'.[24]

HEATHS, MOORS AND MOUNTAINS

In 1780 there was an estimated 2.42 million ha of 'waste' land in England and Wales. By 1880, a reduction to 1.48 million ha had taken place.[25] While at that time such lands existed to be improved and tidied up (even, in 1803, to be 'conquered' in the same way as France), today we find much of the 'waste' land prized as open space, and it is clear that some changes in the cultural valuation of heaths, moors and mountains have taken place.

Scotland as microcosm

During the eighteenth century many changes in Scotland had their impacts on the land, as we have seen above. Since moorland and mountain are so pervasive, it is not sur-prising that these environments should represent the zone where differing economic, technical and social conditions are reflected. Transhumance had produced sheilings at 400 m in the twelfth century and most were in the 300–500 m zone, with only those on high ground left by the seventeenth century and in decline by the eighteenth century.[26] The upper limits of cultivation in Scotland can be quite accurately mapped from the Military Surveys of 1747–55 and again from the Ordnance Survey sheets of 1860. Between those two datum lines, the area of crops and grass increased from 1.2 million ha to 1.7 million ha, though this conceals quite complicated movements of the limits of cultivation. In general, there seems to have been a high limit of culti-vation in 1750 from which there was some retreat in the shape of abandonment of holdings high in the sheltered glens but a corresponding increase in reclamation of moor at middle altitudes. In the Lammermuir Hills of south-east Scotland, there was cultivation at 360 m and even 400 m in the 1750s and there ensued the creation of 80 new farms in the period 1770–1825, when agricultural prices were high. These new farms took in 4000 ha of moorland never before reclaimed and 3500 ha of land reclaimed earlier and then abandoned. So the 1790s saw the high tide of conversion of moorland into land set to turnips or clover for sheep, and for oats. On the other hand, farm amalgamation saw the abandonment of 145 units and 480 ha of land reverting to the wild. At 1800, therefore, the balance sees the limit of cultivation at its highest since the fourteenth century.[27] This state of affairs lasted only until the collapse of prices that came after the defeat of Napoleon at Waterloo: by 1816 there was much debt.

Such a history puts the price of agricultural products in the foreground. When high, then it is profitable to reclaim land, providing that there are no obvious climatic barriers beyond the occasional bad year, and also that capital is available. This proved a problem in some parts of Scotland since the price of fencing the intakes (to keep the sheep from uncontrolled access to the oats, turnips and legumes) was a major cost.

Mountains and moorlands in England and Wales

In the eighteenth century, about one-third of northern England was uncultivated mountain and moorland. The dominant economy was pastoral and cattle were being replaced by sheep. Transhumance was dying out. The edges of the moorland were subject to reclamation in piecemeal fashion by tiny enclosures due to squatting by smallholders wanting to add to their holdings, and by more planned intakes of from 4 to 20 ha being added to solid farms. Lands taken in were subject to an intensive treatment of paring off the peaty top, followed by burning, draining and liming before a crop of oats might be sown. The changes were generally seen as evanescent; in some uplands the enclosures were exempt from the normal requirement to fence the intake land by a certain date: this was called 'permissive enclosure'.[28] Investment in fencing might not prevent the reversion of land if prices fell, as after the Napoleonic Wars, though in the North York Moors it was not the permissive enclosures that reverted but those compulsorily fenced.[29] On the Vynol estate in the mountains of Snowdonia, which is enclosed by five mountain peaks including Mt Snowdon ('a rough and cold country: a tract of bogs and rocks'),[30] poor management had persisted until the eighteenth century. Then, more rational estate management was introduced in which the farms had access to reasonable proportions of the different land types of valley and hill. The new management even extended to the enclosure of the mountain sheepwalk, so transhumance became impossible as well as unfashionable. The organisation of the estate, however, meant that the essential balance between the number of livestock and the valley grassland's (**inbye** land) production of winter fodder for that stock could be optimised. There was additional income, too, from the substantial development of slate quarrying, with urban expansion beginning to provide a market for roofing slates. Even so, the expansion was probably greater than the land could support and out-migration caused the population to fall by the beginning of the twentieth century.

Heaths

Lowland areas of heather, grasses, sedges and scrub of birch and gorse are an obvious target for reclamation. That they were sources of important fuel and grazing for the rural poor was rarely of concern to those able to promote a private Act of Parliament, as became frequent after 1750. The availability of the turnip meant that sheep could be brought in to manure the soil and also to help tread light soils into a better texture. The result was the enclosure and metamorphosis of great tracts of heath into cropland and improved pasture. Areas like north Norfolk and both the Lincoln Wolds and Edge were transformed from rough sheepwalk or rabbit warren to farms which might be as big as 650 ha. To intake the heath, a tenant had to spend from £8 to £9 per acre: gangs of men lifted the turf and then burnt it, dressed the soil with chalk and then with 60 bushels per acre of bones while the first crop of turnips grew. Once cropped for cereals, every four or five years there had to be bones at 20 bushels per acre and farmyard manure in liberal quantities.[31] There was a limit to the penetration of such changes, nevertheless, and large areas of the Breckland, Dorset, Cannock Chase, and Suffolk remained for some time in their wild (though certainly not natural) state.[32] On the sandy periphery of London, Hounslow Heath on the Bath Road, a favourite with highwaymen, has now been replaced with Heathrow Airport where European airlines also practise extortion.

Sensibilities and attitudes to environment

The Romantic perspective

The late eighteenth century is the time of the stirring of **Romanticism**, which came to fuller flower in the early nineteenth century. Part of the alterations of sensibilities brought about by the poetic movement was a revaluation of places like the Highlands, Wales and even the southern heaths. From being wastes and indeed frightening and horrid places where the Devil's imprint could be seen (and is still there in a number of feature and place names), they became places to be sought out and protected. Even if still fearful, then it was a tractable fright, like a horror movie, and the adjective 'sublime' was applied. The impact of poetry and painting on landscape is very marked for the eighteenth century and will be developed further, but for the mountains especially it was revolutionary for their environmental appraisal in later centuries.

Fresh water and fresh-water wetlands

Another reserve of land was the peat mosses of river valleys and other low areas. Large domed bogs were common in the Forth valley, for example, where they covered over the post-glacial marine clays. In 1767 Lord Kames began the reclamation of a large moss near Blairdrummond, allowing tenants to pare off the peat and cultivate the mineral soil beneath. To bridge the income gap between preparing the ground and the appearance of a crop, they were to sell the peat as fuel. In fact, they had to get rid of much of the peat by floating it down the river in large chunks and a system of ditches had a controlled water level to make sure there was enough water to float away the unwanted black stuff.[33] Also in the late eighteenth century, similar peatlands in Lancashire and Somerset were being reclaimed, though no record exists of any peat being sent out to sea as a sort of inverted iceberg. In the Somerset Levels the period 1400–1600 added the most dry land, until the years after 1770 when all the gains were consolidated and extended[34] (Fig. 5.2).

As if to emphasise that gains of well-drained land from earlier reclamations need not be permanent, there is the example of the East Anglian fenland in the eighteenth century. After its successful drainage in the previous century, the peat began to shrink

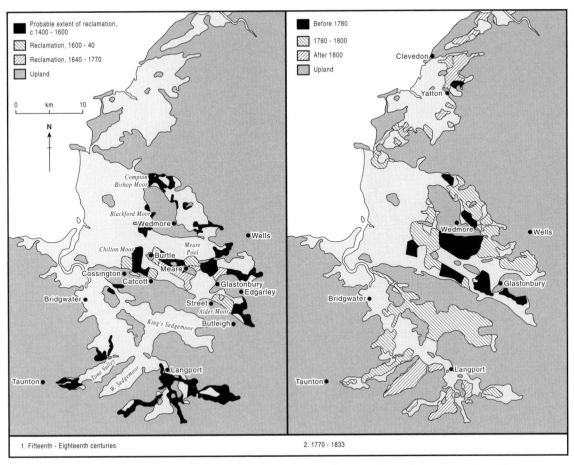

FIGURE 5.2   *Reclamation on the Somerset Levels. (1) Fifteenth–eighteenth centuries, (2) 1770–1833. The piecemeal nature of the drainage compared with the fens of East Anglia is apparent, especially in (1). All the reclamation in (1) is subsumed into the 'Before 1780' category in (2). On both maps the light stipple is clay and silt land that did not need radical measures.*
M. Williams, The Draining of the Somerset Levels, *Cambridge: Cambridge University Press, 1970, p. 84, Figure 12 and p. 129, Figure 16.*

and in spite of the building of hundreds of windmills to lift water up to the main drains (Plate 5.3), the region stood on the brink of reversion back to fresh water or, more frighteningly, inundation from the North Sea. It was saved by the steam pump. The problems of the Bedford Levels did not, needless to say, prevent other adventurers trying to drain the remaining fens of the region. Deeping Fen, between Peterborough and Bourne, was the largest undertaking of the century in East Anglia.

RURAL SETTLEMENT

We must not take the notion of tidying up too far. Although the rural scene may have been more prosperous-looking and clearly infused with a lot more money, the landscape

PLATE 5.3 *Once the drained fenland peat began to shrink, resort was had to the Dutch solution of windmills. This example is preserved by the National Trust at Wicken Fen, north of Cambridge, photographed in the 1960s. At one stage there were hundreds of such mills, harnessing the energy of the wind to lift water into the drainage system which gradually rose above the shrunken but valuable agricultural land.*

was in some areas scattered with many more structures than hitherto. In particular, areas of relatively late enclosure had their farm and related buildings on the holdings and not within a village, as had often been the case when common fields existed. The intaking of moorland and heath, too, encouraged dispersed settlement amid small plots. In the settlements themselves, there was often a difference between those villages where a landlord exerted strong control and those where such power was divided or absent; the former are often planned and exhibit a regular layout on the ground.[35] Sometimes the dwellings may be sited in relation to a great house or indeed neatly out of sight of the lord's residence. In Scotland, this is especially true of many of the reorganisations of the eighteenth century that went with the improvements in agriculture, and also with the removal of tenants from the higher parts of estates as sheep rearing became the main source of income. In both kingdoms, the rich built or amended large country houses and castles and to many of these they attached parks and gardens. All such changes favour a different mix of species from those of the pre-existing times.

URBAN SETTLEMENT

After 1730, the picture of urbanism which had emerged from the Middle Ages changed rapidly. London was dominant throughout, but in 1600 the next five cities in size were Norwich, York, Bristol, Newcastle and Exeter, with from 10,000 to 20,000 inhabitants. By 1800 only Bristol held its position in the face of burgeoning centres like Manchester–Salford, Liverpool, Leeds and Birmingham, all with populations over 50,000. Some towns and cities which experienced rapid growth did so for explicitly environmental reasons, as with the spa towns such as Brighton, popularised by the Prince of Wales after 1783. Sea-bathing centres like Weymouth (a favourite of George III in and after 1789) and spas which became seaside towns, such as Scarborough after 1750, also grew. The queen of all was Bath, which had a population of 32,000 in 1801.

Naturally, none of these towns and cities was without environmental impact in terms of the building-over of land, though they did not rewrite the cover of the land on the scale of the nineteenth century. Many were ports and so quays were built that in turn regulated the flow of the rivers; harbour walls started to influence the movement of sediment along coasts. A few ports reclaimed intertidal areas in order to build the large bonded warehouses made essential by the volume of colonial imports. Some harbours high up estuaries lost their trade because they could not dredge channels efficiently enough. Chester, for example, was the major port of embarkation for Ireland until the late eighteenth century. In 1742 it was still important, for while held up there en route to Dublin for the first performance of his oratorio *Messiah* George Frederick Handel made some alterations to the score. The racecourse now occupies the site of the quays.

London in 1801 held nearly one million people and as such constituted a big market. In its immediate vicinity it drew in vegetables and fruit from market gardens and hay for horses and the milk cattle kept within the urban fabric. The demand for bricks and tiles during the eighteenth century was such that it was said that the kilns surrounded the capital with a ring of fire and smoke.

INDUSTRY

In 1780, about 25–30 per cent of the adult males in England and Wales were occupied in mining and manufacturing. Most of them however were not far from their origins in rural occupations, for industry in the eighteenth century was often a by-employment of farming and smallholding so that industry went to the people; the nineteenth century would reverse that relationship. Nevertheless, even in the eighteenth century there were some industrial landscapes[36] and each of them can be seen to have had some environmental consequences (Table 5.2).

TABLE 5.2 Industry and environment in the eighteenth century

Major factors of development	Material phenomena	Environmental linkages
Sources of raw materials	Minerals Agricultural products Imported materials	Holes, heaps, flashes, toxic waste dumps
Processing plant	Furnaces, kilns Warehouses Mills and manufactories	Waste heaps, emissions to water and air Water withdrawal
Power sources	Water power Tidal power Steam power	Mining wastes Land conversion
Secondary industry	Clay and brick-making Bleaching and dyeing	Heaps and holes, emissions to water and air
Accommodation	Light in houses Mining 'barracks' Transport nodes	Building over farmland, heath for workers' housing
Transport	Packhorse and carrier routes Early railways Canals	Canals require water supply from existing streamcourses.

Sources of raw materials

The increasing overseas trade of Great Britain is contextual to much of the eighteenth century, much of it with the expanding empire. In the framework of industry, the period 1790–1800 witnessed the first doubling of the imports of raw cotton (they doubled again in the next decade) and which by 1800 had overhauled the woollen industry as a feature of the landscape and led to the rise of Manchester and Liverpool. The woollens of Yorkshire, however were beginning to mark out the regional pre-eminence of the Leeds–Bradford zone. The diversion of water courses to power mills was a major consequence of this transition phase to full factory-based production. By 1788, over 120 weirs had been constructed on relatively large streams for use in the cotton milling industry. The increasing tendency to regulate rivers would see conflicts develop between users of low-energy river flows such as corn millers and demanders of high-energy flows in the burgeoning extractive and manufacturing industries of the late eighteenth and early nineteenth centuries.[37]

Inorganic materials included salt: Tyneside was strong in sea salt and Cheshire in rock salt. The industry produced coal-fired salterns by the sea and water-filled subsidences ('flashes') inland. Sandy deposits were opened up for glassworks (as at St Helens in Lancashire after 1773) and clay pits for pottery, thus adding to the heaps and holes produced by metal working, as always a key feature of a regional economy. Among the

PLATE 5.4 *'An Afternoon View of Coalbrookdale', by William Williams. The incipient indus-*
trialisation is perhaps not as strongly dramatised as in some illustrations of the time and is the
more effective for that. As well as the ironworks with their smoke plumes, the harnessing of water
to drive the various mills and machines can be seen.
By permission of Shrewsbury Museums' Service.

non-ferrous metals, the landscape of Cornwall shows most obviously the environmental impact of tin and copper mining. Iron in the eighteenth century was demitting
its older centres in the Forest of Dean and the Weald where the last ironworks was
more or less defunct by 1794. The newer centres were on a larger scale so that both
their demands for raw materials and emissions were correspondingly greater: the
paintings of Coalbrookdale encapsulate the new scale of human activity. The Severn
Gorge near Telford (Shropshire) is a defining place in the history of industrialisation,
for it was here that Abraham Darby successfully smelted iron with coke and freed
the industry from its dependence upon charcoal[38] (Plate 5.4). Even in the eighteenth
century, the Coalbrookdale valley's streams were strongly manipulated to provide
water for the furnaces and mills (Fig. 5.3). About 100 years after the arrival of Darby
in Coalbrookdale in 1709, the transition to coke was complete. This meant that the
locations changed to the coalfields, with iron being transported to the coal where
necessary, as on Tyneside and in the Don valley near Sheffield. In South Wales,
Merthyr Tydfil grew to be the largest town in Wales and held that position for over a
century.

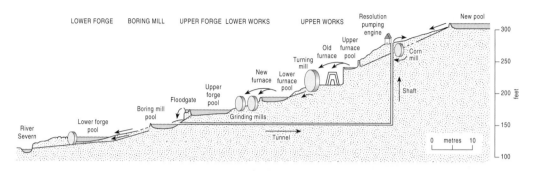

FIGURE 5.3 *Even at an early stage of industrialisation, the manipulation of water flow could be considerable. This schematic section through the Coalbrookdale Valley in c.1801 shows the many pools, dams and wheels which were installed, together with a pumping engine which returned some of the water to the head of the valley.*
J. Alfrey and C. Clark, The Landscape of Industry: Patterns of Change in the Ironbridge Gorge, *London and New York: Routledge, 1993, p. 64, Figure 4.3.*

The processing plant

In addition to the mill, the manufactory shed, the furnace and the kiln, the adapted home was still important in this century: bigger windows to let in light to hand-looms may still be seen in some woollen districts of Yorkshire and Gloucestershire. In 1770–1800 the mill was likely to be small, with only about 200 examples of more than two storeys and over 100–200 employees. It is the larger units that bring with them environmental alterations. Wastes were usually led into watercourses or simply piled up until overgrown and forgotten so that small-scale pollution of air, land and water might be found. The need for building materials was usually satisfied locally, so that a zone of quite intensive hole-making might accompany a new mill.

Secondary industry

Coalmining gave access to clays, so that brickmaking might be carried on as a subsidiary enterprise; when the coal seams gave out, the brickworks might still carry on. Where there was cloth manufacture then there was likely processing of the cloth such as dyeing and this produced wastes which were shed into watercourses with rather obvious results, though not as intense as the chemical dyes of the nineteenth century. But all added to the heaps and often stagnant water holes of the period. It is perhaps a surprise that malaria (still prevalent in fens and marshes in Lincolnshire and Kent, for example) did not spread to these pools.

Accommodation

As people began to move to the mill and factory, then accommodation had to be provided. Barrack-like blocks for male workers who would walk home for Sundays were one response. As well, rows of workers' houses, so much a signature of the nineteenth century, began to appear in a highly dispersed pattern. Land formerly in agriculture might be bought for its site, a precursor of today's practices. Families brought in might be allowed to colonise 'waste' for their own subsistence: in Cornwall between 1798 and 1842, a total of 2000 people were housed on the intaking of 160 ha of heathland. The last open part of Birmingham Heath was enclosed and built over for workers' houses in 1799.

Transport

The eighteenth century was the time of the packhorse, and important carrier routes continued from earlier times. The proliferation of cutting and embanking technology in the eighteenth century soon saw the modification of natural watercourses to produce navigable rivers, of which there were 2250 km by 1750: the Aire and Calder navigations connected Leeds to the Humber quite early in the century. The canal was a new element, as was the railroad. They had in common the need for an absence of gradients for most of their length, coupled with ways of tackling short, steep stretches. This meant earth-moving on a scale which was grand for their time, though the cuts and embankments were made by gangs of labourers, not by machines. Long summit pounds and reservoirs were constructed at the high point of each navigation, 150 existing by the end of the eighteenth century. These supplied the necessary storage to work the locks that allowed the canals to pass over the undulating countryside, with 56,000 gallons needed per lock-full. Feeder channels to these reservoirs and the capture of streams as feeders in lower sections of canal caused silting, however.[39] This silt problem would be transferred down-canal to their confluences with rivers, such as at the works at Knottingley which connected the Calder to the Humber.[40] Landowners frequently complained of both the increased flood control that would prevent flooding of water meadows and the possibility of unexpected torrents inundating other of their lands. The 'diversion' of the Ouse between Goole and Newbridge saw the corresponding decline and silt-up of its Trent and Aire outlets, as well as a reduction of river depths in the upper parts of the Don. Similar channelisation in estuarine river reaches such as that of the Aire and Calder could reduce water flows in tideways to mere trickles, reducing their navigability. The response in some cases, such as on the Severn, was to dredge and embank them to allow the introduction of large steamships in the nineteenth century. Where river outlets were beyond saving, silting reaches were locked and converted into harbours.[41] Throughout the 1700s, piling, cutting, enlarging and widening, shoal dredging and removal of natural obstacles became commonplace in river reaches. Ultimately, disequilibrium resulted in many of these newly navigable rivers, necessitating their continuing maintenance thereafter.

The canals were emblematic of the age and many of them remain in the landscape today, though mostly converted to recreational use. Their florescence began with the Duke of Bridgewater's canal in 1761 which connected coal mines at Worsley with Manchester, halving the price of coal in that city. The next 30 years saw the inland transport of bulky goods transformed, for one horse could move 50 tons of goods, compared with 2 tons in a wagon. Their effects on the spatial economy (and hence in promoting environmental change) were regional, for they were essentially best in lowland country and over distances up to 50 km. But as well as coal, bulk goods like lime, stone and timber could be moved with astounding rapidity and low cost.[42] Both rail- and waggon-ways and canals could affect drainage. The canals needed topping up and so diversions of streams were made and no doubt laden with silt during construction periods. Small reservoirs were also constructed from the 1760s onwards.[43] But all economic and ecological changes were small beer compared with the oncoming of the steam-powered railways.

Energy

During the eighteenth century, the source of power was a prime factor in the location of industrial plant although, in aggregate, human muscle power was still the major source of industrial energy. In 1800, water powered three-quarters of all cotton production (a cotton mill could run on 10 horsepower) but it was overtaken by steam in 1830. The bringing-in of water by leats from rivers or reservoirs was a matter for negotiation with owners in a free market. In upland areas like Dartmoor, peat was still popular for smelting tin and a few tide mills were still useful: Woodbridge Mill in Suffolk was rebuilt in 1793.

The story of the development of water-power and its role in the rise of the factory system of production is very important in the history of the British environment. Yet it is secondary compared with the replacement of charcoal by coal. In the eighteenth century Britain pioneered the use of coke as a replacement for coal and at the same time it was becoming the least forested country in Europe .[44] The obvious conclusion is therefore that neo-classical economics operated its 'invisible hand' and brought about a transition of fuels. Some think that the woods were exhausted by the eighteenth century: having used up the supplies in The Weald by the sixteenth century, the industry moved out to South Wales, the Forest of Dean, Furness, Scotland and even Ireland. Such a simple metamorphosis has been challenged by economic historians. Calculations suggest that the domestic industry could have been sustained on 650,000 acres of coppice woodland. The price of charcoal did indeed increase towards the end of the eighteenth century but there was at that time increased competition for it and so the transition to coke may have had more to do with the relative price of coal and coke, of the costs of transport and the level of wages than with the exhaustion of what was probably a carefully managed fuel supply.[45]

This interesting but unresolved environmental turning point should not obscure the argument that the development of coal mining and the use of steam power generated from coal is without doubt the central, binding narrative of the eighteenth century. The stationary steam engine enabled water to be pumped from shaft mines and so the search for coal (and tin, lead, and copper in other places) could be taken downwards as well as outwards. The Newcomen engine and then the Watt engine were progressively introduced: there is a record of the first in a colliery in Dudley (West Midlands) in 1712 and there were at least 140 of them in Northumbria by 1778. The Watt engine was more efficient and was an excellent winding machine for lowering and raising cages in shafts; it came to be common after 1784. The bell-pit was giving way to the shaft mine. The spoil heaps from bell-pits were sometimes put to fast-growing trees for pit-timber, of which the deeper mines used proportionately less. Below ground the longwall system of coal extraction took out the whole seam so there was more subsidence. From the middle of the century, increasing confidence allowed owners to introduce underground furnaces to improve the ventilation at depth. The greater capital demands of the shaft mine began to change other relationships: whereas early mining was scattered and combined with agriculture, from the late eighteenth century mine owners began to build housing for their workers and so a segmentation began to take place. Interestingly, many rural improvers were also catalytic in developing industry, not least because they (and not for instance the Crown) held the mineral rights beneath their estates.

It is important not to exaggerate the impact of steam at that time. The number of horses in London and Middlesex in 1800 far exceeded the aggregate horsepower of all the steam engines then in use. But a new and growing environmental influence was abroad. The steam engine itself needed water and transport connections, and produced wastes. The coal or metal ore mine it served enlarged those throughputs. The wider economy that was then made possible showed the pump-priming nature of it all: a relatively few joules into a coal mine produced surplus energy 30- or 40-fold, to create multiplier effects through both ecology and economy.

THE RICH AND THEIR PLEASURE

One way in which the gentry exhibited their taste for pleasure was in their houses and the parks and gardens with which they surrounded themselves, especially in rural areas. In the late seventeenth century, the taste was for formal gardens and parks in which the influence of France was strong. The great park at Versailles, designed by André le Nôtre and begun in 1662, was a powerful example. From 1712, English writers began to pour scorn on that style and to advocate parks which drew upon the paintings of Salvator Rosa and Claude Lorrain in having wild elements and a more 'natural' look altogether. The poet William Shenstone (1714–63) coined the term 'landscape garden' and the idea was taken up by fellow poet Alexander Pope (1688–1744), who was influential among the rich in a way not usually associated with poets today. It has been argued that in the late seventeenth century and early eighteenth century, it was the poets and painters who provided a theory of gardens which was put into practice from the middle of the eighteenth century.[46]

Thus evolved the great age of the country house and its park. The park was designed if possible by one of the three great men of the art, William Kent, Lancelot 'Capability' Brown and Humphrey Repton, who between them spanned the years 1685–1818. Into the existing landscapes they introduced numerous characteristic elements: curving streams and serpentine lakes, belvederes and classical temples, grottoes, colonnades, avenues of trees, clumps of trees and even a sunken fence (the ha-ha) which enabled the park to come right up to the house. Special walks or drives might recreate scenes from Vergil or induce a *frisson* of emotion upon suddenly spying a specially created Gothic ruin or, as at Fountains Abbey-Studley Royal, a real

PLATE 5.5   *The epitome of pleasure in gardens is perhaps Stourhead in Wiltshire, where a totally artificial scene has been created, full of grace and classical allusions. Preserved by the National Trust, even they cannot stop the trees (seen here in 1978) growing and perhaps unbalancing the proportions of nature and culture. They are unlikely to be allowed to provide nesting sites for woodpeckers, though.*

ruin taken over as part of the new landscape scene. Villages and cottages might be relocated if they spoiled the outlook, or moved to the background if they could be made to seem, literally, picturesque (Plate 5.5).

New species of tree from overseas were often used, and the years 1736–63 saw the introduction of the dreariest of newcomers, the rhododendron. In all, some 445 new species of trees and shrubs were introduced into England during the eighteenth century. The first book on oriental gardening was published in 1772 and set a fashion for *chinoiserie* still to be seen in the form of the pagoda at Kew Gardens. Urban areas might also benefit from such movements: Regent's Park in London was 'improved' at this time, as were the former hunting parks of St James's and Hyde Park, which were by then open to the public.[47] Woodlands were central to many of the designs since they might also be a source of profit as well as pleasure. The symbolic importance of trees, both native and exotic, is noted by many interpreters of the scene, for they emphasised the power of the landowner in changing nature, delimited the estate from the surrounding countryside and emphasised its exclusivity.[48] (Destroying ornamental trees was a hanging offence after 1722.) It all added up to design on a large scale. Brown is said to have undertaken 188 commissions, which included big estates like

Blenheim and Chatsworth. The park might also function as a larder: in the late eighteenth century, cattle and then sheep became more popular. 'Wild' animals like deer were also eaten, as were hares, partridges and pheasants. Other species (for example, otter, badger and fox) were hunted for pleasure. The generic term of the eighteenth century was 'improvement' and the parks constituted a considerable aesthetic dimension to that trend, spurred on by painterly and poetic examples as well as the demand to plant coverts for pheasants. The parks did not usually contain fox earths so that the farmland surroundings became important sources of animals to chase; arguably, however, the park was the seminary of the fox-hunting which developed in the nineteenth century. In the late eighteenth century the demand for foxes exceeded the supply so they were imported from France and Holland: Leadenhall Market in London handled 1000 foxes a year at that time.[49]

A second pleasure, again not for the poor, has been prefigured in the discussion of the towns: the rise of mineral waters as a source of health. The ubiquitous source of these is the sea but terrestrial springs were also deemed beneficial for many ailments, including the marriage prospects of daughters. The nobility and gentry followed their perception of the habits of Classical Antiquity and so in the first half of the eighteenth century, new spas were established, and established centres like Bath, Epsom and Buxton expanded considerably. The monarchy and its satraps set a powerful example, making the 1750–93 period an even greater period of growth since it was aided by improvements in the roads and the enclosure acts which made expansion of housing possible around larger towns.[50] Hence 39 new spas sprang up, as it were, between 1750 and 1790. The overall success can be seen by the numbers of visitors (Bath peaking at 10,000 in 1812–13) and the steady growth of more permanent residents even if some of them and their descendants were not cured of their irritating ailments, as at Tunbridge Wells. There seemed to be few rivals to the spas except the sea-bathing places like Scarborough, which combined both functions, but the cessation of hostilities against France in 1815 and the coming of the railways meant that there would be changes here, as in so many places, in the nineteenth century.

SACRED SPACE

The actual designation of land as sacred was unthinkable unless it was safely confined to the established church or one of its recognisable imitators. Thus churches, cathedrals and cemeteries were clearly different in purpose, though less determinedly apart from the rest of life than now. The revival of old styles, like the Classical inspiration of the seventeenth century and the Gothic Revival after 1750 did not bring with them the cultures which had inspired them: Greek porticoes were one thing, Orthodox religion would have been another. The Protestant view that prayers could be said anywhere allowed a view of woods as places for meditation and a certain reverence for old trees, for they might have been around soon after the Flood and thus constitute a link with a biblical past. But most of this is tenuous, compared with prehistoric and medieval times.

THE EXTENSION OF THE ECONOMY

The population of Great Britain reached for resources into areas outside their territorial limits in magnitudes which expanded in the eighteenth century. The main ways in which this had happened were firstly the use of the material resources of the seas in the form of fish, and secondly the import of materials won by trade with equal partners or from colonies in a dependent relationship.

Fisheries

Throughout the period of the dominance of agriculture, artisanal fisheries existed round the shores of these islands. 'Artisanal' means that they were small-scale and generally confined to either inshore waters or the relatively enclosed basin of the North Sea. These were supplemented by larger enterprises which systematically followed the migrations of a stock such as the North Sea herring or made expeditions to faraway waters such as those of Newfoundland. The eighteenth century witnessed an unusual development in the way in which the development of fishing was encouraged in Scotland as a way of adding to the resources of the crofting communities there. In Shetland, deep-sea fishing for ling and cod became successful and encouraged the subdivision of crofts, for fishing took up most of the male labour in the summer. In the West Highlands, it was less successful and the herring also began to decline after the 1790s, perhaps as large-decked vessels from the Lowlands dominated the catching fleet. Planned villages like Ullapool, Tobermory and Lochbay were laid out in the 1790s as centres of a fishing industry, with herring curing facilities adjacent. Lochbay was never very successful but the other two grew and led to some imitators, such as Pultneytown, grafted onto Wick in Caithness.

A notable industry of the eighteenth century was whaling, to provide lamp oil and soap. It dates from the seventeenth century in Great Britain but there was a government-subsidised expansion into the grounds off Greenland after 1750. The limitations of the 200–300 ton sailing ships (an example is portrayed in J. W. M. TURNER's paintings of the whaler *Erebus*) and of hand-harpooning meant that the oil was expensive but that the populations of whales were not greatly diminished by the culling until the nineteenth century.[51] The technology of whaling spilled over into home waters with the establishment of a fishery on Canna for the liver oil of the basking shark (*Cetorhinus maximus*) soon after 1760. A few other places on the west coast of Scotland took it up but the impact ended *c.*1830 when whale oil became freely available at the end of the Napoleonic Wars.[52]

Overseas trade

A major product of the seas was trade. The expansion of rates of production in Great Britain and the development of an empire able to supply raw materials brought about a shift in emphasis to everyday materials, of which cotton is a good example. Imports included sugar, timber, tobacco and coffee, and tea had become the main drink of the poor by 1800. The lesson of such lists is that the environmental impact of Great Britain began to extend well beyond its shores. To meet the demands of

British economy and society, large areas of the world underwent environmental change, as plantations were established out of forest or savanna, for instance. As slaves were taken from Africa, it is likely that the forest reclaimed areas then in agriculture. In turn, exported iron goods made it easier to fell forests and make guns to shoot animals. To speak of the globalisation of environmental impact would be a great exaggeration but to conjure up a world-system incorporating all those influences is not fanciful.

A long-term if spatially limited consequence can be found in the dumping of ballast in the estuary of the Tay between 1750 and 1850. This now forms a suitable substrate for mussels (*Mytilus edulis*), which in turn support a large (20,000 birds) overwintering population of eider duck (*Somateria mollissima*). As a last and minor example of English maritime connections, we can mention that in the Scilly Isles. Not far above present sea level at Big Pool on St Agnes, there is a sand bed, sandwiched between peats whose dating has allowed the interpretation that the sand bed results from a tidal wave or tsunami caused by the Lisbon earthquake of 1755. If so, it is a reminder that natural events along our coasts are not always predictable, as borne out in 1953.

A century's trends

Wholes and parts

The eighteenth century can be interpreted as a time when we can see the shape of a tension between the whole kingdom and its constituent parts. In the category of wholes, there is the Union of the Parliaments of England and Scotland in 1707, with some rebellions in 1715 and 1745. The nation which had emerged in the seventeenth century now had a professional government and a stable relationship between Parliament and the Crown most of the time. Not very much legislation which directly affected environmental matters was enacted: cleaning up water and protecting seabirds had to wait until the nineteenth century. Many other government actions, and in particular those affecting trade, fed back into the rural economy and hence into crops, soils and woodlands. Woodlands were of national importance only as sources of ship timber for the Navy and the merchant fleet; the underwood was a private matter and in any case of diminishing importance as the price of coal fell.

Against these wholes, there were more regional perceptions of importance. The improvement of communication and of banking meant that specialisation of economic activity (and hence of environmental impact) could burgeon. A region might specialise in making nails and know that it would be fed with food bought, and brought, in. Self-conscious descriptions of counties were undertaken in the seventeenth century, and the Royal Society commissioned the collection of information about provincial areas from 1645. These adherences of local and regional identities were little diminished in the eighteenth century even though their largely rural base had in a few places become secondary to an industrial economy. The county reports on agriculture (commissioned by the Board of Agriculture and General Improvement from 1793 to 1815), supplemented by the work of William Marshall,[53] provided the first attempts at maps of soil and land use regions. These in part reflected, as was realised by the authors of the time, the physical characteristics of latitude, altitude and soil, but also the area's place in systems of national and international markets. It is not too much to say that some of the regional characteristics then recognised are still units with which people identify today. The way in which map-making and statistical surveys helped make a national identity for Scotland in the eighteenth century has been charted by Charles Withers: he makes the case for regional identities based on forms of self-knowledge,[54] which is not entirely irrelevant to the devolutionist mood prevalent in the late 1990s.

No trend begins and ends with the century to which it is often principally assigned. This is true of both material changes and those which are so intangible as to be called 'sensibilities' or the 'spirit of the age'. Distillations of both are plentiful for the eighteenth century.

Eluding Malthus

THOMAS MALTHUS (1766–1834) is surely one of the most enduringly famous Britons of the eighteenth century. He was convinced that population would outrun its resource base but seems, so far, to have been proved wrong by the story of the nineteenth century and thereafter. In an economic and energy-based interpretation which parallels the layout of this book, E. A. Wrigley suggests that the transition to an industrial economy is the reason that the British outstripped the horsemen of famine and disease.[55] He argues that in the seventeenth century England possessed an 'advanced organic economy'. The energy flows of the society were more efficiently organised than in a 'simple organic economy' with animals taking the place of humans in many jobs requiring effort, and water power extensively employed. Poor communications ensured that most material requirements had to be produced near their sites of consumption. The eighteenth century was the time of transition to a 'mineral based energy economy' which taps the stocks of inorganic energy; in that case coal, and now oil, natural gas and nuclear energy.[56] The use of underground supplies firstly reduces pressure on the land: coppice can be converted to cornfield, as eventually can grassland formerly used to feed horses. A good transport network unites centres of production so that they can become more efficient and thereafter raise the per capita levels of production of almost everything. This required good roads and the railways, for canals were basically tied to regional economies.

That transition covers the metamorphosis of the sixteenth century into the late nineteenth century. We must not exaggerate its progress in the eighteenth century. Bairoch[57] suggests an index for per capita levels of manufacturing industry (not mining, nor construction and utilities) with the UK level at 100 in 1900. Then: 1750 = 10, 1800 = 16, 1860 = 64 and 1913 = 115.

So the eighteenth century may have been a decisive step along the road to an energy-intensive, fossil-fuel-based society but it had not yet gone all that far down it. A fantasy of what it might have been like had we not done so (and had avoided the Reformation into the bargain) is in Kingsley Amis's novel *The Alteration*. If as in earlier sections one particular event or process is to be singled out as pre-eminent environmentally, then for the eighteenth century it must be the concern for trees and wood supplies: something with both domestic and overseas connections, a common enough theme before and since.

Finding romance

It is a commonplace observation that as 'raw nature' became tamed, its appreciation increased among people now sheltered from most of it. The poetry canon, for example, is full of names from the eighteenth century who turned the wild into the elevated and the numinous, whereas before it had been at best an ambiguous place and more often one full of threat. A list containing names like Coleridge, Wordsworth, Burns, Smart and Pope entrenches the view of nature as delight. Even Blake, who was not above taking radical views, endorses the rural idyll, as we have seen in his *Jerusalem*. There is little in their output which reflects, as has been suggested, that the Lisbon earthquake of 1755 was influential throughout Europe in increasing a sense of vulnerability to natural hazards.[58]

A common thread to much of this is the depiction of nature. Following the introduction in the seventeenth century of the word *landscape*, referring to a particular genre of painting, there was a search for the picturesque in words, in oils and watercolours, and in the very materials of the land itself.[59] There was, in Constable especially, a coming concern with the artist as Self as well as being a craftsman at others' service like a carpenter or gardener. So the new consciousness of humanity induced by the French Revolution (Wordsworth thought that wildernesses were 'rich with liberty') complicates the analysis. The changing tastes of the eighteenth century and their representation are far more complex than simply a reaction to the retreat of the wild; however, they ripened into a much fuller set of fruits when faced with the more thorough development of industry in the following century. The eighteenth century, though, was a time when the correspondences between humans and nature (lions as strength, four-leaved clovers as lucky, goats as lustful) were being replaced by a separation, in which scientific observation was paramount and of which the key emblem was giving Latin names to plants.[60] The London Society of Gardeners' *Catalogus Plantarum* of 1730 marks an irrevocable stage in that transition.

CHAPTER SIX

Industrial growth: material empires, 1800–1914

This period demarcates the transformation of the British economy to the world's first fully 'mineral-base energy economy', based first on coal (Plate 6.1) but which by the 1950s had incorporated other hydrocarbons and developed the possibility of nuclear power. This all showed itself at first in a mushrooming technology epitomised by the railway, the steamship and a myriad of devices designed to harness the energy of nature and the ingenuity of the British into making and selling everything that the world could be persuaded to need. Some of the energy went into population growth and some into the acquisition and maintenance of the largest empire ever known.

Economic historians frequently describe the nineteenth century in Britain as a time

PLATE 6.1   *Blaenavon, Glamorgan. A small colliery, in 1998 a museum pit (the car park gives it away), with only a small number of buildings round the head-gear. The tip heap in the background has been landscaped and grassed and the road access smartened up. But many nineteenth-century collieries were about this size and this isolated.*

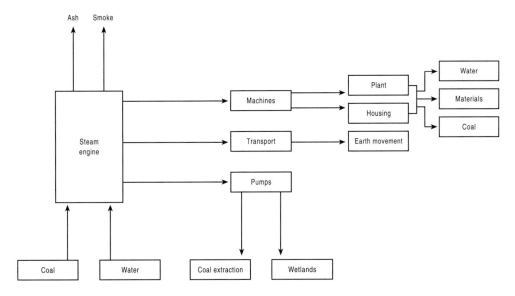

FIGURE 6.1   *Some of the environmental linkages of the steam engine on the local or just extra-local scale. In fact some of these comprise positive feedback: better pumping equals more coal extraction, so more coal can be mined . . . etc.*

of progress towards a 'mature industrial economy'. In some degree this is summarised by Bairoch's data[1] in which an index of manufacturing set at 1900 = 100 gives us 1800 = 16 and 1913 = 115. So in the period of concern here, industrial production was multiplied by seven times. No change of that magnitude can fail to have environmental consequences, as subsequent sections will describe. The clear integrator of much of the change in the environment and all its denizens is the growth of the human population and the changes in its distribution. The dimensions of growth are measured by a doubling between 1780 and 1831 with a rise to just over 20 million in 1831, which doubled again by 1911. Within the 1851 figure, 62 per cent of the population of Great Britain lived in England, 50 per cent of whom might be considered urban. The proportion of people engaged in agriculture, forestry and fisheries fell from 36 per cent in 1801 to 9 per cent by 1901. These are data with a dependable degree of accuracy, since the first national census of population took place in 1801. Some 'shafts of sunlight' become reliable even if they illuminate a rather smoky scene.

The key symbol of this era is the steam engine in its various forms, both stationary and locomotive. The application of some long-standing theoretical knowledge by some practically-minded men had many linkages (some of which are plotted in (Fig. 6.1) to environmental changes as well as transforming economic and social conditions. Most city people were moved away from the close interaction with nature that farmers experience but they were not immune to weather, nor to a particle-laden atmosphere nor to contaminated groundwater. More and more sought contact with 'nature' via the acquisition of scientific knowledge, for example, or by railway excursions to spa towns (Matlock and Buxton in Derbyshire come to mind) or to coastal resorts. 'Nature' might have been built over in a great many places but we can be sceptical about claims that it was bricked out of people's minds.[2]

The atmosphere

In the field of climatic change, the nineteenth century gives us one greater certainty: the data come from direct measurements with instruments of reasonable precision. They enable us to plot the end of the Little Ice Age in the 1850s, when temperatures, especially those of the winter months, began an upturn. The annual averages rose by about 1°C between 1850 and 1950, with winter temperatures more like 1.5°C in the same period.[3] The last frost fair on the Thames in London was held in 1814, though changes in the river's confinement by embankments also contributed to its failure to freeze over thereafter. The amelioration of climate was not steady, though, and there were a number of bad years in the nineteenth century: 1879 and 1880 were especially bad for agriculture and are credited with having been a major factor in the decline of agriculture which started about then. The summer of 1826 was very hot and dry and there was a shortage of water for mills and canals, which led to layoffs of workers and a bout of road building in Lancashire to alleviate the subsequent poverty. More dry summers in the 1850s led to the inception of the British Rainfall Organisation in order to collect data. Less directly economic was the prevalence of volcanic dust in the period from 1807 to the 1830s (for example, from Tambora in Indonesia which caused the 'year without a summer' of 1816) that may well have produced the kind of skies immortalised by Joseph Mallord William Turner (1775–1851) in some of his most famous pictures.[4] The atmosphere of Great Britain in the nineteenth century was, however, more notable for its human inputs than its natural change. The burning of coal in homes, workplaces and railways, together with the by-products of the chemical industry, were led off into the air in a more or less uncontrolled fashion until enforceable legislation began to take effect after mid-century. The major con-tributor was the domestic hearth, which provided about 85 per cent of the smoke. This source was not subject to control until the Clean Air Acts of 1956 and 1968, whereas industrial emissions were tackled about 100 years earlier. The consumption of coal provides one key to sources of emissions. In 1800, the UK consumption of coal was 10 million tons, in 1856, 60 million tons; 1900 saw the burning of 167 million tons and 1913, 189 million tons. This rise is not directly related to pollution because of technological changes and, regionally, the kind of industry. A textile mill with steam engines working at a constant speed could achieve relatively smokeless combustion, whereas an iron-works requiring a burst of power for, for example, rolling, could only achieve that by a blast of very smoky firing. (In 1856, the Dowlais ironworks at Merthyr Tydfil in South Wales had 63 steam engines.) The beehive coke oven was another famously smoky device. The great inorganic pollutants of the air were hydrochloric acid and hydrogen sulphide produced in the manufacture of alkalis. The former rained out immediately; the latter's rotten-egg smell might travel at least 25 km from its point of origin. Places like Runcorn and Widnes in Cheshire and St Helens in

Lancashire were particularly known for the acrid smells and acid soils. Acidification of soils seems to have extended to the Pennine uplands, where a change from bog-mosses of the genus *Sphagnum* to the more acid-tolerant cotton-sedge *Eriophorum* took place over whole stretches of peaty moorland. Ways of precipitating the acids down into their emitting chimneys were devised but the resulting liquor was then piped into the nearest watercourse. Smells were exacerbated by those of sewage from cesspits. Piped water without piped sewage proved a difficult combination: Leeds in the 1870s had three times as many cesspits as connections to sewers. Coupled with ashpits for other organic material and the smells produced by, for example, tallow rendering and bone boiling (to make phosphates), the nineteenth-century city was not a place of high nasal amenity.

In London, as Table 6.1 shows, smoke rose from 122 µg/m^3 in 1800 to 201 µg/m^3 in 1900, and sulphur dioxide (SO_2) from 488 µg/m^3 to 603 µg/m^3 over the same period. The thick and persistent London fogs, whose frequency peaked in the 1890s, were clearly associated with smoke levels and in 1881 there was even a Smoke Abatement Exhibition. Most devices failed to catch on because they did not recommend the continuance of having a domestic open fire; it was even proposed that 'Biovitric rays' from an open hearth were necessary for human fertility.[5]

The early history of pollution control depended a great deal on the zeal of the local inspectors who naturally were faced with industries who resented the rising costs of controlling their wastes. Policies today may still be at the mercy of the quality of the implementation at the local level and the variations to which this can be subject. There may be a considerable distance between the setting of standards and their enforcement: between Vienna (HQ of the IAEA) and the Irish Sea for radioactivity, for instance, which is always going to diminish the effectiveness of the implementation. The decision taken in 1998 to close the Dounreay plant in Caithness because of the mishandling of radioactive wastes in the plant is a good example.

TABLE 6.1 Air pollution in industrial London

Date	City width (km)	Population (thousands)	10^3 tonnes of wood fuel	10^{3}t of coal	Smoke in µg m^3	SO_2 in µg m^3
1750	4.5	675	—	700	76.0	340
1800	5.0	1114	—	1350	122.0	488
1850	9.0	2235	—	4000	135.0	471
1900	16.0	6581	—	15,700	201.0	603

Source: adapted from P. Brimblecombe, 'Early urban climate and atmosphere', in A. R. Hall and H. K. Kenward (eds), *Environmental Archaeology in the Urban Context*, CBA Research Report 43, London: CBA, 1982, pp. 10–25.

Steps were, however, taken to produce environmental amelioration. There was a Smoke Prohibition Act as early as 1821, which encouraged some local by-laws, and London had its own legislation against smoke in 1853. The Alkali Act of 1863, pushed forward by further laws in 1874 and 1881 (when enforceable standards were applied), was the sign of real progress.[6] The local application of the law was sometimes spotty: in St Helens the inspectorate managed to control acid emissions but had much less success with hydrogen sulphide.[7] Yet not until the 1950s were such far-reaching measures equalled: the Alkali Act can be seen as a defining moment in the history of environmental management in Great Britain.

LANDFORMS

Though the British Isles are not generally regarded as earthquake-prone, there are perhaps six tremors per year of Richter 2.5 and above. The strongest known (though it was not recorded seismologically so that its true strength is unknown and there were no permanent landscape effects) happened on 22 April 1884 at 0920 with an epicentre south-east of Colchester in Essex. Its effects were felt over an area of 130,000 km^2, with the most serious damage within a radius of 250 km, where 1200 buildings were largely destroyed and hundreds of people made homeless. There was only one definitely attributable death, however.[8] The next largest shock was in 1990, centred on North Wales and measured at 5.2 on the Richter scale.

The century, however, saw the creation of many human-made landforms, which are dealt with in appropriate places: they might be classified as excavation, attrition, subsidence and accumulation. The coasts were also affected, as were water bodies and perhaps above all the sites of cities. Cataclysmic events were rare but the extraction of rock salt from under the Northwich area of Cheshire started in 1670 and intensified thereafter, being pumped out with steam after the 1850s. On 6 December 1880, a 90-foot chimney collapsed into a hole, followed by two small streams and then the River Weaver. Geysers of mud and gas followed, and then a large water-filled depression or 'flash', which went on sinking for a month. The flashes were then used for lime-waste dumping from 1907 until the 1930s, when such products were injected into the salt strata. Lime waste at the surface was an environmental novelty since it was a slimy semi-plastic mud that was inimical to all plant growth and did not dry out.[9]

WOODS AND FORESTS

In 1800, England and Wales were the least wooded areas in Europe with an estimated total of less than 2 million acres (810,000 ha) of woodland.[10] Shipbuilding timber had disappeared from the south coastal counties and pitwood from the north-east coal districts. Only the Midlands remained well-wooded and big inroads were made into that resource as the canal network reduced the costs of carriage. Even here, forests like Wyre (on the Worcestershire–Shropshire border) had been largely converted to coppice by that time.[11] In Scotland, forests like those on Speyside were felled for charcoal and timber, with log rafts being floated downstream on rivers dammed to provide a good head of water for that purpose.[12] There were some reversals, as in the Chilterns, where monocultures of beech replaced coppice and wood-pasture in the late eighteenth century. These stands have long been associated with the furniture industry around High Wycombe, though Rackham is sceptical about the closeness of this association, asserting that wooden chair seats can only satisfactorily be made of elm.[13]

Since at least John Evelyn's time, there had been pleading that a maritime nation could not afford to neglect its supplies of native oak. Eventually there were results in the form of new plantings in the Royal Forests under the supervision of Commissioners of Woods and Forests. Although the 1850s and 1860s saw the conversion of some woodland to arable and grassland, the reverse was happening as well, with large estates taking the lead in planting up big areas (often with conifers) as well as small-scale copses and shelter belts. The agricultural depression of the later part of the century allowed poorer soils to be sown and from 1871 onwards the area of woodland began to increase. This trend was reinforced by a new appreciation of the value of amenity woodlands (Burnham Beeches and Epping Forest were acquired for the Corporation of London at this time) and by the extension of planting around upland reservoirs.[14] The decreasing demand for coppice products meant that in areas like south-west Yorkshire, conversion to high forest included coniferous species.[15] Two factors changed the forestry scene. The first was entirely predictable, namely that imported species of trees began to demonstrate how useful they might be, particularly in the cooler and wetter parts of the country. So in 1827 there was the import from North America of Douglas fir (*Pseudotsuga menziesii*) and Sitka spruce (*Picea sitchensis*), followed in the 1850s by Western hemlock (*Tsuga heterophylla*) and Lodgepole pine (*Pinus contorta*). Lodgepole and Sitka survive as major commercial species today. The second factor was perhaps less obvious: the battle of Hampton Roads in 1862. The clash of two iron-clad vessels during the American Civil War presaged the demise of the wooden wall and the dominance of the iron dreadnought. Thus afforestation could concentrate on short-cycle timber for immediate use. A truly commercial approach could be formulated in which national interests (articulated by a Royal Commission) suggested in 1908 that 9 million acres (3.6 million ha) should be under trees. Lectureships at universities, demonstration forests and the import of German foresters' methods were all encouraged.

AGRICULTURE

Any environmental perspective has to acknowledge the poor years of 1879 and 1880. In the former, rainfall was very high in spring and summer and, in the latter, it was high in summer and autumn, with an average value of 14 per cent above the 1875–82 levels. This mean value conceals extreme values in the eastern counties and Midlands: Berkshire rainfall was raised by 35 per cent, and Lincolnshire 33 per cent in those years. Hence, cereal yield in 1879 was only one-half to three-quarters of the 1873–7 levels, with wheat being especially affected. Livestock was hit by a poor hay crop in

1879 (too wet in early summer) and in 1880 (too dry in spring); liver fluke followed the rain. Bad conditions returned again in 1892 and 1893 when the spring and summer were abnormally dry. Hay and roots were badly affected and stock became thin; cereal yields were mostly low though locally variable.[16] These natural influences exerted their grip at a time when overseas competition was becoming serious and prices were falling. Wheat prices, for example, fell 51 per cent between 1871 and 1898, wool by 50 per cent and beef by 29 per cent.[17] The move from arable to pasture from the 1870s encouraged the retention of hedgerows to the point where in, for example, Essex and Suffolk they encroached onto roadsides and fields to depths of 6 to 8 m.

Even though farming was still much influenced by the weather and by economics, it was in part busy absorbing the new world of science and technology. The Royal Agricultural Society of England was founded in 1838, and in 1840 Justus Liebig demonstrated that plant growth was often held in check by the short supply of mineral nutrients and that the application of the chemicals made possible by heavy industry (what we would call 'artificial fertilisers') would allow surges in productivity. His book led to the foundation of the Rothampstead Experimental Station in 1843, the first of a series of British scientific agricultural experimental stations. Developments in agriculture mirrored the world: mechanisation became more common, with inventions like the mole plough for the remote installation of undersoil drainage, mechanical reapers (40,000 of these in the 1870s) and steam ploughing on large farms. One farm with such a machine reduced the number of fields from 36 to 9 to allow its use, which is a theme we shall encounter again. There were improved strains of seeds which were encouraged to crop heavily by guano from Peru, nitrate from Chile and potash from Germany, together with the home production of superphosphate and basic slag; the Cambridge Greensand, high in phosphates, was dug out for fertilisers from 1851, turning to raw material for shells in World War I and being a symbolic precursor of cement and fertiliser plants in today's landscape.[18]

The application of the science, helped by a rapidly growing market, saw in the system called High Farming in which there was no longer any fallow. In 1870, there were about 6 million ha of arable land which went through a wheat/root crops/green crops/rotation grass sequence, and which with a flexible attitude to rotation and good leases on farm tenancies was highly productive. Some 4.5 million ha of permanent grassland supported both meat and milk producers. In both, the railways were helpful: beasts could be got to market in better condition and more cheaply and rural milk could replace the insanitary urban dairies which added to the disamenity and waste disposal problems of the cities. High farming, though, occupied only about one-third of the arable area: there were still lots of small farms with older practices. In nineteenth-century Wales, for example, the nutrient cycles were kept up by manure and locally-produced lime only.[19] Turnips were still rare, their dissemination lagging some 50 years behind that in England. Mechanical reaping of hay replaced the scythe only in this century and in general mechanisation was significant only after 1880 and then only slowly. Welsh sheep were thought to be of poor quality, in contrast to the horses, which were universally praised. Modernisation brought about a great deal of protest:

gangs pulled down fencing on enclosed commons in the 1860s and poached salmon on the Wye in the 1850–80 years.

An example of the ecology of a mixed farm in England during the 1820s is given by T. Bayliss-Smith.[20] This farm had arable land with no fallow period and Chalk downland under sheep. These animals were the key to the ecology: they would walk from 4 to 6 km to feed during the day and at night were folded on enclosed fields where they dropped a great deal of dung, enabling cereals to be grown. The ratio of energy produced to energy consumed is about 14: 1, with each farm worker responsible for an output of about five times what he and his family consumed. Note that this was done with a direct fossil fuel input (Fig. 6.2) of a very low order but supported by indirect use of coal in the shape of 400 kg/year to the blacksmith. The scope for increases in productivity once mechanisation was further advanced (this farm already had a threshing machine which allowed the cost of threshing to fall two-thirds from the level for hand-threshing) and full fossil fuel 'subsidies' were applied was immense.[21]

The overall consequences for the biota can be deduced: where farming advanced and intensified then many bird populations either retreated or changed their composition. Where the rank grasses and hedges grew back then there was a greater diversity of flowering plants and birds. The great pests of the more intensive systems of this century were the wood pigeon, rook, skylark and house sparrow, for example. Until the seventeenth century, the wood pigeon had been a thinly scattered woodland species but had exploded by the mid-nineteenth century to the point where bounties were paid for corpses. In similar fashion, skylarks became gourmet food in London markets, 20,000–30,000 at a time, though arsenic was used to deter them from eating shallow-sown grain. Increased attention to drainage meant that snipe, dotterel and woodcock declined, and remaining lowland raptors like the kite became extremely scarce.[22]

HEATHS, MOORS AND MOUNTAINS

In 1800, there was still a great deal of unenclosed land in Great Britain, and much of it was in semi-natural vegetation as heaths, moorlands and mountains. The final surges of enclosure associated with the nineteenth century often resulted in the fencing of such lands and the assignment of former common lands to private ownership. This did not necessarily bring about a change of use, for many of the upland areas were incapable of conversion to arable land. It did mean the installation of miles of fencing and stone walls, often in straight lines, which can still be seen today in upland areas. In Wales, upland enclosures resulted from the high prices of the Napoleonic Wars coupled with the landowners' desires to prevent piecemeal enclosures by commoners. This deprived ordinary folk of resources such as peat. Even by the 1890s however, large tracts of upland remained unenclosed.[23] In northern England between 1815 and 1850, 30 thousand acres (12,000 ha) were enclosed in the North Riding of Yorkshire, and 23 thousand acres (9300 ha) in County Durham. An example of massive transformation occurred on Exmoor from 1818 when 4000 ha of the former Royal Forest

Environmental Linkages

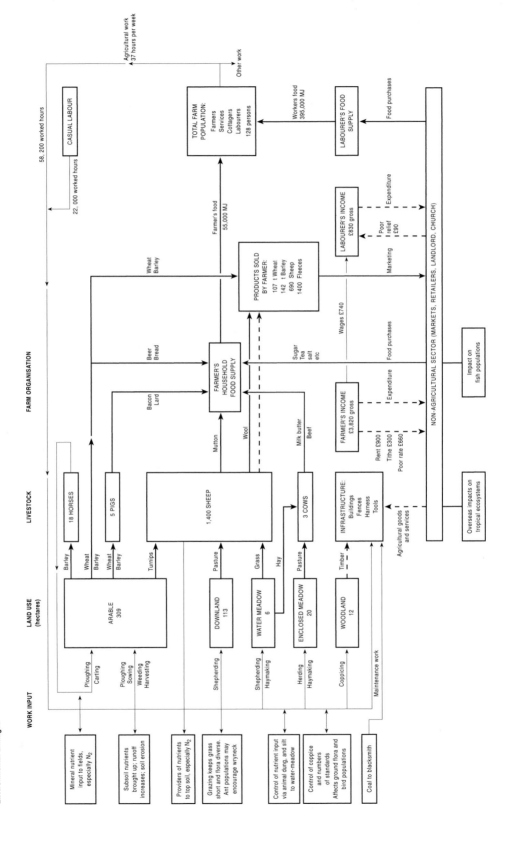

came into the possession of one family, which thereafter developed much of the moorland into cattle and sheep range and cropland. Fencing, paring, burning and liming are familiar sounds in such a history; on Exmoor in 1855 there was as well the creation of a village (Simonsbath), road-building, and the creation of an ecclesiastical parish.[24] A similar transformation was effected slightly earlier on the Mendip Hills in Somerset where 25,000 acres of rough grazing was enclosed and improved on the initiative of one landowner in the economic context of the wars with America and France. Some 1650 miles (2640 km) of field boundaries were thus created, predominantly of dry walls. Had they been quickset hedges, the gain to wildlife would have been enormous.[25]

In both the eighteenth and the nineteenth centuries, the Peak District and the northern Pennines were the main regions of lead mining and in both places an earlier industry was greatly expanded (Plate 6.2). In the case of Derbyshire there was a rapid decline after 1850 except at Mill Close Mine in Darley Dale which persisted until the 1930s along with some fluorspar mines.[26] Environmental relations included very long drainage tunnels ('soughs'), which emitted mine-water into a main watercourse perhaps 5 to 7 km away, the heaps of black slag from stamping mills and the waste lead from refining plant emitted from chimneys. Extraction of the base ore from the major veins and their side branches ('rakes' and 'scrins') was often carried on from a series of surface pits, giving rise to linear strakes of overburden in the landscape, still visible today.

In the North Pennines, flues were used from 1778 onwards. They conveyed the poisonous fumes to chimneys from which dispersal over a wide area could be combined with the recovery of lead which condensed on the inside of the structure. They might well be 1 to 3 km long and usually ran uphill, not always in a straight line: the flue from the Keld Heads mill at 200 m ASL in Wensleydale ran through several bends before discharging from a stack at Cobscar Mill at about 380 m. The Rookhope chimney north of Weardale, though, had only one significant bend in its 3.5 km length and 220 m of ascent.[27]

The development of 'heavy' industry in the uplands during the nineteenth century now adds to our confusion about their environmental management. We see them as havens of rural peace and wildness and the remains of, for example, lead smelting mills as somehow romantic. Yet to replace the smelt mill's chimney (after a gap of 80 years) with a wind farm provokes considerable controversy and policy indecision.

FIGURE 6.2 (opposite)   *Annual energy, material and cash flows at Fyfield Manor Farm, Milton Lilbourne, Wiltshire, 1826. The linkages to the local environment are added to the quantified flows. T. Bayliss-Smith,* Ecology of Agricultural Systems, *Cambridge: Cambridge University Press, 1982, p. 47, Figure 4.7.*

PLATE 6.2 *Nenthead, Cumbria. Lead mining and smelting were the great environmental manipulators at Nenthead in the nineteenth century and scarcely a square centimetre of this scene was unaffected by it: if the substrate was not actually formed of mining and processing waste, then the atmosphere would have toxified most of the ecosystems within sight. Since 1998, use as part of the museum has tidied up the landscape a great deal.*

The new money made in the early nineteenth century, combined with the easy access to rural areas made possible by the railways, created a demand for country sporting pursuits. What landowners needed was a sport which appealed to the new rich, did not interfere with existing patterns of land use to any degree, and made money. The answer was grouse shooting on moorlands. This bird (*Lagopus scoticus*) had long been netted or shot from behind as it was flushed from the stands of heather which are the staple food of adult birds.[28] It was discovered that if the drier moors of northern England and Scotland were burned in 30–50 m strips every 10–15 years (Plate 6.3) then the density of heather and the grades of the plant from young shoots to dense bushes would allow the territorial bird's densities to rise.[29] On the opening day of the season (12 August), beaters would drive the birds across a line of turf or stone butts; in each butt was a sportsman and a loader. The gun was of the breech-loading shotgun type, another introduction of the early nineteenth century. In these circumstances, it was difficult for even inexperienced financiers to miss and good shots might pile up the carcasses at a very high rate. One moor of about 4000 ha in Yorkshire yielded 2843 birds to the guns on 27 August 1913 and about 5000 birds over the whole season to 10 December. One day in the late nineteenth century, Lord Walsingham shot 1070 grouse.[30] The environmental impact of this sport has been quite strong: the low concentrations of nutrients in the acid upland soils are

PLATE 6.3   *A grouse moor in the Pennines (north from Arkengarthdale) showing the patches of heather of different ages: the darker the patch the longer since burning. The sheep emphasise that the moors are grazed as well as shot and probably also that they are accustomed to supplementary feeding, given the densities of the mid-1990s.*

mobilised into smoke and runoff, soil erosion is initiated, and attempts to encourage heather by digging drains ('piping') exacerbate peat erosion and make the streams more prone to flood. Stretches of heather monoculture are prone to accidental fires which then burn underground in peaty topsoils. The strongest effect, however, was caused by the perception that predatory birds and animals were an efficient restraint on the density of grouse. Keepers therefore waged war on raptors and anything else with claws (Table 6.2) and continued to do so until legislation provided some kind of check. By then, however, Scotland had been depleted entirely of the sea eagle (*Haliaetus albicilla*), gyrfalcon (*Falco rusticolus*), osprey (*Pandion haliaetus*) and goshawk (*Accipiter gentilis*).

In Scotland, the sheep bonanza of the Highlands was short-lived: there were many grazings that would not support a high density since the species composition of the pastures passed to the unpalatable very quickly under the intensities of animals initially deployed, and cheap refrigerated meat from Australia and New Zealand became available in the 1870s. Landowners therefore tapped the same rich vein of demand that fostered grouse shooting: land over 600 m was managed as 'forest' carrying red deer and attracted rich industrialists and trendsetters like Prince Albert. By 1850 there were already 42 such estates and 118 by 1885. Some 1 million ha of such land in the 1870s became 1.5 million in 1912, though perhaps as much as 0.4 million ha were also used for sheep. Many of the deer were fed cereals on the low ground in winter and so a density of 1 deer to 6–8 ha was produced; the 'natural' carrying capacity is more like 1 deer to 16 ha. Before World War I, ample labour meant that an annual cull of hinds could be carried out (sportsmen were only interested in the stags and then mostly in those with the largest spreads of antlers): war and other employment opportunities changed that ecological relationship for good. The deer population of the Highlands has rapidly become much too dense, to the detriment of other land uses and of the deer populations themselves; the attitudes of the 1880s have, however, coloured the management of the deer populations right up to the present day.[31]

TABLE 6.2   Wild animals destroyed on one Scottish estate 1837–49

Common name	Number destroyed
Pine marten	246
Polecat	106
Kestrel	462
Kite	275
Rough-legged buzzard	371
Common buzzard	285
Wild cat	198
Raven	475
Merlin	78
Goshawk	63
Short-eared owl	71
Hen harrier	92
Badger	67
Golden eagle	15
Peregrine falcon	98
Osprey	18
Sea eagle	27
Hobby	11
Gyr falcon	6
Marsh harrier	5
Honey buzzard	3
Stoat	301
Otter	46
Fox	11
Magpie	8
Hooded crow	1431
Domestic cat	78

Source: League Against Cruel Sports, *Shooting Birds*, leaflet, n.d.

FRESH WATER AND FRESH-WATER WETLANDS

The rivers could hardly have been immune from industrial and urban effluents. The deference to industry and trade was such, however, that Parliamentary enquiries into the disappearance of salmon from the Thames in the early nineteenth century concentrated on the mechanics of salmon fishing and not on the water quality. In 1858 a hot spell caused such a smell of sewage that Parliament was adjourned for a week. The levels of pollution had become apparent as early as the 1840s and 1850s in Yorkshire's West Riding when the watercourses had the appearance of ink (and indeed letters could be written in Calder water in 1868)[32] and the Bradford Canal could be set on fire. Sewage was just one pollutant: coal ash and cinders, coal washings and trade effluents from dye works, tanneries and distilleries all found their way into the nearest watercourses. The organic materials depleted the water of oxygen and the particulate materials prevented any form of photosynthesis, so that rivers became lifeless. Attempts to clean up rivers have been particularly slow in Great Britain; the Rivers Pollution Prevention Act of 1876 was the main piece of legislation until 1951.

The northern cities polluted water with abandon but getting it was a more difficult task. When local boreholes could not supply enough water and the rivers' quality had become too low, then the local upland catchments were tapped. The high rainfall of the uplands, the shape of the valleys and the cheapness of the moorland made them the obvious sites. When the nearby uplands had all been carved into territories by local authorities, then the tentacular reach for resources stretched further: the grasp of Manchester upon Thirlmere in the Lake District is well known. By 1904, the gathering grounds (mostly moorland) in England totalled about 33,000 ha and in Wales a further 2500 ha were in this use.[33] Moorlands are subject to soil and peat erosion and during this century the loss of capacity of reservoirs due to sedimentation is in the order of from 15 to 20 per cent.[34] In drier parts of the nation, boreholes could now be sunk to 300 m using steam power in drilling equipment. Adits were also drilled and dug to intercept water-laden fissures in rocks such as the Chalk.[35]

In 1805 the main areas of the Fens drained by the Bedford Level were deteriorating fast. The desolation betokened the fact that hundreds of windmills were failing to take the water off the wasting peatlands fast enough. The great salvation came with the installation of the first steam pump at Littleport near Ely in 1819–20. By 1848, such pumps were extensively used in the Fens and along with further improvements in the river channels removed the threat of flooding for a further period even though the drained peat went on shrinking. The effectiveness of pumps meant that other wetlands were made less quaggy: Martin Mere in Lancashire and the Humber wetlands were reclaimed at that time. Confidence in the permanence of reclamation meant that the remaining parts of the great East Anglian wetland (East, West and Wildmore Fens) were drained in the first two decades of the century. The last great areas of wetland wildlife habitat in fresh waters, other than the Norfolk Broads, disappeared and along with them the hope that birds such as the bittern and the reed bunting might ever be plentiful. In fact, the double impacts of drainage and shooting made some birds finally extinct in the Fen country during the nineteenth century: examples are Savi's warbler (*Locustella luscinoides*), the black tern (*Chlidonias niger*), the ruff (*Philomachus pugnax*) and kites (*Milvus milvus*). Species brought to the verge of extirpation in the region included the great crested grebe (*Podiceps cristatus*) and the marsh harrier (*Circus aeruginosus*).[36]

There was some stirring on behalf of the wilder wetlands. In 1895, the first suggestion was made that the remnant of unreclaimed Fen at Wicken near Cambridge should be a nature reserve and most of it was given to the National Trust in 1910. The complex relations between wetland conservation and environmental history can be seen in the various uses and patterns of land tenure of the Fen. In the seventeenth and eighteenth centuries, when it was a common, there were a large number of management units and so the biota were diverse. After enclosure in the nineteenth century there were only three different areas: peat digging (4 ha), 125.5 ha of sedge fen and 30 ha of drained land (Fig. 6.3). The demand for sedge declined and so cutting of it ceased. The reserve thus consisted of a low-diversity set of ecosystems in a particular state of succession: the part drained in 1846–9 was the last Wicken Fen station of *Senecio paludosus*; the uncut sedge areas were invaded by scrub which shaded out the food plant of the swallowtail butterfly.[37] But until this kind of environmental sequence has been unravelled, management for species perpetuation is unlikely to be successful.

FISHING AND WHALING

Though not restricted to the nineteenth century, British whaling was then at its height. The quarry was mostly sought in Arctic waters (off Spitsbergen and in the Davis Strait 1733–1824) and in the southern Pacific and Atlantic oceans from 1790 to 1843. British vessels never bought in more than about one-fifth of the oil consumed in Britain. (Its use until *c*.1815 was mostly for lighting and for cleansing in the textile industries.) In 1812, there were 138 whaling vessels registered in the UK, a number reduced to 18 by 1842. Between 1815 and 1830 there was a slump in demand as rapeseed oil took over the cleansing function and coal gas that of lighting, but the industry was saved by the growth of the jute industry and by demand for whalebone corsets: the price of baleen rose from £500 per ton in 1870 to £3000 per ton in 1902.[38] By 1859, steam whalers were being built and better killing technologies adopted: the industrialisation of whaling was largely the achievement of the Norwegian Sven Føyn who patented a steam-driven catcher with gun and explosive harpoon in 1873. The northern bowheads began to decline from the 1830s onwards and, by the turn of the century, Britain was no longer a major whaling nation, though some attempts to compete in the Pacific and Antarctic were made. By the time of World War I, world-wide declines in nearly all whale populations were apparent. Peterhead in Scotland remained a long time in the trade but was outshone by Dundee, whose boats also caught large numbers of seals for their skins.[39]

The industrialisation of fisheries out at sea begins in 1860 when a steam tug from Sunderland towed two fishing smacks some 8–16 km offshore. In 1881, purpose-built steam trawlers were built which caught four times as many fish as a sail-driven vessel; in 1883 there were 181 steam trawlers operating out of British ports and in 1901, 1573. Hull and Grimsby especially profited from this development since they had easy access to cheap coal. (But 5887 sailing smacks remained in 1902 and the last worked out of Lowestoft as late as 1938.) Scottish fishing was dominated for almost the whole century by the pursuit of the herring, which although having its boom periods in the

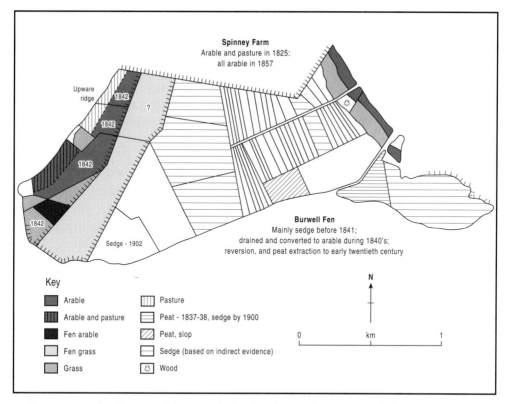

FIGURE 6.3   *Land use at Wicken Fen during the nineteenth century. At that time, the demand for sedge was at its height but thereafter much of the Fen was colonised by bushes. The point is that the value of such a habitat for conservation of, for example, plants and insects may well be dependent on recent use and management.*
T. A. Rowell and H. J. Harvey, 'The recent history of Wicken Fen, Cambridgeshire, England: A guide to ecological development', Journal of Ecology 76, 1988, p. 82, Figure 6.

twentieth century foundered on too great a degree of specialisation. When markets in Germany and Soviet Russia were closed, then the industry could not adapt.[40]

As the trawler developed, the North Sea was progressively exploited, adding to the pattern developed by English smacksmen. The impact on stocks was considerable and very soon felt, for there were numerous government enquiries (including two Royal Commissions) in the later nineteenth century. Laws about mesh size and minimum fish landing size existed but were not enforced and no comparative statistics existed. The President of the Royal Society, Professor T. H. Huxley, addressed an international Fisheries Exhibition in London in 1883 and gave it as his view that at sea,

> the multitudes of these fishes is so inconceivably great that the number we catch is relatively insignificant; and, secondly, that the magnitude of the destructive agencies at work upon them is so prodigious, that the destruction effected by the fishermen cannot sensibly increase the death rate.

He used as the basis for his argument the situation of the Norwegian Arctic fisheries, which was, as it happened, not comparable with the North Sea. Thus he was able to go further and say,

> I believe, then that ... [cod, herring, pilchard, mackerel] ... probably all the great sea fisheries are inexhaustible; that is to say that nothing we do seriously affects the number of the fish. And any attempt to regulate these fisheries seems consequently, from the nature of the case to be useless.[41]

Yet by 1893 a House of Commons Select Committee was faced with good evidence that catches had declined since 1875, due to the introduction of steam trawlers using beam trawls.[42] The Select Committee recommended the adoption of minimum landing sizes, the extension of territorial limits to protect immature fish and the establishment of a Sea Fishery Board to collect statistics. No action was taken on any of these suggestions. The general under-attention paid to pollution and over-fishing also resulted in the decline of oyster production; oysters were the food of the poor in the nineteenth century (consumption in the 1850s was 500 million per year) but, in contrast to France, their subsequent rarity has confined them to the rich.[43]

ENERGY

The nineteenth century was the time of the full transition to a fossil fuel-based economy. In practical terms, this meant the burning of coal to produce heat. The main route of application was via a steam engine, so that their distribution and power along with maps of mining areas are maps of economic changes likely to be reflected in the ecology of the region (Fig. 6.4).

Coal output in England and Wales rose from 16 million tonnes in 1815[44] to 49 million tonnes in 1850 (Plate 6.4). In the latter year there were at least 1704 collieries in England, though that number probably omits a number of small shafts. In 1800, very little coal was hauled overland for more than from 8 to 16 km but a great transformation came with the railways, with the first coal being taken to London by that means in 1845. In the second half of the century, coal was still a major factor in industrial location, with the production rising to 183 million tonnes in 1913, of which one-third was exported. By then, some electricity was being used to power manufacturing industry, but it too was generated locally from coal. In Wales the dramatic change was in the second half of the nineteenth century, with the Principality's output rising from

FIGURE 6.4 (opposite)   *Coal output by region from 1800 to 1911. This shows the early lead of the north-east and the relatively late arrival of South Wales, so that local environmental transformation in the latter region lacked some of the small-scale extraction plants of the north-east. Beyond the local effects, the abundance of coal made possible the wholesale transformation of the economy and led to much environmental change.*
J. Langton and R. J. Morris (eds), Atlas of Industrialising Britain 1780–1914, *London: Methuen, 1986, p. 73, Figures 8.1–8.4.*

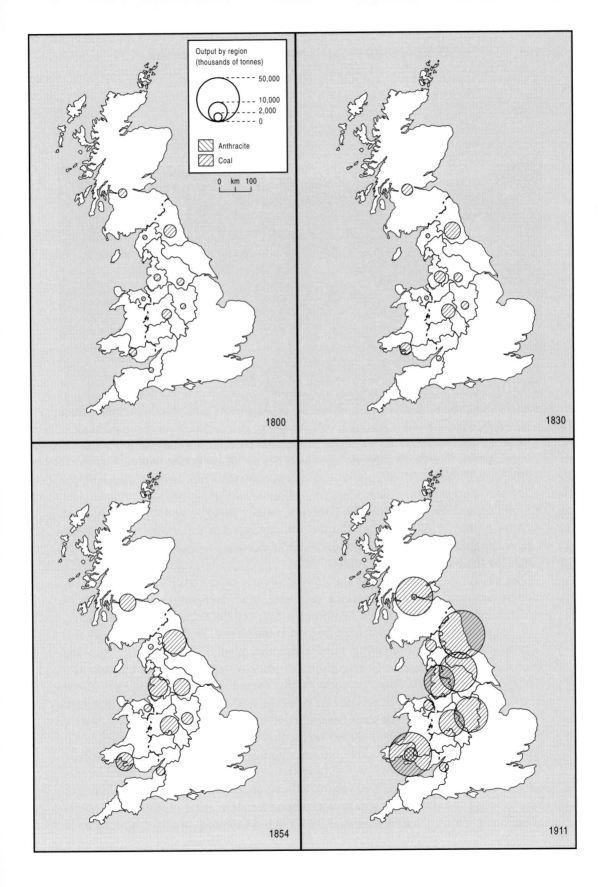

Output by region
(thousands of tonnes)

50,000
10,000
2,000
0

Anthracite
Coal

0 km 100

1800

1830

1854

1911

PLATE 6.4 *Although this is a museum (Beamish in County Durham in 1983), it conveys the sense of environmental disintegration that accompanied the development of coal in the nineteenth century. The mine, the transport links and the housing seem to be placed almost randomly, which was not actually the case.*

8.5 million tonnes in 1854 to 39.3 million tonnes in 1900 and the labour force increasing from 29,100 to 147,600 in the same period. The expansion was down the valleys towards the coast from the earlier iron districts, and saw the emergence of Cardiff as the focal city for the trade.

The environmental impact of this massive development is difficult to overstate, though not much quantitative work has been done, compared with the mass of statistics extracted by economic historians.[45] Between 1800 and 1875, for instance, coal that could not be sold was either burned at the pithead or flushed into rivers and streams; slack and dust was simply piled into heaps. Small coals to be used for coking were usually washed and so huge volumes of black water were fed into brooks and rivers alike. At times of flood, meadows were covered in coal dust sedimented out from the water. In the air, a mixture of carbonic acid, sulphuric acid and nitrogen oxides checked the growth of some trees and destroyed others. Crops of all kinds were injured and the wool rendered useless before it even came off the sheep's back. Even more pervasive than pitheads were domestic chimneys, where the soot made the incidence of sunlight 20 per cent lower in towns when the first systematic measurements were made in 1920. In 1918, it was estimated that an average urban household might have to spend an extra £1 per year on washing materials in order to combat the effects of pollutant fallout, at a time when a skilled coal hewer would earn 38 shillings a week gross. Measurements in Attercliffe (an industrial district of Sheffield) in 1914–16 reported a soot deposition of 9.6 tons/square mile per month, to be compared with eight stations in London which averaged 5.9, Manchester at 4.3 and the spa town of Malvern at 0.4. (9.6 long tons/sq mile is about 38 kg/ha.) As Clapp puts it, 'To the inhabitants of Attercliffe it meant eight ounces of dust falling every year on every square yard of the district' (8 oz ≈ 0.23 kg).[46]

Some idea of the implications for other systems comes from the data that in 1870, the 100 million tons of coal used produced the same quantity of calories needed to feed 850 million adult males for a year or that the steam engines in use that year were the equivalent of 6 million horses or enough wheat for the food of 30 million men, which was three times the actual population of males in 1870.[47]

INDUSTRY

In this period, the transition to a fully industrialised economy was completed as far as manufacturing is concerned. A complex of processes which include geology, land ownership and the means of transport focuses on the industrial process itself, which then has downstream effects like pollution, discarded by-products, subsidence, quarrying and ephemeral phases which leave no later trace. The changes were profound in an ecological sense, for the relationship with the land and the seas was changed in many ways. The symbol of it all is the Great Exhibition of 1851 in Hyde Park, housed in a great crystal palace of iron and glass (4500 tons of the former, 293,665 panes of the latter), which was above all a showcase for British industry. It was a larger version of the glasshouses at Chatsworth House in Derbyshire, which had been designed by the Duke of Devonshire's architect, Joseph Paxton. They are still there.

The pillars of this new economy, in which Britain led the world for a few decades, were made of iron which had been smelted and worked using the energy from coal (Plate 6.5). Apart from a few remaining charcoal-powered furnaces in the Furness district of north Lancashire, iron mining was also concentrated mostly on the Coal Measures. In the north-east of England for instance, the Jurassic ironstones of Cleveland met the coals of County Durham at the river Tees and there sprang up an iron producing region which by 1854 possessed 54 blast furnaces at Eston and Skinningrove. The most manipulated region of all was the Black Country: an area west of Birmingham of some 26 km². Nowhere else was so transformed by mid-century and by then the thickest coal seams were already exhausted.[48] Local iron also became scarce in some of the producing regions and by the 1870s the quarrying of ironstone in the Jurassic belt from Wiltshire to Lincolnshire was under way, leaving roads high if not dry above the worked-out pits either side.

Whereas most of the iron went into manufacturing or capital goods, the coal powered both industry and the domestic hearth (Table 6.3). Coal and iron met in the form of the steam engine, a device undergoing continuous improvements in fuel efficiency during this century. By 1880, steam engines supplied c.80 per cent of the power in the cotton industry and 85 per cent in woollen manufacturing. Textiles used about one-third of the total industrial horsepower and it has been suggested that a

PLATE 6.5 *Blaenavon ironworks in 1798. Most engravings and paintings of ironworks in the nineteenth century show huge billows of smoke and this installation seems relatively clean. Note the detail which shows the impact of the plant on the surrounding area's landforms. An engraving from a drawing by Sir Richard Colt Hoare. © Cadw (Welsh Historic Monuments).*

great deal of water-powered industry remained.[49] Where the steam engine was undisputed king was on the railways. Once reliable and reasonably efficient locomotives had been developed then an explosion of railways took place. The signal event, if we may use that term, was the opening of scheduled passenger services on the Stockton and Darlington Railway in 1825. Even in 1850, though, the average journey of coal by rail was from 50 to 65 km: sea transport was still very important until the 1890s when the production of the Yorkshire–Derbyshire–Nottinghamshire coalfield overtook that of the north-east. Railway mania (Fig. 6.5) added new elements to the nation's landscapes: raw cuttings and embankments, viaducts and bridges, stations and marshalling yards.[50] Drainage might have to be diverted, or occasionally a railway junction's triangle of lines would create yet another pond of stagnant water. At

FIGURE 6.5 (opposite) *The growth of the railways c.1840–1900. Two environmental consequences stand out: first, the whole new sets of habitats that were created in and around the tracks, and, second, the movement of materials that allowed urbanisation and the location of industry to expand.*
J. Langton and R. J. Morris (eds), Atlas of Industrialising Britain 1780–1914, *London: Methuen, 1986, p. 89, Figures 9.10–13.*

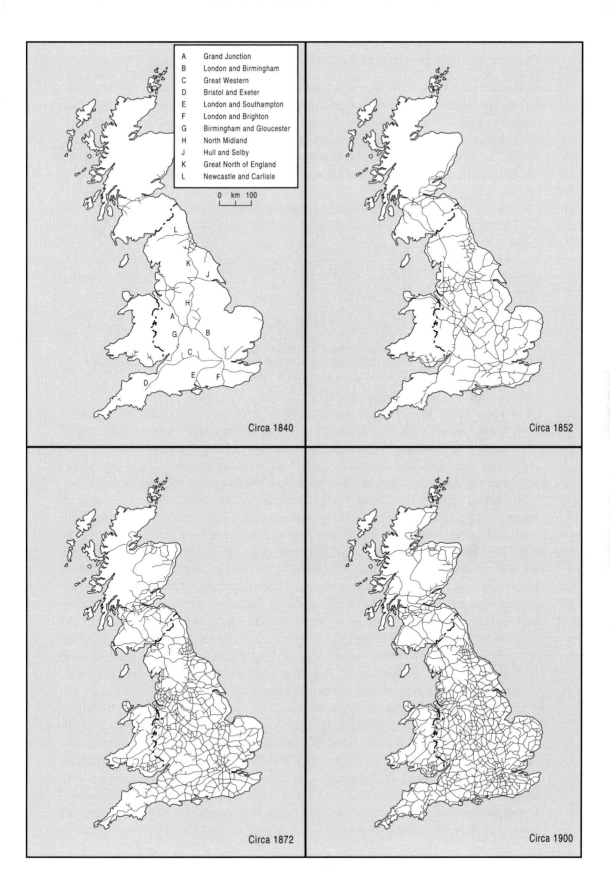

A	Grand Junction
B	London and Birmingham
C	Great Western
D	Bristol and Exeter
E	London and Southampton
F	London and Brighton
G	Birmingham and Gloucester
H	North Midland
J	Hull and Selby
K	Great North of England
L	Newcastle and Carlisle

0   km   100

Circa 1840

Circa 1852

Circa 1872

Circa 1900

Hinksey, near Oxford, the wet pits from which the gravel was taken for the ballast for the tracks became a source of drinking water and then of recreational use. Land formerly in agricultural use was swallowed up in new railway towns, designed to serve the industry or at important junctions: hence the growth of Swindon, Wolverton and Crewe, for instance. The railways made it necessary to synchronise time over the whole kingdom, a hitherto unnecessary uniformity.

TABLE 6.3   Use of coal in the UK (Percentages by use)

	1840	1913	1929	1955
Iron industry	25.0	11.0	10.0	12.0
Gas and electricity	1.5	8.0	11.0	32.0
Manufacturing	32.5	22.5	22.5	28.5
Domestic	31.5	13.5	15.5	16.5
Exports	5.0	32.5	30.0	6.0

Note: Not all categories of use are included and so the columns do not sum to 100 per cent.
Source: P. Deane and W. A. Cole, *British Economic Growth 1688–1959*, 2nd edn, Cambridge: Cambridge University Press, 1967, p. 219, Table 55 [This is about half to two-thirds of original table].

Two facets bear mention: the first that, when new, the railways created storms of protest and some aristocratic disdain that led to curious routing decisions but once established and bedded in to the landscape became objects of affection; second, that the construction of them (as many contemporary paintings and engravings show) was largely by human labour without the help of steam-powered machines.

Important though the railways were, the nineteenth century was also a time of improvement of road transport. The government paid for many new roads in the 1820s and 1830s, a time which was the greatest age of road improvement before the motor age. The environmental linkages included some notable engineering works, of which the Shrewsbury to Holyhead road through Wales is an excellent example. Coaching towns had their great apogee and some had to designate special places (usually a part of a river bank) for dispersal of the horse dung. Sea transport was still important for bulk goods so that the mines on the Isle of Man could be at their height in the 1820–90 period, producing large quantities of, especially, zinc. The great water wheel at Laxey was built in 1854 to pump out those mines, which created large spoil mounds in the valley, known as the 'deads'. Their size was later diminished when they were used for hardcore in airfield construction on the mainland.[51]

If we follow Wrigley's analysis of the stages of industrialisation then transition to a mineral-based economy reduces pressures on the land: 1 million tons of coal provided as much heat as 1 million acres (405,000 ha) of forest.[52] Even more graphically, Wrigley envisages this as an addition of 15 million acres (6 million ha) of cultivable land between c.1560 and 1820. Equally, food could now be bought from abroad with the profits from selling manufactured goods. Mineral stocks of fuel, too, are suited to rapid economic growth, in a way that living resources are not: a 6 per cent per annum growth rate in coal production was achieved in the 1850s. Positive feedback is an undisputed result: energy extracted from coal gives a surplus to extract yet more and to divert into intensifying the production of other manufactures and into fertilisers for agriculture: the throughput of the whole economy is moved up several notches.

The environmental effects of this energy production and use are taken up in various other paragraphs, but they were indisputably profound. Many writers and travellers of the period remark upon the new densities of settlement that they found, as well as the distressing levels of inquination of air, water and land. As early as 1835, a canal in Manchester was declared to be as black as the Styx and 'pestiferous' from gas and effluents. Manchester attracted much of the comment on both environmental and social conditions at the time: it was on every traveller's itinerary. The most famous passage is from FRIEDRICH ENGELS, who lived there, relating to 1844:

> The view from this [Ducie] bridge . . . is characteristic for the whole district. At the bottom flows, or rather stagnates, the Irk, a narrow, coal-black, foul-smelling stream, full of *débris* and refuse, which it deposits on the shallower right bank. In dry weather, a long string of the most disgusting, blackish-green slime pools are left standing on this bank, from the depths of which miasmatic gas constantly arise and give forth a stench unendurable even on the bridge forty or fifty feet above the surface of the stream. But besides this, the stream is checked every few paces by high weirs, behind which slime and refuse accumulate and rot in thick masses. Above the bridge are tanneries, bone mills and gasworks, from which all drains and refuse find their way into the Irk, which receives further the contents of all the neighbouring sewers and privies.[53]

The actual living conditions of people in small back-to-back houses recently erected without basic facilities but at high densities appalled an observer (A. B. Reach) in Leeds in 1849:

> Conceive acre upon acre of little streets, run up without attention to plan or health – acre upon acres of closely-built and thickly-peopled ground, without a paving stone upon the surface, or an inch of sewer beneath, deep-trodden sloughs of mud forming the only thoroughfares – here and there an open space, used not exactly as the common cesspool, but as the common cess yard of the vicinity – in its centre, ashpits, employed for dirtier purposes than containing ashes – privies often ruinous, almost horribly foul – pig-sties very commonly left pro tempore untenanted because their usual inmates have been turned out to prey upon the garbage of the neighbourhood.[54]

Even in such surroundings, the enthusiasm for the new economy was certainly strong. A German visitor of 1844 recorded that the graffiti on the desks at Manchester Grammar School were of canals, railroads, ships and locomotives. Yet there was some precariousness: many towns had railway tracks, embankments and bridges that had never borne traffic since the company had failed before it had even got started, or had been bought out by a competitor at the embryo stage. The Ketley ironworks in

Shropshire was the second biggest in Britain in 1804 but by 1817 was declared to be an appendage to the Wellington workhouse. None of these events retarded the growth of an industrial lifestyle but they do provide a reminder that being based on finite supplies of a mineral at any scale always carries with it the possibility of running out. None of these considerations prevented the agglomeration of pollution-producing industries in areas deemed most suitable for them as around the estuaries of the Mersey and Tyne (Fig. 6.6).

URBAN SETTLEMENT

Along with the industries that made them grow, Victorian Britain is above all else associated with the expansion of towns and cities. As the population of the nation rose, the proportion of it that lived an urban life grew spectacularly (Table 6.4). Some of the new industries were land-intensive: 1 ha of cotton mill produced as much as 25 ha of spinning jennies, and 1901 levels of output using pre-Victorian technology would have left much less of the country looking rural. Shaft-mined coal was relatively less consumptive than bell-pits but nevertheless in 1901 there were 60,000 ha of colliery workings, which was about one-third the area of Oxfordshire. The railways in that year were responsible for the use of about half the area of that county, since they needed land at the rate of some 3.0 ha/km for all their purposes in both town and country. The total area of urban land in England and Wales in 1901 was of the order of 162,000 ha, the size of Bedfordshire.

TABLE 6.4   Urban growth in the nineteenth century (England and Wales)

Date	Total population (millions)	Urban population (millions)	% urban	Index of urban population (1801 = 100)	Number of places with population > 100,000
1801	8.8	3.0	34	100	1
1851	15.9	9.6	54	322	10
1901	32.5	25.4	78	843	33

Source: H. Carter, 'Towns and urban systems 1730–1914', in R. A. Dodgshon and R. A. Butlin (eds) 1990, *A Historical Geography of England and Wales*, London: Academic Press second edition 401–428, p. 403, Table 14.1 [This is about 25% of the original].

The environmental connotations are mostly well known. The towns expanded concentrically but patchily, taking in agricultural and other extensive land uses as they

FIGURE 6.6 (opposite) *Chemical plants around the Mersey and on Tyneside in 1882. These maps were compiled from an early official survey under the Alkali Act of 1891. Some are in fact acids rather than alkalis but, since most were led out to the nearest watercourse, the subsequent cocktail could scarcely be imitated on Saturday nights nowadays in Newcastle or Manchester. R. Pope, Atlas of British Social and Economic History since c.1700, New York: Macmillan; London: Routledge, 1989, p. 38, Figures 2.13 and 2.14.*

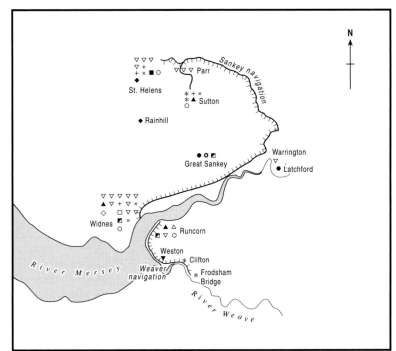

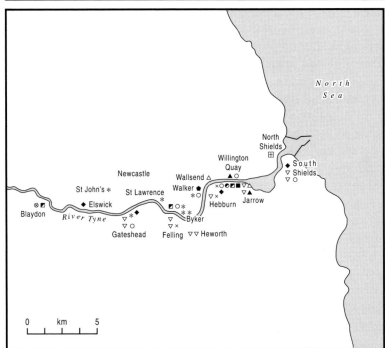

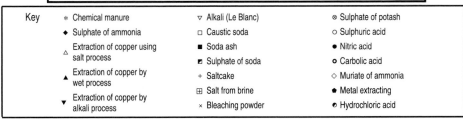

Key			
* Chemical manure	▽ Alkali (Le Blanc)	⊗ Sulphate of potash
◆ Sulphate of ammonia	□ Caustic soda	○ Sulphuric acid
△ Extraction of copper using salt process	■ Soda ash	● Nitric acid
▲ Extraction of copper by wet process	◪ Sulphate of soda	◎ Carbolic acid
▼ Extraction of copper by alkali process	+ Saltcake	◇ Muriate of ammonia
	⊞ Salt from brine	⬟ Metal extracting
	× Bleaching powder	◓ Hydrochloric acid

went, but also creating more open areas as well. Parks, asylums and hospitals, cemeteries, workhouses and waterworks were all part of an urban fabric. Parks in particular
were seen as emblems of civic pride and provision, though they were in some cases
thought to be efficacious against urban riots: the first examples were in Derby and
Birkenhead, in 1839 and 1843 respectively. Pride might also spur the development of
civic grandeur, with the building of neo-classical town halls or the ceremonial
sequence of Admiralty Arch, the Mall, the Victoria Memorial and Buckingham Palace.
In many cities, the tidiness decreed of the countryside in the eighteenth century was
now required of the towns and especially of their inhabitants: most Victorian city
fathers would have loved to have had CCTV at their disposal. They did have the villa
with a large garden, part of the same process as the enclosing, privatising and frag-
menting of the countryside that had gone on in the previous century.

The city had for some time been the object of censure, especially in the 1830s and
1840s, when the health of the working population was affected by poor conditions,
which led among other illnesses to outbreaks of Asiatic cholera and typhoid. Motivated
by either philanthropy or loss of profits among important denizens, civic authorities
began to improve the cities, aided by national legislation and local bye-laws that, for
example, regulated the density of housing, allowed the building of sewage collection
and treatment systems, and the construction of tramways along public roads. One
result was that housing became much more land-hungry: fewer dwelling units per
hectare were built after the Public Health Act of 1875. Birkenhead ran trams in the
street in 1860: thus the suburb was initiated and walking to work was no longer the
automatic lot of the working classes.

The environmental effects of the totality of these progressions have not been
studied in any detail. The removal of sewage from large cities as water-borne waste in
pipes is analogous to building higher factory chimneys in one sense: it shifts the
problem somewhere else. So while London's sewage was collected in wide-bore
underground sewers but discharged into the River Thames downstream of the city,
then the inhabitants of Barking and the fauna of the river were subjected to the risks.
Not until there was treatment of the sewage could there be any ecological improvement
and even today the sludge is taken to sea and dumped. (It is, however, argued that the
productivity of the North Sea would be even lower without the additional nitrogen
from that source.) Better quality water usually meant new sources and so more rivers
were tapped and their levels lowered at critical times, especially in the summer;
hence they became depleted of flora and fauna, especially fish. The development of
boreholes lowered water-tables and hence the flow rates of streams dependent upon
underground aquifers; short bores allowed some subsidence where extraction rates
where high, exactly as happened with shallow coal mines and with salt. The Victorian
sense of cleanliness found one outlet in the attempts to turn sewage into profit by
getting it (in dry or wet forms) to farms. This was in general a failure and so a land-
intensive, urban-based model of processing was adopted in which there was no desire
to recycle urban wastes, simply to avoid more river pollution. But the term 'sewage
farm' has persisted until quite recently.[55]

The city also called for the transmutation of land via the resources which it sucked in. The lower Medway valley was altered beyond imagination after 1850 since it provided cement for London: 15,000 tons per week in 1898. South of Peterborough the landscape became one of pits, at first dry, then wet, from the brickworks that were filling up wagons bound for London. North Wales was host to some huge quarries where the slate for roofing the cities of the nineteenth century was produced. In the urban areas, these materials were turned into huge stores of brick, stone and wood which created many habitats for plants, animals and micro-organisms. Aberdeen quarried its own granite pillars of Victorianism: 6 million tonnes of it from the Rubislaw quarries, 60 m deep and within the city boundary. They also changed the local hydrological cycle by accelerating runoff and evaporation from impermeable surfaces, as well as contributing all manner of substances to the water chemistry. Particulate pollution in the air often produced higher rainfall totals downwind. Towns and their suppliers alike were generators of solid wastes, starting a set of processes that were to be magnified every time that affluence became greater. The brickfields west of London that had stretched from Southall to Slough became even more valuable as holes to receive the solid wastes of the capital.

Lastly, land was taken up for the building of new urbanisations by those few industrialists determined that their workers should be healthy. Several examples exist, but Saltaire (built 1853–63), now part of Bradford, can represent them. It catered for all grades of worker in the woollen mill and offered a 'wholesome' environment in both material and spiritual terms. 'Far be it from me', said Titus Salt, 'to do anything to pollute the air or the water of the district'.[56] We might wonder if he ever had an environmental audit.

RURAL SETTLEMENT

A number of changes may be noted in this century. The completion of enclosure meant that more buildings than ever before were erected away from villages and upon the enclosed lands themselves: farms were often named after some event of the time, like Waterloo. Farm workers' cottages often went with them but at the same time an estate owner might fill in gaps in a village. Generally speaking, though, this was a century of rural depopulation, as families emigrated from a countryside with recurrent periods of agricultural depression. So the resulting rural pattern is in general a fragmented one.

A very noticeable addition to the village came when coal was mined in the vicinity. In some cases a settlement devoted largely to the rural economy might have grafted onto it several rows of miners' dwellings, in addition to the industrial gear of the mine itself not far away. In other places, a mining hamlet was built at a remove from the original village but near to the shaft and appeared as an island of terrace housing in a largely open landscape. The same is true of mining in the uplands, as with lead in the Pennine Dales. It was at one stage common for the lead miners to live in villages and walk many miles to their work, staying during the week in a barrack-block or 'shop'. More paternal companies, such as the London Lead Company, preferred to keep family units intact and so terrace rows were built in quite remote parts of the upland valleys. To improve the workers' health, small patches of land would be granted as smallholdings and so the margins of cultivation (albeit of an especially intensive kind) were extended beyond the economic level associated with the climate of those regions.

Observers of 1900 in the countryside who were able by some miracle to compare their experience with 1800 would notice first that there was more poverty. The condition of the rural poor was often as bad as their cousins who had gone to the towns. Secondly, that there was a filling-up of the countryside, with more buildings and more activity, in a way probably not parallelled until after the later intensification of farming beyond 1950. A small place might develop an industry and grow rapidly: Leiston in Suffolk claims to be the birthplace in 1825 of the production line for steam engines. More people and more activity, in both cases, mean more environmental impact.

LEISURE AND PLEASURE

The onset of industrialisation produced new opportunities for leisure. The rich made more money and hence could hire more people actually to do the work. The 'labouring classes' shared very little in this until the railway companies began to see that cheap fares on excursion trains to resorts would provide profits. In environmental terms, one important manifestation turned out to be a continuation of old habits and another almost an entirely new phenomenon: namely hunting and shooting, and the seaside town, respectively.

Hunting and shooting

One of the problems with the great glasshouse built for the Great Exhibition of 1851 was the house sparrow. A population established itself in the structure, with unfortunate results for ladies' dresses, hats and sensitivities. Queen Victoria turned to the ever-resourceful Duke of Wellington for ideas and in a moment that combined ecology with aristocracy, he is reported as saying, 'Try sparrowhawks, ma'am.' That encapsulates the response of the (especially) English ruling classes to changed environments, at home or abroad: what is there to be killed?

As woodland shrank in area and enclosure spread, the popularity of deer as a target for rich hunters fell away, though there were still 16 packs of staghounds in 1895.[57] After 1750, however, the chase of the fox rose to be the dominant sport, at its height in the nineteenth century. A mounted hunt followed a pack of dogs which chased a single fox that had been forced out of its earth in a small copse or thicket of rough ground. The sport was well adapted to a post-enclosure landscape, for the fox is an animal of edge habitat and could be encouraged by planting small copses of trees or of gorse bushes. The horses were bred to be able to clear a five-bar gate and the hunting country would be managed so as to provide an open aspect (areas with a lot of permanent grass were ideal, so that Leicestershire, Northamptonshire and Rutland became the most famous regions) with well-kept hedges and ditches, few wire fences, marginal grass gallops to arable land and rides cut through woodland. In 1895, there were 153 packs of foxhounds in England and 10 in Scotland. There were also 44 beagle packs in England, whose quarry was followed entirely on foot.

The pre-eminence of the Midlands was confirmed by the railways, for a man might come up from the City for the day and his horse might also be boxed by rail to the meet. The novelist Anthony Trollope was a keen rider to hounds and a hunt features in many of his stories:

> It was not customary with them to draw the forest . . . and they trotted off to a gorse a mile and a half distant. This they drew blank, – then another gorse also blank, – and two or three little fringes of wood, such as there are in every country . . .

gives the flavour of the hunting country of the Brake Hunt, as depicted in *Phineas Redux* (1874) which was 'easily reached by a train from London . . . was near to two or three good coverts, and was in itself a pretty spot'.[58]

The chances of 'drawing' (that is, causing a fox to bolt from its earth) were always at least 2–17 against, so that the whole pursuit can have something of a farcical affair, especially when the hunt claims that it is engaged in necessary pest control;[59] Oscar Wilde's famous epithet of 1893 sums up that side of it when he talks of the English country gentleman galloping after a fox: 'the unspeakable in full pursuit of the uneatable'.

Perhaps there is little need to point out how difficult it is to eliminate hunting with hounds from Great Britain. Bills such as that promised by the government in 1999 tended to be sidelined by government business managers because they knew that the measures will get savaged in the old House of Lords. Leftish administrations fear too that the removal of killing for pleasure will lose them what friends they have in the financial and landowning communities and possibly the Royal Family also. But the moral shift which pushed governments in the 1990s into dealing with the cruelties of live animal transport seems bound eventually to be reflected in the demise of fox, stag and hare hunting with dogs. Such hunting was the (negative) subject of one of the first Bills in the new Scottish Parliament of 1999.

No such uncertainties have attended the shooting of pheasants. After Waterloo (1815), a number of farms on marginal soils were planted to mixed woodland and these provided a good density of pheasants.[60] This density could be improved by rearing the chicks in cages, free from predation, and then releasing them as young adults into the woods. Beaters then drove the birds out of the woods across a line of guns with loaders; the practice is analogous with the management of grouse. Large bags could thus be obtained, sometimes competitively; the best account of the combination of country-house party and shoot (set in, ominously, 1913) is Isobel Colgate's novel *The*

Shooting Party. The social pinnacle came with the royal purchase and development of Sandringham in Norfolk in the 1860s; it is significant that it had its own railway station. Enclosure also improved the habitat for partridge, and the planting of taller hedges and belts of pine encouraged large coveys of that bird. After 1850 these too were driven across lines of guns: at Weston Colville near Newmarket in four days of January 1860, nine guns shot 1077 brace of partridges. John Betjeman's poem on the death of King George V in 1936 distils the essence of identification of the country house, royalty and shooting:

> Spirits of well-shot woodcock, partridge, snipe
> Flutter and bear him up the Norfolk sky.[61]

Beside the seaside

The 1909 popular song 'I do like to be beside the seaside' (written by John Glover-Kind, d. 1918), draws attention to the phenomenon of mass excursions to the sea. Though the benefits of sea-bathing were already known, and had occasionally been combined with spa waters as at Scarborough, the attitudes of the railway companies in providing workmens' tickets for travel into cities and excursion tickets for escaping them brought about a leisure revolution, especially when the cultural hurdle of running trains on Sundays had been overcome. One result was the development, mostly after 1870, of towns which had little purpose other than to please in this way (Blackpool is the first example that springs to mind) or which had grafted a very substantial leisure component onto fishing or port functions, such as Margate in Kent. Interestingly, most of the leisure time of the working classes was spent in urban environments; cycling and hiking were more activities of the lower middle class.[62]

Just as there were company towns for manufactures, there were single-agent developments for this purpose. An example is Skegness, on the coast of the Lincolnshire marshland, an insignificant farming village of 134 people in 1801. The ninth earl of Scarborough held estates here and decided to diversify his landholdings during the agricultural depression of the late nineteenth century. With the aid of the railway (extended from Wainfleet in 1873), he founded a resort for the working-class populations of cities like Leicester, Nottingham and, to some extent, Sheffield. The permanent population expanded to 2140 in 1901. A pier was opened in 1881 to signify the status of the new resort and early photographs[63] show the encroachment of urban features onto the dune-slack system above high water mark. The town never extended southwards as far as Gibraltar Point where the dune and marsh system is still more or less complete and managed as a nature reserve. The density of visitors (106,000 'trippers' came by rail in 1880[64] and in the 1890s the Bass brewery in Burton-on-Trent sent its works outing to Skegness in 12 special trains) nevertheless ensured that the town's dunes underwent blow-out as the vegetation cover was damaged. Thorn bushes were even 'planted' along the foreshore in 1881 to encourage new dune formation in front of the first concrete esplanade walls, in a very modern-sounding piece of environmental management. In general, though, the new resorts acted like towns everywhere: they covered up the pre-existing terrain and piped their sewage as far as was essential and as near as was affordable: in these places it often came back to cause dis-ease to its creators.

SACRED SPACE

Most histories of the nineteenth century emphasise the desacralisation of nature which came to its height then. Counter-arguments are difficult to produce but it could be noted that this was a period of church and chapel building in town and country alike and that, if a burial ground was attached, then a small haven for wildlife was created. This was especially true of the monastic and conventual foundations which grew up at the time, often with a teaching or caring role. In both cities and rural areas an enclosure was desired and preferably one with a large garden. We might think that black and brown habits led to more blackbirds and thrushes.

WILDLIFE AND ITS PROTECTION

The better killing technologies of the nineteenth century, coupled with obsessive collecting in the name of 'science' eventually produced some reactions. One early practice which eventually caused a great deal of outcry was the 'battue' of the 1830s at Flamborough Head on the east Yorkshire coast where parties of from 12 to 30 men were rowed out to the cliffs where they fired guns indiscriminately into the roosting and nesting bird colonies. This finally led to the first piece of national legislation for species protection: the Sea Birds Preservation Act 1869 which protected 33 species between April and August; egg collecting was still allowed where the local economy was accustomed to gathering gulls' eggs. Further legislation in 1880 and 1881 was followed by a series of minor amendments to existing legislation.[65] These later laws also protected less obvious species: the Dartford warbler (*Sylvia undata*) had undergone a dramatic decline in the nineteenth century at the hands of egg collectors, for example. Many wild flowers, too, were collected for amateur herbaria and mountain species for rockeries. It was extinction due to to 'building, drainage, disafforestation or in consequence of the cupidity of collectors' which led to the founding in 1912 of the Society for the Preservation of Nature Reserves (SPNR). At first essentially a ginger group for the National Trust (which had by then acquired 13 sites of special interest to naturalists), its destiny was to lead eventually to state involvement in the protection of wildlife.[66] Parallel efforts to protect plants as individual species were to some extent eclipsed by the rise of ecology with its emphasis on plant associations.[67] The impact of human activity on these latter was rarely understood in the case of ecosystems which appeared to be wild.

Thus while the nineteenth century was mostly a bad time for wild nature, it also exhibited the stirrings of an attitude to environment in which protection of the biota for their own sake, rather than as food species or game, was emerging. An impetus was given by the foundation of the Plumage League in 1885 which led to the foundation of the Society for the Protection of Birds in 1889. This was occasioned by the use of feathers in womens' hats: the London commercial feather sales of 1902 processed the equivalent of 192,960 herons killed for their feathers[68] and the great crested grebe suffered as it were grev/biously. (A conservationist in New York in 1906 counted 40 native species of birds whose feathers adorned three-quarters of his sample of 700 hats.) Between 1900 and 1910, the UK imported 6000 tonnes of feathers (not all for hats) per year; legislation to control this trade eventually came in 1921.

INTERNATIONAL LINKAGES

The outreaches and spinoffs of a great city are but a microcosm of empire. Just as London drew in materials, transformed them and exported them beyond its bounds, the British Empire acted as a source of raw materials which were brought to the mother country, processed and stored, and then exported to the world. In parallel with the material flows, there were movements of people and ideas. All of these were immensely accelerated by the steamship and the telegraph. The money in circulation could also be used to buy goods outside the empire, so that wheat could come from the USA or meat from Argentina. The repercussions on the environment of Great Britain were usually at a remove (if we except the enlargement of ports at the expense of salt-marshes and mud-flats), with impact upon agriculture being an important theme. Cheap wheat and cheap meat from abroad obviously undermined the profitability of British agriculture and helped contribute to the depression in that occupation in the later nineteenth century when large parts of Essex, for example, were recorded as being under weeds and secondary vegetation (Fig. 6.7). The term used for all the imports of organic materials, like cotton, cereals and fish might be 'ghost acreage': control over lands overseas and over the oceans added to the effective land surface of the kingdom, just as the use of underground fuel resources released land at home. The environmental relations of the British population have to be seen therefore outside a strictly local 'people–land' framework.

> The question has been with us until quite recently as a policy issue: to what extent ought the UK to be self-sufficient in basic resources such as food? Membership of the EU has submerged the question except for the most rabid Europhobes but a world with twice its present population may just force the item back onto the agenda during the next century.

Increased trade and travel had many effects, not least of which were some of the changes in outlook dealt with in 'Sensibilities', below. At a more material level, there were many imports. Rarities and exotica were the special concern of zoos and botanical gardens; neither were necessarily founded at that time but they usually expanded greatly as science rather than the gratification of curiosity became part of their mission. The cathedral of this redirection of collecting from all the world is the Natural History Museum in London, built somewhat appropriately in Victorian Gothic. Many

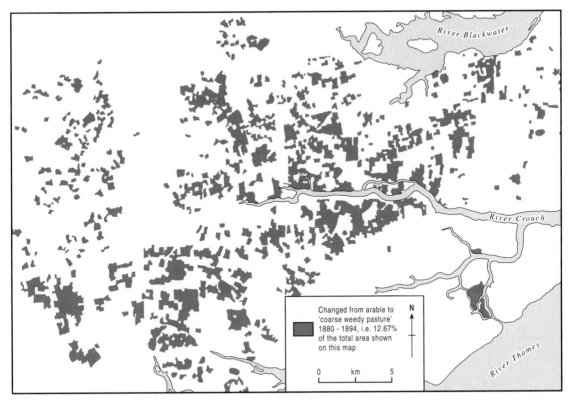

Changed from arable to
'coarse weedy pasture'
1880 - 1894, i.e. 12.67%
of the total area shown
on this map

0   km   5

FIGURE 6.7 *Agricultural depression and land reversion: the change from arable land to 'coarse weedy pasture' in late nineteenth-century Essex. A bonanza for tall 'weed' species like* Senecio *and for seed-eating birds like goldfinches.*
J. T. Coppock, 'The changing face of England: 1850 – circa 1900', in H. C. Darby (ed.), A New Historical Geography of England after 1600, Cambridge: Cambridge University Press, 1976, p. 310, Figure 63.

new species were brought from empire and foreign lands: economic species of conifers from North America have been mentioned and there were many other plants which have become part of the 'natural' scene: some of the large rhododendrons, for example, have in 100 years become naturalised to the point of becoming the understorey vegetation in woodlands. Animals from this period include the grey squirrel (*Sciurus carolinensis*) around 1900, where releases were made on some 32 occasions for the pleasure of landowners who liked to have an exotic species on their estates.[69] Other nineteenth-century introductions of more than passing significance include Japanese *sika* deer which after 1860 escaped from park herds to establish a number of feral populations, and the little owl (*Athene noctua*), succcessfully set free into Kent in 1874.[70] Plants which now appear to be native include Policeman's Helmet (*Impatiens glandulifera*), the third touch-me-not to arrive in Great Britain in the nineteenth century; this species is a native of the Indian Himalayas and very common in damp and shaded sites. The Sicilian species of ragwort (*Senecio squalidus*) escaped from the Oxford Botanical Garden in *c.*1800 and spread discontinuously during the nineteenth century; Kew Gardens made a similar contribution with the Peruvian *Galinsoga parviflora* which in the 1860s eventually became Soldiers of the Queen after a period when it was called Gallant Soldiers. In this century, both became common in open habitats: the former on bomb sites in cities and the latter in suburban gardens.[71]

The rather forced military pun seems to fortify the view that, both in empire and outside it, coercion was a normal way of extending the flows of people and goods: 'trade follows the flag' is capable of ironic as well as patriotic interpretations. The possibilities for exercising power conferred by the development of machines were central in the developing sensibilities and world-view of the century. As if to emphasise the concern with overseas, the 1890s were witness to the first big purchases of land by the War Office for military use, especially for training. The first was a large tract of land on Salisbury Plain; next came heathland in Surrey and Hampshire, arrayed round Aldershot. Both centres have persisted to the present day, though the ecology of the tank is rather different from the troop of cavalry.

Sensibilities and attitudes to environment

One difficulty in discussing the ideas which inform a society's view of its environment in this or any age is the gap between what is said and what is done. Legislation is an obvious example: Parliament passes laws which are then ignored or subverted on the ground if the enforcement procedures are inadequate because not enough public money is spent on them. This problem becomes more acute when we discuss ideas which are placed in the public domain and may or may not have had some influence in determining actions at some future time. So a sense of caution is necessary when thinking of contributions to the ways of thinking (as expressed in the German word *Zeitgeist*) in any particular epoch.

Science and technology

Two strands, among many, can be identified as important for our theme: one pragmatic and the other theoretical. The first is the triumph of the engineers in producing machines that nearly always worked, even if not always at first. Locomotives, steamships, bridges, forges, machines for making almost everything: all were confidently hailed as successes. They were roads to profit and perhaps equally important to 'the conquest of nature'. Failures on a large scale were relatively few and were quite quickly brushed aside. The celebration of the machine (epitomised in the Great Exhibition of 1851) was nowhere greater than in the business community, and in many sources it is apparent that the interests of 'trade' were paramount above those of health and amenity, which together formed the main way in which the environment presented itself at that time. (The notion of resource scarcity was little discussed except in the matter of coal.) In accounts of some of the gross pollution in industrial cities, there seems to be a collective sigh, 'Of course it's awful, but the interests of trade are so strong.'

Nevertheless, there were some powerful reformers who did not subscribe to this mainstream thinking and who were convinced that the mass of the people deserved something better. The Victorian instinct to reform was strong and so there were powerful voices at individual and civic level for a reining-in of both pollution and disamenity. There was a National Health Society founded in 1873 which inspired a Smoke Abatement Committee to mount an exhibition in London in 1881–2, and in 1877 the Kyrle Society was founded 'to bring beauty home to the people'.[72] The leading figures in these groups were members of the establishment: Lord Aberdare, Lyon Playfair and Sir Frederick Leighton (a President of the Royal Academy) were founders and even the patronage of the Duke of Westminster was granted. The Smoke Abatement Society is still in existence, as the National Society for Clean Air, and continues to have a full agenda. But its very title of 'abatement' suggests that there was in the nineteenth century no equivalent of a precautionary principle in which there is an attempt to avoid problems: 'trade' acted and society in general tried to shift the worst effects at the margins. The results came in a series of pieces of legislation, which tried to reduce the toll exerted by bad health on the working population and to provide for their leisure. In an ideal world this would have been disinterested philanthropy but it is hard to escape the idea that the real purpose of it was to get more work out of the inhabitants and enhance profits.

The role of 'pure' science's relations with society in the nineteenth century is interesting. The key figure is CHARLES DARWIN (1809–82) whose concepts of evolution and competition were so powerful at the time. His ideas of the descent of humans challenged those of religion; those on competition and predation were taken up into a form of 'Social Darwinism' whose main protagonist was HERBERT SPENCER (1820–1903). Spencer coined the phrase 'the survival of the fittest' and his laissez-faire attitudes centred on the view that if you were poor then that was the working out of an iron law of nature. Darwin never approved of this use of his work but successive Conservative governments have never quite been able to disavow it. Science in general, though, flourished in the nineteenth century: where conflict over, for instance, water quality existed, then sharp differences in attitudes at local level provided a fertile field for innovations in knowledge and technology.[73] Science helped to bring together the rural and the urban in ways other than the obvious. The pioneers of agricultural chemistry pointed out that the 'modern agriculture' of the time might reduce soils to exhaustion, just as industry might run out of capital. The factory as the source of profits for investment became an important standard against which farming was to be judged. Another subtle change also resulted from science: the colonial experience had revealed all kinds of linkages between human activity and the environment. The relations between land cover and runoff, for example, and the effects of deforestation upon local (and even global) climate were commented upon, especially by foresters, agriculturalists and governors of tropical islands.[74] It is possible that here we find some of the origins of today's environmentalist thinking in Europe, for the same was true of the colonies of other European imperial nations. Lastly, it was this period in which it became acceptable to argue that a true picture of the world might best be obtained by observers who were detached from immediate social pressures. So was cultivated the 'objective' view of the man (sic) in the white coat.

PLATE 6.6   *The Italianate fantasy of Portmeirion in North Wales (although a twentieth-century development) illustrates a continuity of fascination with the Mediterranean scene from the paintings of Rosa to the 'villages' of today's shopping malls like the Metrocentre in Gateshead. Photographed in 1999.*

Arts and humanities

The most important influence in changing the sensibilities of the nineteenth century was romanticism. The origin of this movement can be traced at least to the middle of the eighteenth century (Plate 6.6) but its dominance was from the 1790s to the 1840s with inevitably a lag effect producing romantic ideas long after that time. In many ways, romanticism was a reaction against the mechanistic world-views which came to practical dominance in the nineteenth century. Spurred on by the revolutionary temper of the times, it came to prize the individual over authority, the imagination over the industrial and the re-positioning of nature as central to human concerns. The Romantics' perceptions were different, too. There was a considerable focus on their depictions being unique to a particular time and place.

In Britain perhaps the biggest change was in the revaluation of the wild places. Before the late eighteenth century, the heathlands, moors and mountains were regarded as objects of fear, scorn and avoidance.[75] Some diarists and travellers had vied to pen the most off-putting descriptions of, for example, the Surrey-Hampshire heathlands, Dartmoor and the Lake District. The shift to our present-day reverence and valuation of such places started when others tried to revalue these places: Gray and Gilpin were both influential, as were (at a remove, so to speak) painters like Claude, Poussin and Rosa.

Sources of quotations to back this up are easy to find, especially when taken out of context. The idea of a holistic continuity between humanity and nature is often cited as the centre of the new world-view, as with William Blake (1757–1827), whose *Jerusalem* has already been quoted. Less direct in meaning but replete with significance for the incipient Romantic movement are his lines from *The Marriage of Heaven and Hell* of 1790:

> If the doors of perception were cleansed, everything would appear to man as it is, infinite.
> For man has closed himself up until he sees all things thro' narrow chinks of his cavern . . .[76]

The most intense expression of this type of vision comes in William Wordsworth (1770–1850) of whom the novelist Margaret Drabble writes, 'Never since Shakespeare has scenery been made to bear the weight of such significance.'[77] This came about because Wordsworth saw the natural world not only as a significant force upon humans but also as an entity which lived in its own right as an active, living universe. This impulse breaks out in the famous Tintern Abbey poem,

> And I have felt
> A presence that disturbs me with the joy
> Of elevated thoughts; a sense sublime
> Of something far more deeply interfused,
> Whose dwelling is the light of setting suns,
> And the round ocean and the living air,
> And the blue sky, and in the mind of man;
> A motion and spirit, that impels
> All thinking things, all objects of all thought,
> And rolls through all things.[78]

Beyond that, the inexorable processes of nature are of a piece with human survival: both need each other but perhaps we need nature more: at funerals (and high-profile deaths) we console ourselves with flowers.[79] This leads to the reading of Wordsworth as a radical in the sense of scepticism about any spiritual benefits of economic growth and material production. To that extent, therefore, Wordsworth has a significance beyond the Romanticism of his time,[80] on into modern environmentalism. It might be argued that the Romantic idea was most influential in its declining phase, when it was taken up by the Court and led (with Sir Walter Scott at the head of the procession) to the infatuation with a mythologised image of Scottish Highland life and to paintings such as Landseer's *The Monarch of the Glen*.[81] The artist (and stag) as lonely individual coincided, too, with the Social Darwinism that aggrandised the notion of individualism.

The equally Romantic idea of the individual's importance in the face of the oncoming factory system and of mechanical world-views is often associated with the painter JOHN CONSTABLE (1776–1837). Not only did his paintings aspire towards grand statements of the relations between birth, growth and decay, but also he typified the 'artist' as a creative individual ('Painting is but another word for feeling.' [The sky was] 'the chief organ of sentiment')[82] in contrast to the journeymen painters of landowners and their estates such as Gainsborough, who died in 1778. That Constable's paintings have become so popular in our time (being used as icons by many preservationist movements) shows that, like William Wordsworth's poetry, there is a depth which can move people greatly removed from the temper of the times of their composition.[83] The passing of the old solar-powered world and the thrust of the new and coal-fuelled empire is made visible by Turner's paintings. The most celebrated expression is the depiction of the hulk of the ship of the line *Téméraire* being towed up the Thames to the breaker's yard at Rotherhithe in 1838 (Plate 6.7).

To show us what to think, there is a brilliant sunset even though for this purpose the sun is setting in the east. In the original, though difficult to pick up in most reproductions, the hull and masts of the *Téméraire* are a ghostly white, in contrast to the browns and blacks of the steam tug providing the motive power.[84] Almost as famous is his *Rain, Steam and Speed – the Great Western Railway* of 1844. Here the new age is celebrated in the ability of the train to forge unimpeded through bad weather. As Daniels points out, the itinerary is significant for this was the London to Bristol route on which Prince Albert had travelled to launch the iron ship *Great Britain*, such a potent symbol of maritime power. Part of this line was as well indeed the scene of Queen Victoria's first railway journey, from London to Slough en route for Windsor.[85] But here also, the past is with us: the painting has an abundance of symbols of an older order: a road bridge, a plough team, a rowing boat and a fleeing hare.

The scenes, products and values of the new era naturally enough provoked reactions. JOHN RUSKIN (1819–1900) coupled aesthetic values with social justice and was a firm advocate of medieval values which he claimed were lost; the revival of Gothic was one of his enthusiasms; the Renaissance and industry were not. In a more domestic sphere, WILLIAM MORRIS (1834–96) equated beauty with simplicity and being hand-made. He implemented this in the fields of furniture, stained glass, fabrics and wallpaper and (like Ruskin) championing the socialist movement of the later nineteenth century, making a great deal of money decorating the homes of the wealthy. Both Ruskin and Morris lived into the time of the onset of Modernism in the arts, which was to embrace more fully the economy of the age than Romanticism; Turner might be seen as an early Modernist and the Impressionists, in their concern for the mundane (albeit mostly the rural everyday scene), the first identifiable phase. Impressionism is mostly associated with France though Monet and Pissaro both painted in England from time to time, producing work that smells of wet summer grass rather than golden-dry hillsides: see, for example, Camille Pissaro's 1871 painting *Lordship Lane Station, South London*. The growth of suburbs in response to the availability of commuter railways comes across in raw brown houses roofed in Welsh slate, seen through a frame of grassy field and embankment; the railway is not fenced, which may draw attention to its Promethean role.[86]

The years to 1914 had many creative men and women who dealt with the themes of land, water and the human presence. There were writers who encompassed the

PLATE 6.7   *J. M. W. Turner's 1838 picture,* The Fighting Temeraire, tugged to her last berth to be broken up. *It is generally interpreted as depicting the end of the old order of solar power (the sailing vessel) and the oncoming of the new, fossil-fuelled world. The sunset may be influenced by volcanic eruptions in Asia at that time; it is also in the east, which suggests a world turned on its head by industrialisation.* © National Gallery, London.

broadest world-views like Wordsworth but also those who dealt with smaller-scale subsets of concern. John Clare (1793–1864) wrote a great deal in anger about unenclosed land and made of the word 'common' a place of untidy, wild and bleak character but one which also conveyed preciousness which was being lost by enclosure, often symbolised in both its natural and human aspects by the trapping of moles.

> And spreading lea close oak ere decay had penned its will
> To the axe of the spoiler and self interest fell a prey
> And crossberry way and old round oaks narrow lane
> With its hollow trees like pulpits I shall never see again
> Inclosure like a buoanparte* let not a thing remain
> It levelled every bush and tree and levelled every hill
> And hung the moles for traitors – though the brook is running still
> It runs a naked stream cold and chill . . .

*A reference to Napoleon and his conquests in Europe.

Which runs almost without a break into the more aesthetic Gerard Manley Hopkins's evocation of a Scottish burn (in 'Inversnaid' from *Poems 1887–89*),

> What would the world be, once bereft
> of wet and of wildness? let them be left –
> O let them be left, wildness and wet;
> Long live the weeds and the wilderness yet. (11. 13–16)

We are close to the rural idyll,[87] yet the nineteenth century provides an example of a much cooler appraisal of country life and its relation to natural (indeed cosmic) phenomena at the pen of the most consummate all-rounder since Shakespeare, namely THOMAS HARDY (1840–1928). He is most celebrated for his imaginative transformation of a real region of England (centred on Dorset but extending to Devon and Somerset, Salisbury Plain and Oxford) into a Wessex with which his characters are deeply engaged. After his death, however, his poetry gained in acclaim, in a just reflection of Hardy's own priorities, for he wrote novels for income and poetry for love. In both poetry and prose, his delineation of the interaction between nature and humans is complex: his most famous piece of landscape is 'Egdon Heath' (the model for which is now mostly afforested) of which he wrote that it was unconventionally beautiful and thus in tune with the age, 'a place perfectly concordant with man's nature – neither ghastly, hateful, nor ugly; neither commonplace, unmeaning, nor tame, but like man, slighted and enduring; and withal singularly colossal and mysterious in its swarthy monotony'.[88] He wrote the novels out of a countryside suffering agricultural depression and in which not to adapt meant dispossession, as happens to Henchard, one-time Mayor of Casterbridge when a harvest fails. This seems to be the influence of Darwin percolating into creative writing. An economy trembling on the brink of industrial revolution is neatly shown in *Tess of the d'Urbervilles* (1891) where a threshing engine was used, and the engineman,

> was in the agricultural world, but not of it. He served fire and smoke; these denizens of the fields served vegetation, weather, frost and sun . . . The long strap which ran from the driving-wheel of his engine to the red thresher under the rick was the sole tie-line between agriculture and him . . . the environment might be corn, straw, or chaos: it was all the same to him. If any of the autochthonous idlers asked him what he called himself, he replied shortly, 'an engineer'.

Tess contains a great variety of rural scenes and Tess herself carries the weight of the imagery that contains the relationship between the humans and nature. Alfred Alvarez writes of the remarkable way in which the landscape is a sounding board, 'in order to deepen and intensify whatever it is that Tess is experiencing'.[89] The world is, in the last analysis, more than a little indifferent to its human inhabitants, something that derives from a cosmic connection:

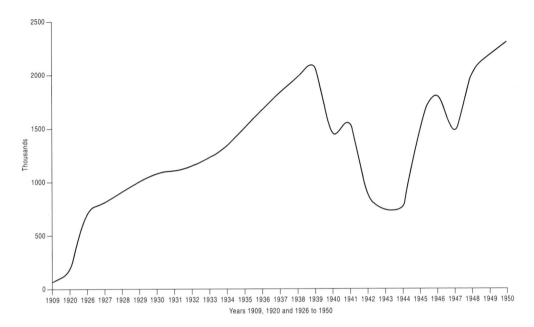

FIGURE 7.1 *Private cars from 1909 to 1950 in thousands. Note that the extreme left of the horizontal scale is not linear: it jumps from 1909 to 1920 and then to 1926.*
Department of the Environment, Transport Statistics of Great Britain, *London: HMSO, 1977.*

its use in each of the two major conflicts by the imposition of fuel rationing. A growth in the total registered from 8465 in 1904 to 2,257,973 by 1950 must, however, bring about some environmental changes.

TABLE 7.1   Cars and total motor vehicles 1904–90

	Motor cars	All motor vehicles	Cars as percentage of all motors
1904	8,465	17,810	48
1920	186,601	650,148	29
1939	3,148,600	2,034,400	65
1945	1,468,600	2,552,500	58
1950	2,257,873	4,409,223	51
1960	5,525,828	9,439,140	59
1970	11,328,000	14,950,000	77
1980	14,772,000	19,132,000	77
1990	19,742,000	24,673,000	80

The percentages are rounded to the nearest whole number.
Source: British Road Federation, *Basic Road Statistics*, London: BRF, 1993.
Note: In terms of traffic created, all motor vehicles travelled 422 billion vehicle kilometres in 1994, of which 345 bn v-km (82 per cent) were in cars and taxis; only 2.5 per cent was contributed by the dreaded articulated lorries. The use of bikes was about 1 per cent of that of all motor vehicles.

In broader terms, there is now more discussion of control systems than hitherto, for the inter-war period's demand to produce 'a fit country' led to some early legislation aimed at what we now call development planning. National consensus during World War II resulted in a much more comprehensive system of control of land use and built environment. In both instances, war (itself hitherto only a minor manipulator of the environment in Great Britain) acted as a catalyst for emphasising particular environmental attitudes.

Climate

The facts of secular climatic shifts are relatively simple: it was cool and dry at the turn of the century, became warmer and wetter to the middle and thereafter cooler and drier again. The warmest decade was the 1940s, with 1949 the warmest year until quite recently; 1933–52 was the warmest 20-year sequence of summer and annual temperatures in the last 300 years in Europe. The years 1918–31 were exceptionally wet just as 1887–1917 were unusually dry, as were 1932–45. The wettest decade was the 1930s, with another peak in the mid-1960s with 1946–55 being about average.[4] But practically none of this matters in terms of anything except the memories of people who experienced, for example, the clear skies of September 1940 in which the Battle of Britain was fought, or who, more frivolously, try to correlate the onset of trendy PVC clothing with the wet years of the 1960s.

One aphorism applicable to the twentieth century calls technology 'the knack of so arranging the world so that we don't have to experience it'. In Britain, this is most evident in the heating of buildings, and the main domestic means in this period was still the open fire, with the creation of smoke from the quality of bituminous coal used, which was still about 40 million tonnes/yr in the post-1945 period. This led to the continuing contest between urban air quality, the health of the population and the biota of towns. A number of attempts to bring in smoke control legislation foundered on problems of the definition of visible smoke and on the intractable demand for open fires in domestic premises. Not until 1946 was cleaner air legislation available to a few cities, when it was adopted relatively enthusiastically in Manchester and Coventry,[5] though national consideration had to await the catastrophic events of 1952 in London (see page 240). Results of all the delay included the continued incidence of various diseases that resulted from inhalation of smoke particles, and also the disappearance of susceptible plants from urban areas and city fringes. Lichens are probably the best indicators of long-term sulphur levels, with inner cities at that time being more or less devoid of that group of organisms. There was response in animals, too. From the 1880s onwards, evolution produced a dark form ('melanic') of the moth *Biston betularia* which grew to prominence in polluted zones of British cities where its dark colouring allowed more individuals to escape predation and to breed.[6]

Landforms

Though industrialisation rather disguises the fact, the creation of landforms is continuous, albeit much modified by human activity. Soil erosion, for example, occurs all the time on agricultural land although its intensity varies with tillage and grassland management practices: root crop areas like sugar beet farms in East Anglia are very prone to wind-blow. Storms around Ely in May 1943 raised smoke-like clouds 50–100 feet (30–60 m) high with a deposition of dust one inch (2.5 cm) deep.[7] Most forest planting before World War II was done by hand, but, after 1945, afforestation in large blocks drained by plough-created ditches in the mineral soil allowed accelerated soil erosion on slopes as low as 2 degrees. Similarly, forest roads and tracks became foci for rill and gulley erosion in ways which peats did not.

Some human-made landforms diminished in quantity: the formation of heaps of waste in, for example, the Lower Swansea valley took place at a slower rate than in the nineteenth century since there were fewer firms operating and the copper smelting which had been a key industry was declining through the twentieth century. The legacy of derelict land was immense and a massive reclamation project was needed, a necessity that had been seen as early as 1912 when the Borough Surveyor presented a report to the Council which advocated the reclamation of tip waste. Nothing happened until the 1960s when the task of improving some 800 acres (324 ha) was attempted.[8] 'More of the same' might, however, be applied nationally to the creation of holes and heaps by successful industries, of which coal was pre-eminent. The growth of pit heaps and the creation of subsidence flashes was characteristic of all major coalfields. Only after nationalisation in 1948 was there a different attitude but its results were not evident in the landscape until after the 1950s. A new element of significance was the winning of coal on a large scale from open pits. Open-cast mining on an industrial scale was undertaken from a standing start in 1941 as a wartime measure and by June 1942 there were 22 working sites in the Yorkshire–Staffordshire belt, a number which expanded in other areas to a total of 419 sites by 1944 and achieved a production of 10 million tonnes/year in 1945, which was about 5 per cent of total coal output. Production was stabilised in the 1950s and 1960s (Plate 7.1) and then began to grow again in the 1970s.[9] A great revolution in output was made possible by access to US technology, with the first walking dragline (a 1200 ton machine) in operation at Bedlington, Northumberland, in 1949. From 1943 onwards, restoration was compulsory, so that there are many areas which were once huge pits but are now in other uses, such as agriculture, forestry, housing and theme parks. The restoration, although very thorough, cannot reproduce the original landscape. Underdrainage may in fact be improved but stored topsoil loses its fertility, runoff may suffer from acidification due to the different set of strata through which it passes, and wildlife has little place in the new, simpler landscape.[10] Archaeological sites are lost for ever. In spite of all the drawbacks, the restoration of open-cast has had a very beneficial effect in showing that the mindless creation and abandonment of land for mineral extraction is not acceptable.

Other land-forming elements happened, as always, sporadically.[11] The importance of extremes of climate needs mentioning, since these may produce landforms: very heavy and intense storms, for example, may saturate soils to the point where landslides occur, such as the bog-burst on the North York Moors in 1945 which covered 4.5 ha. Floods may mobilise riverine sediments along with any contaminants they

PLATE 7.1 *Swadlincote, Derbyshire. As open-cast coal mining became apparently more economic, its scale increased and by the end of the 1950s (this picture was taken in 1960) large pits were being worked with very heavy equipment.*

contain. Hence, lead fines from the Pennine smelters and crushers which had their heyday in 1815–1900 can be re-worked and sent further down the Tyne in flood-time alluvial sediments, causing crop damage such as stunting and chlorosis where they are re-deposited.[12] Similar deposits of iron, tin, copper and arsenic are stratified into the sediments of the estuaries of the Red River and the Fal in Cornwall where a sea-level rise might re-mobilise them to the extent of forming a significant environmental hazard.

One apparently continuous process from the seventeenth century to the present is the shrinkage of peat in the Fens. Drainage in the seventeenth century relied first on gravity and then on the windmill. It was followed by the introduction of the steam pump, succeeded by diesel and electric technology. All these accelerated the loss of peat, since it dries out, becomes oxidised and shrinks in volume. A good measure of the rates of wastage can be seen at the Holme Fen post 10 km south of Peterborough: a cast iron post was driven into the peat level with its surface in 1848, shortly before drainage operations started on nearby Whittlesey Mere, which was one of a number of small lakes that had survived previous bouts of draining and reclamation. In the mid-nineteenth century the level of the Mere was 1.6 m OD and that of Holme Fen peat surface 1.62 m OD. The 'mere' is currently 2.0–2.5 m below OD and the length of exposed post at Holme Fen (which was flush with the surface in 1848, but protruded 3.05 m in 1913 and 3.35 m in 1940) is 3.87 m (Plate 7.2). It looks from accounts of

PLATE 7.2   *The Holme Fen post in 1996. When driven into the peat in 1848, the base of the pyramidal ornament was level with the ground surface. The nature reserve in the background has passed through several phases of land cover and the birch wood is relatively recent.*

the Mere as if wastage accelerated with each new pump, and that sea level was passed in early 1852 (Fig. 7.2).[13] Most of the peat fens of East Anglia are about 3.7–4.3 m below their pre-pump drainage level, with Mean Spring Tides in the Wash at 3.78 m to 1.86 m OD. Wider surveys quote wastage rates of 0.6 cm/yr for the 200 years of wind pump drainage and 2.5 cm/yr thereafter, though the variation in the 1940s–1955

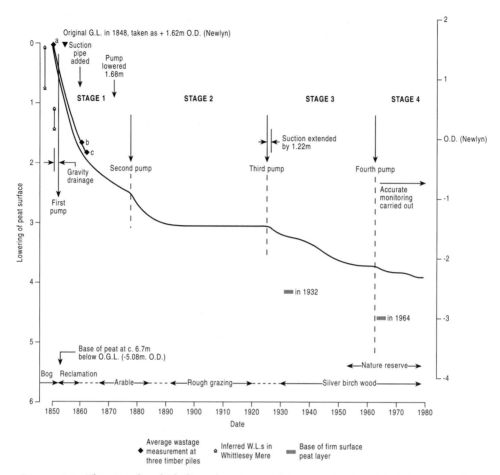

FIGURE 7.2 *The stages by which the Holme Fen post has emerged. The original bog surface in 1848 has shrunk to the 1978 level with the application of a series of pumps and the presence of a number of land use types. The present nature reserve woodland is not 'natural' but has succeeded rough grazing and, before that, arable land.*
J. N. Hutchinson, 'The record of peat wastage in the East Anglian Fenlands at Holme Post, 1848–1978 A.D.', Journal of Ecology 68, 1980, p. 239, Figure 3.

period was between 1.17 and 3.09 cm/yr.[14] The Fens contrast with the Somerset Levels, which stayed as grasslands and shrank very much less.

WOODS AND FORESTS

The interest of the national government in forests and woodlands was strong even in an era of liberal economics, partly because in 1900 some 90 per cent of all timber and forest products were imported.[15] Between 1850 and 1914, six Royal Commissions, Select Committees and Departmental Committees had sat and reported to little effect, though an Office of Woods, Forests and Land Revenues had undertaken small-scale afforestation on estates it had acquired at Hafod Fawr in Merioneth (1899) and

Tintern in the Wye Valley (1900).[16] Nevertheless, national control was very light: it was still possible for an estate to clear-fell its woodlands to pay off debts or for the planting policies to be determined by the head keeper solely in the interests of the pheasant shooting. Nor had there been an accurate estimate of the national inventory since the data published in 1881, though it was clearly realised that the resource was almost all in private hands. Unsurprisingly, the impacts made by the great wars of the century have been profound. The Great War resulted in the felling of 0.5 million acres (202,000 ha) and World War II in the clear-felling of 373,000 acres (151,000 ha) and the so-called 'devastation' of a further 151,000 acres (61,000 ha) by the removal of the best timber. In 1942, 1800 acres (728 ha) were being felled every week, an amount which would likely have been higher were it not for the use of reclaimed wood from packing cases and from bomb-damaged buildings. The data in Table 7.2 summarise many of the trends above.

TABLE 7.2   Data on forest and woodland changes

	Britain 1881–1949 (Million ha)		
	1881	*1924*	*1947–9*
England	0.59	0.65	0.75
Wales	0.06	0.10	0.13
Scotland	0.30	0.40	0.50
Total	0.99	1.20	1.40

Note: these figures are converted from acres and rounded.
In 1924, 92 per cent of these woods and forests were private, 7 per cent belonged to the state and 1 per cent to corporate bodies; in 1947–9, 82 per cent were private and 18 per cent state-owned.

Proportion percentage of land area in forest and woodland

	1881	*1924*	*1947–9*
All England	3.4	3.9	4.8
Upland England	4.5	5.1	5.8
Wales	3.2	5.0	6.2
Scotland	4.3	5.6	6.6
All Great Britain	4.3	5.3	6.1

Percentage growth in forest and woodland area

	1881/1924	*1924/1947–9*
England	11.3	14.4
Wales	55.7	24.9
Scotland	29.5	17.9
Great Britain	20.4	16.6

Adapted from E. J. T. Collins, *The Economy of Upland Britain 1750–1950: An Illustrated Review*, Reading: University of Reading Centre for Agricultural Strategy, 1978.

One effect of the wars was to enlarge immensely the role of the state in forestry and woodland management. The report in 1917 of the Forestry Sub-Committee of the Reconstruction Committee was the most significant, for it led to the establishment of the Forestry Commission.[17] That body was established by the Forestry Act of 1919 and the first trees were planted on 8 December of that year in Eggerford in Devon. They were the first visible elements of an ambition to acquire reserves of standing timber sufficient to carry the nation through a 3-year emergency, an objective not abandoned until 1957 when presumably it was thought that the outcome of a nuclear conflict with the USSR would not depend upon timber supplies.[18] The essential task of the Commission did not vary in the period 1919–84: it was to expand the forest area and increase timber production.

The Forestry Act of 1919 set ambitious objectives of planting up 200,000 acres (80,000 ha) in 10 years, and of having an estate of 1.7 million acres (0.69 million ha) in 80 years. But given the slow growth of trees, action needs a long lead time. In 1924 there were in fact 2.9 million acres (1.2 million ha), which grew to 3.4 million acres (1.4 million ha) in 1947–9. Nevertheless the initial planting programme was not in time for World War II in most respects, when the Commission's forests comprised only about 6 per cent of the cleared and 'devastated' areas; in Scotland, some 92 per cent of fellings were from private woodlands, and the level of replanting was low.[19] As with the Great War, national stocktaking was a socio-political process and there was a 1943 report by the Commission on Postwar Forestry. It suggested that a national stock of 5 million acres (2 million ha) should be established, of which 3 million acres (1.2 million ha) should come from the afforestation of 'bare land' by which it largely meant moorland and heath. The result within our present period of interest was a total stock in 1947 of 3,448,000 acres (1.4 million ha) of which 52 per cent was productive forest, plus another 178,000 acres (72,000 ha) of small woods between 1 and 5 acres (0.5–2 ha) in size.[20] This supplied about 7 per cent of the national demand, with the Commission in *c*.1950 having about half of its total land holding under planted forest, in some 480 units. Some of the largest blocks started out in the 1920s plantings in the Borders.[21]

Complete control of the felling of private woods was only brought in in 1939 but it was continued after World War II and is still in place, buttressed by grants and loans to woodland owners. The concept of dedicated woodlands, where owners would trade total control for grants and loans was also brought to the fore and became established in 1947; in fact, much national policy-making in the reconstruction period of the 1940s was directed at the private sector.[22]

The planting policy of the Commission had been largely dictated by the need to build up timber stocks quickly after the Great War and to keep a reserve of fast-growing trees. Using the expertise of private woodsmen gained in the second half of the nineteenth century, North American conifers were much favoured. In 1954, 122 million trees were planted, and 109 million of these were conifers. The native Scots pine accounted for 25 million but the rest were imported species: Sitka spruce (*Picea sitchensis*, 29 million), Norway spruce (12 million), Japanese larch (*Larix kaempferi*, 16 million), Lodgepole pine (*Pinus contorta*, 10 million) and Corsican pine (*Pinus nigra* var. *maritima*, 8 million). Only about 12 per cent of planting was of native broadleaved species, so the new forests were easily perceived as alien in their ecology, environmental impact and appearance in the landscape.

In the latter evaluation, large units were seen as intrusive on account of their colour and especially their shape when the upper boundary of the planted area was ruler-straight, without regard for the contours of the land (Plate 7.3). Any proposal by the Commission to plant large blocks, especially in the English uplands, was therefore met with a determined campaign led by organisations like the CPRE and the Ramblers' Association.[23] At the same time, as outdoor recreation became important in the post-war period, people in great numbers visited the forests and used the parking, picnic sites, guided walks and nature trails (Plate 7.4).

It is an interesting speculation as to whether it is really the visual aesthetics of the plantation that offend or its foreignness, centred on the feeling that conifer forests are dark, sinister phenomena from central Europe (especially Germany),[24] and which are invaders of open hills supposedly more used to Blake's 'countenance divine'.

The strength of the opposition was demonstrated in the Lake District in the 1930s, for example, when the Forestry Commission proposed to plant 740 acres (300 ha) of upper Eskdale. In the end, the Commission had to be compensated by public subscription (at the rate of £2 per acre) for not planting there and the land passed into

PLATE 7.3 *The Forestry Commission's upland plantations often attracted adverse comment on account of their interruption of rights of way and changed ecology but most frequently on aesthetic grounds. This example in the upper Honddu valley near Capel-y-ffyn in Gwent in the 1990s shows the straight edges so often deplored. Hay Bluff is on the skyline.*

the hands of the National Trust.[25] One outcome was a voluntary agreement between the Commission and the CPRE in 1936 that a central zone of 300 square miles (777 km²) in the Lake District would not be subject to planting by the Commission (Fig. 7.3). At about that time, the Commission decided to accede to demands for public access and so Ardgarten (north of Glasgow) became in the 1930s the first of a number of National Forest Parks (in England, the Forest of Dean was added to the list in 1940) which foreshadowed the post-1950s role of the Commission as providers of recreational resources. At one stage, interestingly enough, it was proposed to limit access to the 'right sort' of people (for example, members of the YHA), reminding us of the character in E. M. Forster's *Room with a View* who wanted all potential (British) visitors to Florence to take an examination at Dover, something which is even more likely today.[26]

The perceived ugliness of many Forestry Commission plantations was ameliorated when in the 1990s the government decided to sell some of them off to a private sector which would not guarantee access and recreational usage in the way the public body had undertaken to do. So they were ugly, but not that ugly, as it were.

One last potential virtue may be mentioned. In his great survey of the human ecology of the Scottish Highlands, FRANK FRASER DARLING discussed the role of cover in rural innovation. He pointed out that the treelessness of the Highlands and Islands discouraged any behaviour which was different from the traditional norms, and hindered the adoption of anything new that might make life better for the inhabitants of the region.[27] (There is a whole new research project for a multi-disciplinary team containing a fancy statistician here.)

AGRICULTURE AND GRASSLANDS

The metamorphosis of Great Britain into an industrial society must not blind us to the reality that most of the land was still rural in the sense of being under crops, grass, orchards, forests and semi-natural ecosystems like heath, moor and mountain. In 1920, about 68 per cent of England and Wales were crops and grass and 14 per cent **rough grazing**; in 1949 the corresponding figures were 66 per cent and 14 per cent: 9.9 million ha and 2.2 million ha respectively.[28] In a long-term view, there was the consolidation of a pattern in which the west is mostly grass and the east mostly cereals. This pattern, now so strongly implanted, has, however, an essential continuity with the last 500 years.[29]

All periods of agricultural change, however, involve shifts (both national and regional) in the proportions of arable and pasture, in the shape and size of fields and responses to technological innovation, including the advent of new chemical treatments for many purposes. In the twentieth century, new forces affected agriculture.

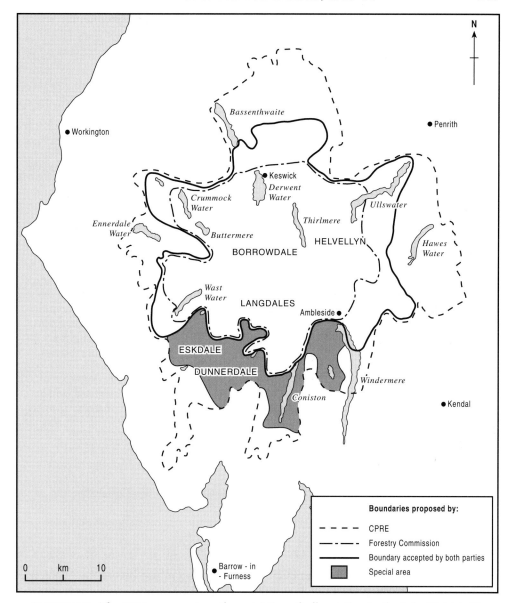

FIGURE 7.3 *The 1936 agreement on the restriction of afforestation in the Lake District. We might wonder whether the negotiations were conducted by face-to-face exchanges or by the iteration of maps.*
J. Sheail, Nature in Trust: The History of Nature Conservation in Britain, *Glasgow and London: Blackie, 1976, p. 85, Figure 16.*

The most important have been firstly the spread of mechanisation and of effective chemicals, and, secondly, high levels of government intervention. We cannot ignore, however, the effects of the world outside Great Britain, especially of cheap imports of agricultural produce and also of import shortages during wartime. Some of these innovations are summarised in Table 7.3; nearly all have environmental consequences.

TABLE 7.3   Agriculture 1915–45: A summary

Period	Economics	Government	Agricultural changes	Environmental effects
World War II	Farms profitable again	Rapid activation of war-time policies, for example, Agricultural Development Act 1939	Control of crop areas; provision of machinery and labour by government	More intensive land use and drainage; wildlife not a consideration
Inter-war period	Depression due to cheap overseas production until government protection	Wheat Act 1932 introduced deficiency payments for wheat and continued them for, for example, sugar beet	Little capital for maintenance of drainage; lower fertiliser use	Scrub advancing in, for example, Berkshire, Essex, land to golf courses, housing; wetlands, heaths available for nature reserves
World War I	Rapid rise in prices until government intervention	1917: government control of prices and production	Move to production of high-calorie foods like potatoes and cereals; also sugar beet	Reclamation of reverted land in marginal districts
1870–1914	Agriculture (especially cereals) in deep depression until c.1900, with some recovery thereafter			

Overall changes

Perhaps the most pivotal environmental change in agriculture before the 1960s was the area of arable, which fluctuated with the prosperity of farming, since in bad times farmers turn to grassland and livestock rather than cereals and roots. In the last decades of the nineteenth century, land went out of cultivation, later to see-saw in and out with the onset of wars and depressions. As a proportion of the total area of crops and grass, the proportion of England under tillage in 1890–4 was 35 per cent, a figure which receded just before World War I to 32 per cent and down to 27 per cent in the 1935–9 years but reached a little over 42 per cent in World War II and was about 38 per cent at 1960.[30] The animal husbandry (of Great Britain) of the period shows an overall increase from 13,800 stocking units/1000 ha of grass in 1873–85 to 18,287 in 1961–3, with a Depression level (1930–2) of 12,124 units.[31] However, the productivity of the industry made it possible to undertake capital investment at times of good profits, especially after World War II. Part of the capitalisation came in the form of mechanisation, for only in 1950 was the number of horses exceeded by the number of tractors. In 1920, for example, there were 10,000 farm tractors in Great Britain and 800,000 farm horses. The shift from horse to tractor between 1920 and 1950 must have released about 1.6 million ha of land for other uses. World War II was a potent accelerator of change: in 1939, there were 52,000 tractors but this increased to 125,000 in 1943; 150 combine harvesters in 1939 became 1500 in 1943.[32] In sum, an index of agricultural output which took 1936–9 = 100 would have 1916 = 75 and 1950 = 145, no mean achievement.

Chronology of modifications

The food supply of an island nation at war is always a matter for concern: imports may fall sharply and the demand for calories to feed soldiers, factory workers and horses will rise. At the outset of World War I cereal prices rose sharply: the demand for fodder for horses was such that oats were 44 per cent more expensive in 1915 than in the previous year. All cereals indexed at 1913/14 = 100 were at 234 in 1918/19. Such was the climate of economic liberalism that only in 1917 did the government try to effect some control with the Corn Production Act, which set minimum prices for wheat and oats, though not for barley because of the temperance lobby. It also gave county committees powers to order the ploughing of grassland on the pain of eventual dispossession (Fig. 7.4), though this had happened to only 125 farms by 1918. Helped by government direction of labour onto the land, but hindered by the lack of imported nitrogenous fertiliser, farming managed to regain the pre-war level of output of about 21.4×10^{12} kcal of food by 1918, having dropped to 19.4×10^{12} kcal in 1916.[33]

World War II was to some extent pre-empted by government action, for the Agriculture Act of 1937 raised deficiency payments for barley and roots (such as sugar beet), provided grants for land drainage and set up veterinary services with the aim of eliminating, *inter alia*, dairy tuberculosis. The outbreak of war brought on the Agricultural Development Act of 1939, with grants for ploughing permanent pasture and for repairing drainage: in essence, government spending to repair some of the ravages of the inter-war period of depression in agriculture. County targets for ploughing-up were set each year, with the aim of growing more potatoes, sugar beet and flax. Livestock was also discouraged by the diversion of more wheat brans to bread rather than cattle feed. The proportion of arable to total crops and grass became 61 per cent in 1945, compared with 40 per cent in 1939.[34] Whereas the cost of living rose by about 30 per cent between 1939 and 1945, agricultural prices nearly doubled so that many farms became profitable (small dairy farms and hill farms excepted) with a taste for machinery and fertilisers and the willingness to raise capital to buy them.

This state contrasted markedly with the inter-war period. In 1918–21 all controls on agriculture were lifted and the freed market brought down prices in wheat and wool. Dairy farming was a better bet than cereals and so development was faster: 1922 saw the first Friesian cow to yield 3000 gallons (13.6 thousand litres) of milk from a single lactation. A significant outcome was the retreat of tillage: scrub advanced on Berkshire hillsides and on fields of Essex clays. Hedges, fences and drains became

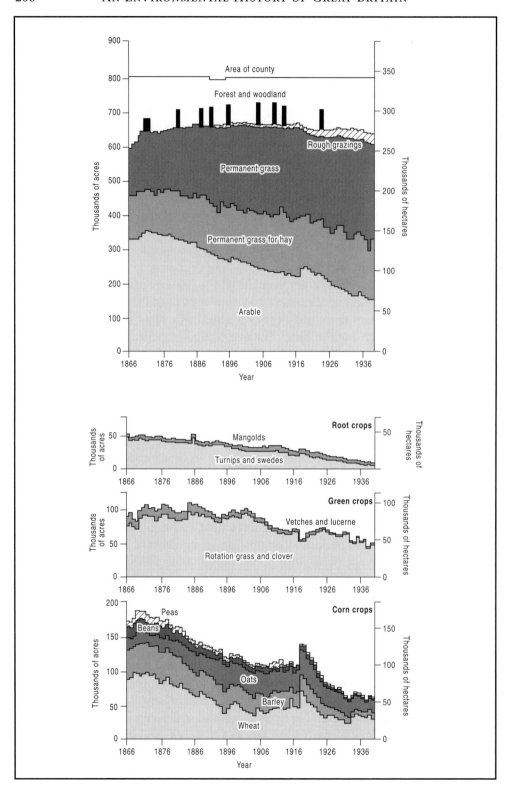

neglected, all symptoms of low investment. When agriculture picked up again from 1934 onwards, its increased output was at the expense of soil fertility, since only manure and cheap basic slags had been used for 20 years. This proved a problem in the wartime hustle for home-grown food.

> So we are launched into the programmes of agricultural subsidies which have underpinned the industry ever since, with avid proponents and opponents both. We might note that subsidies to productive processes are common in other sectors thought of as private: there is no VAT on aviation fuel, for example. The incentives given to manufacturing industries to locate in less favoured areas are also immense.

Themes

The Land Use Survey (discussed below) carried out in the early 1930s provided evidence of the transfer of land from agriculture to other uses. Together with other estimates and later surveys, we can gather that about 1.18 million ha of crops and grass went out of that use from 1916 to 1940, and 0.87 million ha were altogether and permanently lost to agriculture. In the 1918–40 period, 36,000 ha/yr were transformed and, after 1940, approximately 14,000 ha/yr until the mid-1970s. Urban and industrial use claimed 77 per cent of the land thus altered: from 1926 to 1939, some 25,000 ha/yr went to urban use in England and Wales.[35] Knowledge of this shift led to some discussion about the productive value of houses and gardens, with the more extreme claims suggesting that the best way to expand cultivation of fruit and vegetables was to build semi-detached houses. In a suburb, the proportion cultivated is actually at most from 9 to 16 per cent, a level normally achieved only in wartime; allotments and gardens produced about 10 per cent of home-grown food in 1944. In spite of such utility, the inter-war problem of the loss of the ability to produce food at home and the apparent 'urban sprawl' informed and indeed drove the debates which resulted in a comprehensive planning and development control system for the nation. In 1950, the best estimates suggest that 9.7 per cent of England and Wales was urbanised and 2.4 per cent of Scotland. Of the open land, crops and grazing were about 66 per cent and unenclosed 'rough grazing' about 14.5 per cent, in England and Wales.

FIGURE 7.4 (opposite) *Buckinghamshire 1866–1938. Before the great ploughing campaign of World War II, the area of arable cultivation was steadily declining, though the effect of World War I demands for corn is also seen. Not only grassland expanded but also 'rough grazing' which we might infer is unmanaged. Thus the area of intensively used land declines, to the benefit of many wild species.*
L. D. Stamp, The Land of Britain: Its Use and Misuse, *3rd edn, London: Longman Green/ Geographical Publications, 1962, p. 63, Figure 30.*

The rural environment

Since the 1930s were remembered clearly by post-war conservationists, there was a tendency to romanticise a time of run-down farms, unkempt hedges and blocked drains as one of organic wholeness between the farming community and wild life.[36] Any shift to mechanisation and chemicals was therefore seen as a major economic-ecological metamorphosis rather than the recovery of a formerly ailing industry. Changes occurred: if reversion happened in the Breckland, for example, then the stone curlew (*Burhinus oedicnemus*) gained in numbers from its very low levels; as arable land expanded, so did the range and population density of the stock dove (*Columba oenas*). Quite intensive grazing is needed to allow the ant *Lasius flavus* to build nursery mounds and in turn these provide the food for the wryneck (*Jynx torquilla*);[37] the inter-war depression allowed the decline of the wryneck to levels where it is now regarded as a threatened species in Great Britain. Some of the biological interest in the flora of Chalk grassland may also stem from the fact that the grassland had been ploughed at relatively remote periods in its history.

The removal of hedges to allow more mechanised farming is a feature not just of the most recent years but of earlier times too: 15 per cent (130,000 km) of the total length of hedges in England and Wales were removed between 1946 and 1963. The reclamation of moorland affects most aspects of flora and fauna: on Exmoor an area of 59,000 acres (24,000 ha) in 1947 was reduced to 47,000 acres (19,000 ha) by 1976, with 9500 acres (3800 ha) going to agriculture. The annual rate of moorland change was highest in the pre-1960 period and 31 species of plants have become extinct since the 1930s.[38]

The relationship between farming and wildlife in this period was to be transformed yet again by the intensive agriculture of the post-1960 phase. But the general trends were set earlier: as farming becomes more prosperous, most species retreat. The exceptions are those which can take advantage of 'windfall' conditions: higher stocking densities on hills may mean more dead sheep and lambs (and ewe's placentas if they lamb on the hill), to support more ravens, carrion crows and foxes.

The Land Use Survey

In the pre- and inter-war periods there was no national picture of the land situation. That one came about was due to the tenacity and foresight of L. Dudley Stamp, then Professor of Geography at the London School of Economics. Out of small grants and a lot of child labour came a national picture of land use in the early 1930s. Although not perhaps a major aim, it enabled whole sectors of society to see how urban growth had taken place and how complicated the mosaic of land uses had become. Though it never achieved the central place in the national planning which Stamp wanted, it acted as a catalyst in many directions: there had, after all, been nothing like it since Domesday Book of 1086, and King William's survey was a much less complete document.

The survey involved the mapping of the land use of more or less every parcel of land in Great Britain using the OS six-inch scale (1:10,560) maps as the base.[39] The fieldwork, mostly by schoolchildren, was undertaken in the early 1930s and a series of publications of maps at one-inch and smaller scales and of descriptive regional monographs was begun, in spite of wartime losses of records and printing capacities. The result is an unparalleled archive of land use and hence an interesting pointer to environment at the time. Urban and industrial land use was included, with a sub-text of determining the extent of the 'urban sprawl' of the twentieth century. No government department had ever tried to achieve such a synoptic picture of the nation's land resource; neither have they tried again, though an attempt to repeat the LUS in the 1950s was made. The inventory of the 1930s was without doubt an invaluable foundation for the Scott Report of 1942, discussed below.

Heaths, moors and mountains

There runs through the inter-war period a sense of deterioration. Recruitment for the army in the Great War showed that many young men were physically unfit, and the growth of cities and of industry, combined with the Depression, reinforced the mood of national run-down. This even penetrated to the Scottish Highlands, where a perception grew during the inter-war years that there was a deterioration in the quality of the upland sheep grazings of Scotland. This perception was given further approval during World War II, in a set of official reports. As Mather puts it, 'what had originated as opinion ended up being regarded as fact.'[40] Any attempts to provide reliable historical data fall foul of the problems of the data sources themselves, of the increased stocking rates of wartime and of the effects of severe winters like 1941, 1947 and 1951.

The uplands were, without doubt, the scene of depopulation. In one study of upland England and Wales, those parishes which lost people between 1931 and 1951 retained on average 80 per cent of their population at the latter date, which is not perhaps the spectacular decline implied in some discussions. In fact, one parish in the sample (Glyntawe in South Wales) gained population. The number of farms, however, had almost universally declined by 1976 to about 56 per cent of their 1925 level, typically from the 50–60 per cent range but as low as the 30–40 per cent levels.[41]

Heaths

Lowland heaths generally are easy to transform. They are on low slopes and are vegetated mostly with low heathy plants and scrub, their soils are light (though acid) and usually well drained. The trends of the nineteenth century continue into the twentieth, with a further segmentation of lowland heaths, such as those of Dorset, the London and Hampshire basins, and the Suffolk Sandlings. In Suffolk in 1889 there were 19,000 acres (7700 ha) of heath, of which only 40 per cent remained in the 1960s. When agriculture was less profitable, as in the late nineteenth century and the 1920s, then land passed into golf courses: some 1250 acres (506 ha) of the Sandlings have been thus converted. Likewise, the Forestry Commission acquired land when it was cheap in the 1920s. Military use preceded World War II but greatly expanded during it, with airfields, training areas, and defences against the prospect of invasion from the North Sea.

The former system of harmonised land use, in which sheep grazed on the heaths and then on the agricultural land, forming a kind of integrated land use ecology, was thus broken down.[42] As late as 1938 there were 13,000 sheep in the Sandlings area but in 1962 this had dwindled to 4100. The ploughing had calamitous consequences for wildlife since few species can survive the transition from heath to farmland; more can persist (for example, the nightjar, *Caprimulgus europaeus*) in plantations if good-sized rides and openings are left. In the event, the corn bunting (*Emberiza calandra*) was one of the few that appeared to have adapted successfully. Another heathland bird, the black grouse (*Tetrao tetrix*), suffered great reductions in range in the later nineteenth century but stayed on in the New Forest in the 1930s with the help of some stocking; also found on moorland edges, it disappeared from Dartmoor and Exmoor in the 1960–70 period.

Moors

The moorland story of these years has elements of continuity. The grazing value was continued, though with doubts in some areas about the quality of the forage. The sporting interests, too, went on. In Scotland, deer forests were still a vital part of the economy of the land and the ego of the laird, even though the Great War robbed many estates of the cheap (and fit) labour needed to maintain the proper annual cull of the hinds between November and January. As a result, deer numbers expanded greatly, to about 150,000 in the 1960s. This was in spite of the fact that certain forms of poaching were not illegal in Scotland until the Red Deer (Scotland) Act of 1959.[43] The area of deer forest was about 3.2 million acres in the 1950s, a slight fall from the 1912 maximum of 3.5 million acres (1.3–1.4 million ha). Fraser Darling argued that the proper number of wild deer in the Highlands should be 60,000, with an annual cull of from 10,000 to 12,500 per year.[44] The question of the optimal management of Scottish deer continues to the present.

By contrast, discontinuity may be represented by the demands south of the border for increased recreational access to moorlands, which grew especially quickly in the inter-war period. Increased leisure time and disposable income, coupled with the combined appeal of some forms of radical politics and hiking sports, brought many more people onto the hills, especially of the Pennines and of the Lakes.[45] In the former region they encountered two land users who thought their presence inimical: the grouse moor owners and the water authorities. The first were scared that the nesting birds would be exterminated and that frightened birds in summer would never return to their territories; the second were concerned, basically, that walkers would contaminate water supplies. The outcome of these conflicts was never properly resolved since the eventual legislation, the 1939 Access to Mountains Act, was disowned by the outdoors groups and in any case was overtaken by World War II.[46] It was repealed in the 1949 National Parks and Access to the Countryside Act (dealt with elsewhere) but the problem is still not solved, nor perhaps ever likely to be.[47]

The grouse moor was a touchy subject with landowners since the red grouse appears to be subject to cyclic populations.[48] In poor years the sport is inferior and the income likely to be attenuated. In fact the years from 1925 to 1937 in England and from 1922 to 1935 in Scotland were good years for grouse, with lows either side.[49] So the perception that any form of public access to keepered moors would result in substandard sport was more than a little exaggerated and recent research has usually vindicated this view. The water authorities that feared bacterial contamination of their catchment areas rather undermined their case by letting out the moors to grouse syndicates, whose guns, beaters and keepers could not be guaranteed to be totally continent while on the land.

Mountains

Much of what is said above on moorlands applies also to mountains: deer forests in Scotland encompass both types of terrain. As with moors, recreational demand greatly increased and might have been the cause of conflicts had not much of the Lake District, for instance, not come under the control of the National Trust, as had a number of Scottish and Welsh mountains. One undoubted result of easier access to mountains at this period was the stripping of rare plants by enthusiastic gardeners who wanted alpines: a number of species were reduced to very low levels, especially those from nutrient-rich rocks. In terms of recreational use, however, the number of visitors in the period to 1950 must look insignificant beside today's impacts.

In England and Wales, many of the mountains, moors and heaths were common lands in ownership terms, having survived being assigned to individual owners during periods of enclosure. By the 1950s they were often poorly managed, access and use rights were in dispute or forgotten, and there was a feeling that a national asset was being misused. This led to a Royal Commission on Common Lands which reported in 1958 and whose findings will be discussed at an appropriate moment.

FRESH WATER AND FRESH-WATER WETLANDS

Though there were no great changes in these environments in this period, this seems a good moment to take stock. In the lowlands, drainage was very important at times of increasing agricultural production, such as the two major wars. In between, lack of capital rather allowed systems to fall into disrepair. So marshlands where there had been intermittent flooding, as in East Anglia, lost many of their species of plants and animals when they were put to more intensively used grassland or even ploughed. Where once stork, spoonbill and cranes had graced the wet areas, there were now a few pipits. Surprisingly, the otter was still relatively common, compared with many European countries: enthusiasts for the now forbidden sport of otter hunting will claim that this was due to hunting. Small bodies of water were under threat in all the lowlands, for Leicestershire lost 30 per cent of its ponds between 1930 and 1970 and the meres of the Cheshire and Shropshire plain were subject to loss of open water by silting and to eutrophication from agricultural runoff. At times of building boom, especially after World War II, the extraction of sand and gravel created a large number of new lakes which enhanced the populations of, for example, the great crested grebe and tufted duck. Gravel pits and water authorities became in fact the great new providers of open fresh water: by the 1960s there were 19,000 ha of reservoirs in England and Wales and 750 water bodies over 2 ha from sand and gravel extraction

(Plate 7.4). Scotland is home to three-quarters of all the lakes in Great Britain with 25,000 inland water areas, of which 4000 are over 4 ha in size. All these are potential wildlife reserves, the more so since sailing was permitted on water impoundments in England, Wales and southern Scotland only after 1960. The other great expanse of wetland is blanket bog: on the uplands over about 400 m in England and Wales and in the great swathes of Galloway and the Flow country of Sutherland. These peats have been subject to cutting for centuries but also prone to erosion, the causes of which are not accurately known.[50] A combination of climate, the physics of saturated masses of peat from 4 to 5 m deep, grazing and burning has allowed the formation of linear or reticulated systems of channels often cut down to the mineral soil. The result is more rapid runoff of water, occasional trapped sheep and hard work for walkers.

River clean-up began in earnest with legislation in 1951, which started to yield real compliance in the early 1960s. By this time, sewage rather than industrial effluent was the main contaminant of British rivers. Hitherto, poisons from industry created conditions like those of the River Trent in 1936. It then comprised 880 km of river of which 200 km was lethal to all forms of plant and animal life, with a further 80 km that had plant life only. It received, among other noxious tributaries, the Tame from Birmingham plus the Aire and Calder from the West Riding of Yorkshire, to add to its own course through the Potteries. The tidal Thames deteriorated steadily from 1900 to 1950, as London grew and contributed pollutants of many kinds, mostly from 'point' discharges, especially sewage works. Only after 1961 was a serious attempt at restoration undertaken, with some success except in times when drought reduces the freshwater flow to very low levels.[51]

Industry and wastes

One story of industry in this period is that of government attempts of varying success to influence both the location of industry and its environmental relationships. A series of Acts to try and steer new industry into areas of greatest unemployment achieved their greatest success with the Development Areas Act of 1945. Under this legislation, two-thirds of the recorded moves of industry went to Development Areas between 1945 and 1951. In parallel legislation to control emissions was strengthened from time to time (though not during World War II), and improved technology both reduced waste volumes and made possible the recovery of profitable materials from wastes. At the same time, new technology made heavy demands on land: an electricity generating station might take up 200 ha and an oil refinery 500 ha of land.

Modifications of the nineteenth-century pattern

During World War I, nineteenth-century industrial patterns were reinforced for the simple reason that what was required was greater production from existing plant at any cost; in World War II this principle was modified to some extent by the need to locate new industry where possible away from bombing. Nationalisation in the 1940s offered the chance of central control. Thus isolated iron and steel plants like Consett

PLATE 7.4 *Near Rothbury, Northumberland, 1980s. Gravel pits have been one of the major environmental creations of the post-World War II period. Once worked out, restoration conditions may produce good sailing facilities or, as here, a dedicated nature reserve.*

in County Durham had whole supply networks built around them even though their early nineteenth-century resources of iron ore were worked out. The production of derelict land followed in such industries' wake without much amelioration, so that in 1954 about 70,000 ha of land in England and Wales were officially declared derelict, even on a rather restricted definition. In the 1960s the growth rate of such terrain was 1400 ha/yr in Great Britain as a whole. (1400 ha is equivalent to an area 3.75 km square; 3.75 km is about the direct distance from Paddington to King's Cross stations in London.)

Results

Air and water pollution were still bad, right up to the end of the 1950s. If health is excluded, damage to the nation from air pollution during the 1950s was from 1.0 to 1.5 per cent of GDP. Even though smoke was brought under control quite rapidly there was little attempt to deal with sulphur emissions, which had most of their effect as acid precipitation of SO_2. The national fallout of SO_2 in 1900 was about 4 million tonnes, a level which stayed the same in 1939 but rose to 5 million tonnes in 1951 and got even higher with the shift to oil for power generation. Very few power stations were desulphurised during this period. Fulham and Battersea were converted in the 1930s and Bankside was added after the war, though the cleansing process was everywhere suspended during the conflict. The **CEGB** did not in fact agree to a complete pro-gramme until 1987. The one positive outcome seems to have been the inhibition of black-spot fungus on roses.

An outcome of both social and environmental concerns was the development in new locations of 'light' industry powered by electricity. The heavily contaminated air, land and water of places like Rotherham, coupled with their poor employment prospects outside wartime, produced the move to the 'Trading Estate', an area of newer factories and service industries laid out on the edge of a town, with an attempt at landscaping. The air was cleaner though the inevitable presence of a power station ensured that it was not all that clean. The most famous example is at Slough, with the Team Valley in Gateshead and Trafford Park in Manchester other well-known sites.[52] Of the first, a report says that neither Slough nor the trading estate can claim aesthetic merit, but that 'It may even be not without importance for a trading estate to be within a few miles of Eton and Ascot.'[53] These developments were to some extent the forerunners of the small clumps of factories, warehouses and service buildings that were placed in many rural areas to try and retard rural depopulation, a strategy that was copied for some small mining settlements in, for example, County Durham during the 1960s. Most were built on greenfield sites, adding to the perception of inter-war urban 'sprawl'.

Extractive industry

An industrial economy cannot afford to import all its raw materials of low unit value. This has been especially the case with sand and gravel, which is essential for concrete aggregate but which cannot bear the costs of transport for more than about 30 km. In 1950, the national production was 127 million tonnes and in 1960 this rose to 192 million tonnes, with a 1987 figure of 315 million tonnes. (An average house takes 50 t of sand and gravel, and 100,000 t builds 1.6 km of new motorway.) Most of it has come from wet pits in river valleys and thus formed a much-contested intrusion into land use patterns until it was realised that restoration conditions might allow the provision of very popular recreation space for sailing and angling as well as some wildlife habitat (Fig. 7.5). The environmental losses of wet pit extraction such as disappearance of woodland, meadow and farmland together with the noise, dust and transport impacts of the extraction period, were thus set against long-term 'amenity' provision. Such was the quantity of land converted in some areas that in 1973 one geographer called the area of west Middlesex peripheral to Heathrow Airport a 'crisis area', to be set alongside the gravel extraction in the Trent Valley and tin extraction in Cornwall, as well as emerging threats to National Parks like the Peak District's limestone quarries and copper mining in Snowdonia.[54] Not all of these other crises have in fact materialised, and large complexes of wet pits in the Upper Thames can be popular enough to be labelled 'The Cotswold Water Park'.[55]

It is difficult to summarise the environmental linkages of industry simply since the sector is in fact so complex. We can perhaps say that it has become a larger converter of land for production but that the wastes have become less pervasive. But much of this clean-up belongs to the post-1960 era rather than that before 1950.

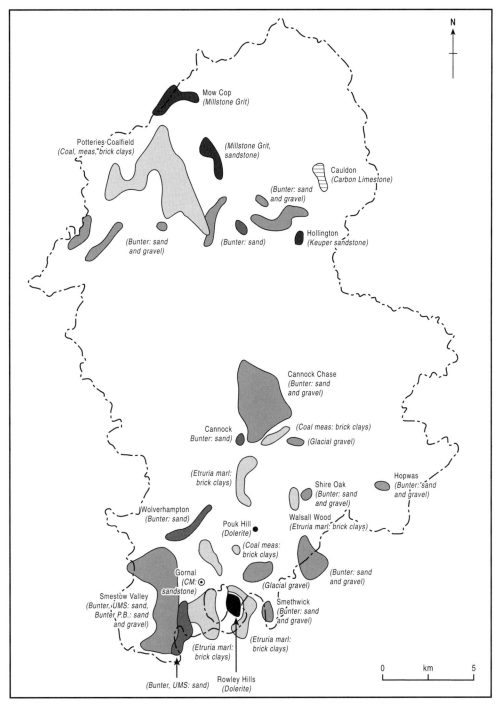

FIGURE 7.5 *An indication of the potential area and diversity of surface mineral workings in Staffordshire in the post-war period. Like many areas in lowland Britain, the demand for sand and gravel was immense at that time. One beneficiary of wet pits was the great crested grebe, which had hitherto been quite a restricted species.*

L. D. Stamp, The Land of Britain: Its Use and Misuse, *3rd edn, London: Longman Green/ Geographical Publications, 1962, p. 233, Figure 140.*

Urban settlement

A synopsis of the environmental impact of urbanisation is seen in Table 7.4. The 'unproductive' figure from the Land Use Survey is an approximation of inner city areas together with industrial zones and waste land. 'Houses with gardens' captures the British (and especially English) desire for a house with a garden. Also a feature of the time was the allotment, fitted into small patches of suburban and rail-side land: from 1936 to 1939, Great Britain held 51,275 ha of them, an area which rose to 72,441 ha in 1942 under the pressure of wartime food demands. Roads in 1937 took up about 185,000 ha and railways 101,000 ha.

Table 7.4   Land use survey: urban areas

	Houses and gardens	Agriculturally unproductive*	Total area
Scotland: area (ha)	155,273	207,726	77,170,442
Percentage	0.8	1.1	
Wales: area (ha)	29,652	47,621	2,063,463
Percentage	1.4	2.3	
England: area (ha)	60,180	432,881	12,964,002
Percentage	4.6	3.3	

*Not including 'rough grazing'
Source: L. D. Stamp, *The Land of Britain: Its Use and Misuse*, 3rd edn, London: Longman Green/Geographical Publications Ltd, 1962, pp. 197–200.

The fact that in the 1930s, some 25,000 ha/yr of 'open land' was being 'lost' is a pervading theme of these years. The growth of suburbs and the by-pass road, immortalised by the cartoonist Osbert Lancaster with his identification of 'By-pass variegated' domestic architecture,[56] the sprinkling of bungalows in seaside areas like Peacehaven in Sussex, and the extension of housing along roads out of town in 'ribbon development', all contributed to a perception that the precious countryside of Great Britain (and especially of England) was being eroded. In a book entitled *Britain and the Beast*, G. M. Boumphrey wrote of:

the march of an inglorious suburbia across our countryside; the wanton sterilization of so much of our productive agricultural and market-garden land; the marring of vista upon vista . . . by the erection of unsuitable buildings, by thoughtless felling of trees, by Philistine methods of road-making and road-widening – in short the blighting touch of the townsman upon the country.[57]

Given this context, it is not surprising that the LUS reports and summary volume give off the distinct feeling that the 'proper' use for land is farming and that all else is a regrettable lapse, often of a distinctly moral kind. The intrusion of morality into British politics is sufficiently rare that it is often followed by legislation, and indeed there were a number of attempts at 'planning' in the 1930s, including the control of ribbon development, which failed, however, because local authorities were unable to afford the compensation promised by an Act of 1935. Positively, there were attempts to found the right kinds of towns in the form of 'Garden Cities' with healthy layouts, good housing and nearby employment: Letchworth in 1914 was the first, with Welwyn in 1920 and Wythenshawe in 1930.[58] Progress towards a Green Belt for London was made in a 1938 piece of legislation enthusiastically promoted by the Labour-controlled London County Council. But even by 1942, in the midst of wartime controls on almost everything, only 3 per cent of Great Britain was covered by any kind of planning scheme. Aestheticism usually failed to notice the quantity of unfit housing that was in need of replacement, a problem exacerbated by war.

All these currents formed part of the great stream of national reappraisal that marked World War II. Once it became clear that the Allies would in time defeat the Axis, the planning for the post-war period began. Stock-taking on an unprecedented level preceded the legislation that a post-war Labour government was temperamentally suited to emplace and which had as its centerpiece the 1947 Town and Country Planning Act. Until reconstruction was really underway in the 1950s, city environments bore many visual reminders of destruction, not least in the populations of rosebay willowherb (*Chamerion angustifolium*) which cloaked so many empty sites. Its American name of 'fireweed' was even more appropriate.

This attitude is deeply engrained. Even though the passage of land from farming to other uses has not been dominated by house building the way it was in the inter-war period, it is still regarded as a major problem at, for example, the level of the serious newspapers. *The Guardian* for example, carried at least two extended discussions of it during 1997. The problems for policy are the demand for more households per thousand people now that nuclear families are less common, increased expectations that the countryside will provide fewer problems than living in cities, and the unknown chemical characteristics of much 'reclaimed' urban land.

RURAL SETTLEMENT

If there can be a single image of rural environments of this period, it is that of the village. Clustered round church, pond, pub or green (occasionally, all four), some of the cottages will have thatched roofs, and the road will very likely bear traffic only of an obviously rural nature. Such has been for some time one construction of rural life in Great Britain.[59] The reality has often been at odds, for agricultural depression sends people away to seek work elsewhere (Fig. 7.6); the result may be the physical decay of church and cottage, and the filling-in of the pond with debris. The Big House and its park become a theme park, a corporate HQ or a residential establishment, a transformation captured by Peter Scupham:

> Time now to stump the hill
> Where stone is dressed to kill
> With bobbled obelisk and dumpy octagon,
> Where some sly toad lifts an unjewelled head,
> Revisiting the mansions of the dead,
> Each son et lumière gone[60]

In contrast, the image of the good life in the country brings retired people, escapees from cities, and, with better transport (especially the private car), commuter families. Since the values of the remains of the farming and allied populations and those of the more recently arrived may not coincide, the latter often attract opprobrious labels. In the LUS reports, for example, the term 'adventitious' is about as value-free as 'immoral', and the subject is still good for a TV documentary today. But even the 'primary' rural population who were presumably allowed to be there adopted rural electrification and allowed it, for example, to replace human labour in the dairy and, eventually, in the indoor animal production units.

Military lands

In the 1930s, many farmers found that the most profitable crop was bungalows, but for a few the acquisition of land by the military was a solution to their income problems. Such a use grew markedly in actual wartime but the premonitions of the 1930s added to the totals. In the years from 1927 to 1934, for example, agricultural land was acquired by the Air Ministry to the extent of 26,225 ha.[61] These were small compared with the 231,488 ha that passed into urban-industrial utilisation. On the brink of World War II in 1938, the services controlled 102,000 ha; in 1945 their holdings were 4.6 million, which was about one-fifth of the area of Great Britain. This land was subject to legislation which allowed the government to hold it in perpetuity (while paying compensation) while it decided upon its future use. No government department was directed to consider the amenity, landscape and wildlife issues involved.

The coasts

Many of the contests of the twentieth century have been manifest on the shores of the country. The battle between protection of the open country and building development of most kinds was fiercest in the south-east: on the Sussex coast and on Canvey Island, for example. It was also fought in Lancashire between the 'popular' developments of the Blackpool type and the more 'refined' attitudes of Southport. The fear of invasion in 1940 led to large areas of the east and south coast being festooned with barbed wire and blistered with concrete cubes and pillboxes, the remains of which can sometimes still be seen. At least one pillbox has become a listed building.

Just after 1950 (but of a piece with the attitudes of the earlier period) there was an example of the old struggle between the forces of the sea and the human desire to win and retain land. This instance occurred on the night of 31 January–1 February 1953.

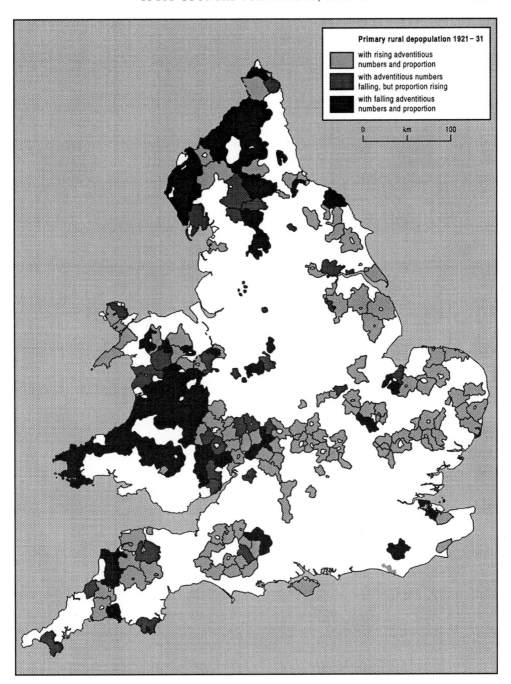

FIGURE 7.6 *The LUS view of 'primary' rural depopulation 1921–31 in England and Wales. Since 'adventitious' people represented changing users of the countryside, it is that category which the LUS has shaded strongly. Who are such people leaving the Lake District and rural Wales? It looks very like the 'primary' families.*
L. D. Stamp, The Land of Britain: Its Use and Misuse, *3rd edn, London: Longman Green/ Geographical Publications, 1962, p. 460, Figure 242.*

On 31 January, the North Sea coasts were experiencing spring tides. In addition, a deep depression tracked south-eastwards from the Faeroes, reaching the North Sea shore of Germany in the early hours of 1 February, bringing Force 10 winds from the north, which pushed water into the narrowing basin between England and the Netherlands.[62] The wind, the storm surge and the spring tides combined to break down sea defences and to over-top them with water levels of, for example, 7.9 ft (2.5 m) higher than expected at Great Yarmouth. Most of the east coast experienced levels of from 6 to 9 ft (1.8–2.7 m) above normal spring tides. The water was driven by the wind into waves of over 20 ft (6.5 m) and so shingle and sand were driven inland to distances not recorded within the living memory of that time: bungalows were partly buried by shingle in Suffolk. Approximately 158,000 acres (390,000 ha) of land along the east coast were flooded by salt water, of which 20,000 acres (49,000 ha) were in Lincolnshire (Figs 7.7 and 7.8). The other major affected areas were Norfolk, Suffolk and the Thames estuary.[63] Some £30–50 million of damage was done to crops, grassland, drainage systems and urban areas, and over 300 people were killed. Temporary repairs were swift and more substantial defences followed: today there is little trace of the event, which was without doubt Britain's worst weather disaster of the twentieth century. It is, however, recalled to memory more often now that the possible effects of global warming upon sea-levels have come into focus and the Thames Barrier seems to be lowered rather more often: on three consecutive tides in the winters of 1996 and 1999, for instance.

Fishing and whaling

Any tendency to be triumphant about the management of the British environment has to stop headlong when it comes to the seas. From inshore controversies over cockle beds to the naval clashes off Iceland, there has been a series of tensions over stocks. Only the withdrawal from commercial whaling can be regarded as an example of rational management. In general, mythologies of various kinds have eclipsed science.

The story of the nineteenth century was of the industrialisation of fisheries and the slow realisation of the finite nature of the stocks. That of the first half of the twentieth century is of the improvement of the technology for catching fish divorced from progress in stock conservation. The total demersal fish landings in England and Wales in 1921 were 9 million hundredweight (0.4 million tonnes) and, in 1937, 13.8 million cwt (0.7 million tonnes), about half of which was at Hull. The only serious check on this raid on waters as far afield as the Barents Sea and Iceland was war: in July 1939, there were 1031 deep-sea trawlers working out of British ports and in December 1944 only 276. Landings fell from 22.4 million cwt (1.13 million tonnes) in 1939 to 4.7 million (0.2 million tonnes) in 1944.[64] Thereafter, government money allowed the modernisation of fleets of all kinds. The results are now familiar, including the unedifying clashes between trawlers, Icelandic fishery protection vessels and the Royal Navy (the 'cod wars') which led to the eventual loss of the Icelandic fisheries after Iceland had declared an Exclusive Economic Zone in 1975.

The UK was involved with whaling throughout the period to the collapse of

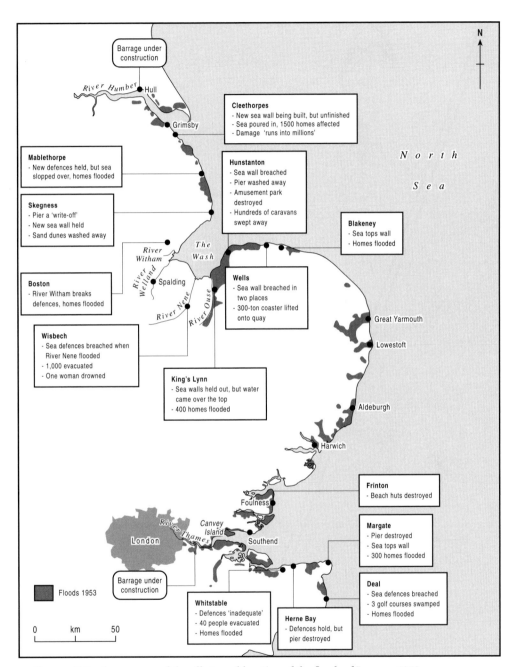

FIGURE 7.7   *A summary of the effects and location of the floods of January 1953.*
A. H. Perry, Environmental Hazards in the British Isles, *London: Allen and Unwin, 1981, p. 78,*
Figure 4.3.

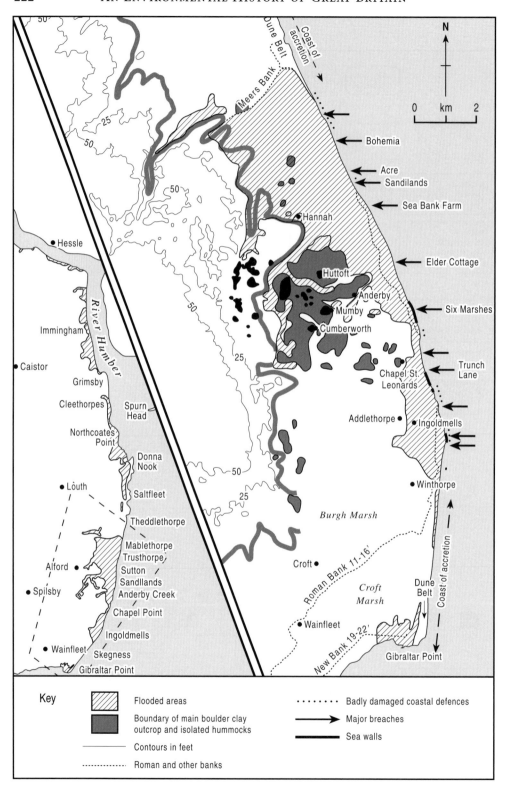

Key

Flooded areas

Boundary of main boulder clay outcrop and isolated hummocks

Contours in feet

Roman and other banks

Badly damaged coastal defences

Major breaches

Sea walls

commercial whaling and finally withdrew from the trade in 1962 after 300 years of activity in various waters.[65] Demand for oil had always been variable and when whales were no longer wanted for lighting and corsets, then soap and margarine took most of the product. Competition here was from palm oil, and the continued fall in whale prices had brought about the closure of the Scottish land-based industry in 1929. In the 1915–50 years, most of the effort was expended in the Antarctic with little to show at home except for the prosperity of the home ports of whaling vessels, such as Leith. The number of whalers declined, however, as the size of vessels became larger, with the factory ship replacing the on-shore station. Attempts had been made to carry out shore-based whaling in Scotland and successful enterprises were set up in the early years of this century at Olna (1904) and a number of other sites including West Tarbert.

The Lever Bros board of directors (representing the major purchasers of whale oil in the UK) had published in 1928, 'the view held in all well-informed quarters that there was no likelihood of a shortage of whales generally'. This was either misinformed or misleading and the economic extinction of whales followed, with a short intermission for a world war. The price dropped from a high of £172 per barrel of whale oil in 1951 to £56 in 1962, the government having failed to persuade even a meat-starved people to eat whale steak. So soap, lighting, stay-bones, soap again and margarine failed to demonstrate that the conservation of wild creatures is best ensured by a programme of controlled use. In this case, far too little control came much too late.

ENERGY

There were no radical changes in this period compared with the shift to coal in the nineteenth century or its abandonment after 1970. So coal continued to supply 90 per cent of energy requirements in 1929, with some 22 per cent of it being used in households, 32 per cent in general manufacturing and 16 per cent in the making of gas and electricity. Energy consumption per head in 1910 was indexed at 612 (with 1800 = 100), at 606 in the nadir years of the 1930s and at 656 in 1950.[66] In absolute terms, consumption per capita rose from 4.3 tce in 1925 to 4.9 tce in 1951. Since most of this came from coal, its environmental linkages are mostly those of coal production and use. By 1931, some 60 per cent of coal was cut by machinery: the output per man rose by one-third between 1924 and 1938, so the trend was towards larger but fewer pits.

The increasing use of electricity meant that more and larger power stations took

FIGURE 7.8 (opposite) *Details of the effects of the 1953 floods along the coast of Lincolnshire north of Gibraltar Point. Notably, the urban area of Skegness itself did not experience much inundation but the areas behind the dunes and seawall to the north were badly flooded. Note that 'Roman . . . banks' in the key is the local term for what is probably a Saxon rather than a Roman feature.*
F. A. Barnes and C. A. M. King, 'The Lincolnshire coastline and the 1953 storm flood', Geography 38, 1953, p. 147, Figure 1.

up land and emitted particulate matter and sulphurous gases. High-voltage transmission was coming into being, with a 132 kV National Grid being commissioned in 1926 and nationalised in 1948. Power stations were still mostly part of the urban fabric, however, until the trend to larger plants began to concentrate them in more rural areas of, for example, the East Midlands and Yorkshire where there were both coal and good supplies of river water for cooling: the series along the Trent is a good example. Between 1950 and 1960, output of power more than doubled (from 15 GWh to 32 GWh), but the number of plants in the UK declined from 338 to 289. Some 43 per cent of the electricity was generated in the English Midlands.[67] The construction of the National Grid meant that the transmission of electric power was a new and highly visible element in the landscape, the more so when a 275 kV grid was constructed in the 1950s, most of which was upgraded to 400 kV in the 1960s. Even the lower voltage power-lines attracted opposition in the inter-war years: many a book on scenery anathematised the 'march of the pylons'. (Much of the same kind of controversy has attended the upgrading and building of new grid transmission lines ever since, and recently some has surrounded wind farms as well.) In the end, though, one house in three was connected by 1920, and by 1939 it was two out of three.

Some of those houses were in the rural areas of Great Britain. In 1926 in England and Wales there were only 20 farms with electricity. By 1941–2, about 27 per cent were wired up and, by 1961, the figure was 85 per cent.[68] As well as lighting, this allowed the mechanisation of, for example, milking and of on-farm dairy processes. This, along with the tractor, meant that the energy delivered on a farm went from 1 hp per person in 1910 to 3 hp in 1939 and 50 hp in 1980. Among other effects, 1.6 million ha of land were released to other uses by the replacement of the horse after 1920.

The electrification of the Scottish Highlands as part of an economic regeneration strategy was an important function of the North of Scotland Hydro-Electric Board, which pursued a development programme through to the 1960s. In this, the Board had ample precedents: electric street lighting was introduced for a while in Greenock in 1885, powered by a 40 hp water-driven turbine; the monks of the Benedictine Abbey at Fort Augustus had an 18 kW generator installed in 1890 and there were a number of other aristocratic schemes. The growth in demand for aluminium (where 1 tonne required 24,000 kWh of electricity for smelting) and the scheme of the British Aluminium Company for the Falls of Foyers from 1895 onwards, with big extensions of capacity in the late 1920s, sparked the movement towards public schemes, and a Grampian scheme of 1928 pioneered this strategy. In so far as capital and labour could be spared, the World War II years carried on the expansion. A comprehensive development plan prepared in the 1950s had the aim of quadrupling the production of 900 million kWh per year of that period. **Pumped storage** schemes were introduced, and in 1962 two schemes were under promotion and four more ready for that stage, when the economics of an integrated Scottish grid swung against the cost-benefit ratio. The Board had successfully tapped the environmental resources of abundant water and a mountainous terrain without too much impact upon scenery and fisheries (its salmon ladders were famous) and had become a part of the fabric of northern Scotland, though electricity proved to be insufficient as the sole catalyst of economic growth. By 1970, the contribution of hydropower to the Board's output had fallen from its apogee of 84.6 per cent in 1963 to 66.8 per cent.[69]

ENVIRONMENTAL PROTECTION AND CONSERVATION

One outcome of many of the processes narrated so far was, especially in the inter-war years, an active and articulate group which was on the one hand devoted to the notion of the rural idyll but on the other hand worried that the 'green and pleasant land' was changing faster than they wanted and in ways that they did not like. The years from 1918 to 1950 were those of the emergence of 'rural England' as a special source of value: an organic place, one of continuity and organic accumulation and of enduring values. It was a great contrast to 'modern Britain' of the cities with their disorder and false values.[70] Something of the same attempt to imagine an 'organic' unity can be seen in the 'Character of England' map published by English Nature and the Countryside Commission in 1996 (see p. 333).

Wildlife and cultural protection

One upshot was the upsurge of interest in protecting the countryside, wildlife and historical monuments, the latter including archaeological remains as well as buildings. The origins of public interest in such matters, along with access to rural land, are often to be found in the latter half of the nineteenth century, in legislation such as the first Wild Birds Protection Act of 1872, amended at regular intervals down to 1925; in the Ancient Monuments Act 1882, which enabled the state to take into its care buildings and other monuments of historic significance; and in the foundation of the National Trust in 1895, with its holdings being declared **inalienable** in 1907. The Society for the Promotion of Nature Reserves was set up in 1912; the Commons, Open Spaces and Footpaths Preservation Society as early as 1865.[71]

The declaration that some species of birds were to be protected against indiscriminate slaughter and casual egg collecting was extended through a series of measures, culminating in the Protection of Birds Act 1954, amended in 1967. The effects of these laws are difficult to assess since they are aimed at direct influences upon the birds themselves, rather than on, for example, massive habitat changes or on pollution. In general, their thrust has been towards making all birds always protected, with exceptions for proven pests and sporting birds, rather than the other way round. This style of protection was deemed to be successful enough for it to be extended to mammals such as grey seals in 1932, though amended in 1970 to allow culling, and badgers in 1973. Plants had fewer friends: at the Cheltenham floral fête in 1933, a prize-winner's entry contained 22 specimens of the orchid *Epipactis palustris*, then found only in one bog of 1.2 ha in Gloucestershire. Wide-ranging legislation came only in 1974.

In the inter-war years, concern focused as well on the impact of land use changes: the growing consciousness of the finiteness of the native flora and fauna and their embeddedness in habitat supported the foundation and growth of bodies like the Society for the Promotion of Nature Reserves (SPNR) from 1912, and gave impetus to the National Trust's acquisition of woods, downs and cliffs as well as historic houses.[72]

Lists of potential and desirable nature reserves were drawn up and a parallel exercise was taken in the hope of the designation of national parks, though the exact role of these was not nearly as clear as that of nature reserves. John Sheail gives us the fascinating precursor example of Charles Waterton of Wakefield who built a 4.8 km wall round his estate in 1813 and instructed his gamekeeper to look after predatory birds and small mammals; he even provided barn owls with new breeding sites to 'supply the places of those which . . . are still unfortunately doomed to death by the hand of cruelty and superstition'.[73]

In general, however, the earliest reserves tended to be on wetland sites or along the coast, which reflected both their biotic diversity and their vulnerability to habitat destruction. Lepidoptera were, for instance, the main reason for the purchase and management of Wicken and Wood Walton fens in Cambridgeshire and Meathop Moss in Cumbria. Individuals were often instrumental either in buying key pieces of land or in running campaigns to raise money. These were not always successful: in 1925 the RSPB failed to gather together the £5500 needed to buy part of Dungeness which was the breeding ground of the rare Kentish Plover (*Charadrius alexandrinus*) and the land was developed for housing. The bird became almost extinct as a breeding bird in Great Britain by 1940 due to shooting and egg collecting for the table and indeed bred for the last time near Rye (Sussex) in 1956.[74] The designation of reserves was impeded by the worsening of the financial climate in the 1930s and private munificence was often the only possibility. However, some municipalities were persuaded to change their policies: the Walthamstow reservoirs of the Metropolitan Water Board were open to shooting parties so that the number of birds fouling the water could be reduced but the RSPB succeeded in having this practice stopped in 1923.

The management of reserves could be difficult. Regulation of people was often a major headache since there were some who killed through ignorance, some for profit and others because they saw the reserves as a protected reservoir of 'vermin'. Early hopes that the designation of a reserve and the erection of a fence was enough to perpetuate plants and animals were often disappointed: voluntary bodies often had to control rabbits and wood pigeons in order to appear as good neighbours. In some cases, intervention had to be severe: when the avocet (*Recurvirostra avosetta*) began to breed again on Havergate Island in the estuary of the Ore (Suffolk) in the 1940s there was a long debate about whether the apparently natural predation by rats and carrion crows should be tolerated. The resolution was in favour of avocets and against rats.

Protection and leisure

Another set of groups was interested in the protection of the countryside for its recreation value: sometimes as the site of activities like hiking and camping, sometimes simply as scenery which contrasted with that of towns and cities. The failure of the attempts to pass an effective Access to Mountains bill in the inter-war period was symbolic of the power of the landowners over the urban dwellers: in the end it was mainly a piece of class warfare.[75] The increasing organisation of the recreationists during that interval is impressive: the Federation of Rambling Clubs was founded in 1905 and had 40,000 members by 1931. It was transformed into the Ramblers' Association in 1935. Nine years after its foundation in 1930, the Youth Hostels Association had almost 300 hostels and over 83,000 members. The key bodies for the look of the landscape were the (separate) Councils for the Preservation of Rural England, Scotland and Wales founded in the 1920s.

One set of aims which united many of these bodies was the desire to follow other nations (and especially the USA) in having National Parks. That Great Britain could not sustain the same kind of wilderness areas as the western USA was obvious but nevertheless the government of the day decided it would examine the possibilities and so set up the first of many committees which were to presage the foundational National Parks and Access to the Countryside Act of 1949. The Addison Committee of 1929–31 saw conflicts between promoting access to the land and the protection of wildlife; its solution was to suggest that there be 'national reserves' for the protective function and 'regional reserves' where recreational interests would be paramount. No action was taken in spite of continued lobbying by the amenity and recreation bodies.

World War II: taking stock in time of war

During World War II land use planning was one area which had high priority: as early as October 1940, SIR JOHN REITH was set to work thinking about post-war planning and he in fact met a deputation from the amenity bodies in January 1942 to talk about National Parks. This eventually led to two more committees, associated with the names of John Dower (1945) and Arthur Hobhouse (1947). Their drive was set in the context of the Scott Report of 1942 which was inspired by Dudley Stamp's work and considered all aspects of the use of rural land.[76] In the field of the built environment, the year 1941 saw the establishment of a National Buildings Record which began the listing and grading of structures of historical significance: a task accelerated, we might imagine, by the fact that so many of them were disappearing nightly.

The work of Dower and Hobhouse took place in a largely encouraging atmosphere. The sense that wild places and beautiful scenery were high in the list of values that underlaid the war effort, the introduction to the countryside afforded to many city children by the evacuation schemes,[77] the equivalent dispersal of many service personnel to rural locations and the sense of the need for greater equality of opportunity that brought in the Attlee government of 1945, all worked in favour of the proponents of legislation. The rural scene was recognised at the highest levels in wartime, indeed: Queen Mary (1867–1953) was evacuated to Badminton in Gloucestershire and is reported as having said, 'So *that's* what hay looks like.'[78]

The Act of 1949

In a general sense, there was probably no more important piece of post-war legislation than the 1947 Town and Country Planning Act. But in this more specific area of amenity and access and attention to conservation, the 1949 Act was fundamental. The National Parks and Access to the Countryside Act 1949 applied only to England and Wales, it is true, but it provided for the establishment of both a National Parks Commission and gave powers to the Nature Conservancy, established by Charter in the same year. The designation of National Parks, Areas of Outstanding Natural Beauty, National Nature Reserves, and their management, was all made possible, though not well financed. The law relating to rights of way was amended and the recording and maintenance of public paths was strengthened. The Act was strong on the maintenance of the rural economy and so both farming and forestry were exempt from statutory development control. Therein lay the seeds of conflict which will be discussed in the post-1950 material.

In 1950, the mood was one of hope rather than achievement. The losses of habitat, species and amenity during the inter-war years, exacerbated by the necessary depredations of World War II, were the starting point for something better: new legislation and a new appreciation of the value of the non-human world as well as the core role of 'nature' in what it meant to be British, meant that the conservationists and ramblers alike set out on a pathway that led upwards and, if anything, got wider. They could not, of course, see round the corners.

Specialised leisure

The increase in demand for recreation resources up to the 1950s made clear a trend that had hitherto been hidden. If the activity is not very specialised then it can often be accommodated in a variety of terrains and alongside other uses. Where there is a concentration of demand this breaks down, as is shown by the history of conflict between ramblers and grouse moor management. It also breaks down where the recreation is so specialised that no other use can be tolerated: in recent years, the boating marina has provided a good example. A longer-standing instance is the golf course. Close examination of the golf course as an environment in Scotland reveals that a number of types of terrain are favoured for conversion. The most common types of course cannot in fact be tied to any recognised landform type (the terms 'undulating' and 'hillside' are used by Price,[79] but some 20 per cent of all courses in Scotland are on modified sand dunes ('links'). By 1930, Scotland had 330 golf courses and this rose to 425 in the 1980s, occupying about 18,200 ha, mostly in the urbanised central belt of the nation. Proud though the municipal tradition may be, 355 of those courses were in fact private and so this form of recreation is segmented off from the rest of the land, just as the woodlands of a pheasant-shooting estate may be. The popularity of golf since the 1980s in Great Britain as a whole has led to an area the size of Dorset being dedicated to this activity. The land ownership may be separate but the environmental management in terms of land sculpting, woodland and heathland clearance (with subsequent changes in, for example, insect and bird faunas), water and biocide usage, are not.

Sacred space

In absolute terms, the fastest growth of this category has been in municipal cemeteries, to absorb the great nineteenth-century growth and concentration of population. But a new outdoor element was the public commemoration of those killed in World

War I. Such memorials, often depicting a military figure on a plinth, are usually set in a small grassy or railed-off enclosure and constitute a strong reminder of the human havoc wreaked by that conflict. They may also have been altered to encompass World War II. The post-1960 years saw something of a diminution in the respect paid to these spaces, but interestingly the keeping of two minutes' silence on 11 November itself, as well as the nearest Sunday, has been revived in the mid-1990s when the fiftieth anniversary came. The military cemeteries of World War II provide another new element. These are the graves of allied service personnel who died while based in Britain, and a large number are air crew from the USA and Commonwealth. The attention paid to the upkeep of these areas (as indeed to those in continental Europe) marks them off in a singular fashion. Lastly, the growth of recreation and of travel in these decades brought many people into the presence of the great religious monuments like abbeys and cathedrals, both ruined and preserved. Cathedrals in particular began to be objects of purely secular pilgrimage, as objects to be seen simply because they were there, just as high mountains like Everest had to be climbed. The conjunction of sacred and secular space in this period is perhaps to be seen as Lords and Old Trafford: the saying goes that the English (*sic*) are not a spiritual people, so cricket was invented to give them an idea what eternity might be like.[80]

ADMINISTRATION AND PLANNING

The inter-war years in particular witnessed the rise of concern about the British scene. The main elements of this have been mentioned already: the poor quality of housing in major cities, the 'sprawl' of urban centres, including the much-execrated ribbon development, and the relocation of population to the south-east as depression closed up employment opportunities in the former coal, iron and steel areas. The Jarrow March of 1936 was in some ways an ironic symbol of what was happening. A first major difference between 1800 and 1950 was the increase in the south-east in both absolute numbers (from 2.4 million in 1801 through 11.7 million in 1911 to 15.1 million in 1951) and in its share of the national total, from 22 percent in 1801 through 28 per cent in 1911 to 30 per cent in 1951.[81] The second was the dichotomy between those rural areas which lost population in the 1930s (a core of northern England, upland Wales and parts of both Wiltshire and Suffolk in the 1920s was expanded in the next decade) and those that gained, which were mostly near large urban centres.

In 1900, planning had been restricted to a few areas of workers' housing set down by paternal companies: Bournville near Birmingham and Port Sunlight on the Wirral are examples. Before 1939 there was legislation with the aim of restricting certain kinds of changes (including an Act in 1935 designed to control the dreaded ribbon development) but in general it was ineffective since it depended on local authorities having enough money to pay compensation for betterment foregone. There was, too, a generally negative approach to national planning in the 1930s which was summed up in its description as part of a 'mechanistic mentality'.[82] One great shift in the inter-war period was the control of London by the Labour Party which designated a Green Belt round the capital, starting with the protection of 25,000 acres (10,000 ha) of land

and culminating in a national Green Belt Act of 1938. In part, this recognition was a response to the increasing pressures for recreation out of doors, with bodies like the YHA and Ramblers' Association functioning then much as the more vigorous Non Governmental Organisations (NGOs) do today.

At all events, the atmosphere was suitable for the ready reception of a Report on the south-eastward drift of population and the need to secure a more even distribution of the new industries that were likely to absorb more of the labour force. The Barlow Commission was set up in 1937 and reported in 1940, which was not the most auspicious year. Yet the exigencies of wartime meant that many of its proposals on the location of new industry were de facto carried out and the national mood of wanting to plan for the future instead of drift into it meant that its effects were far greater than appeared on the statute book. Even as the dust settled from the Blitz of 1940, new evaluations and new mechanisms for urban and land use planning began to emerge, with key individuals carrying forward specific plans (such as Patrick Abercrombie's *Greater London Plan* of 1944,[83] Fig. 7.9) and new ministries coordinating previously competitive schemes.

By the late 1940s, therefore, government and people were supportive of the notion of planning. The planners were seen as latter-day prophets, indeed. This did not prevent them making mistakes, as in the case of the Hull dockers re-housed on a city-fringe estate without transport to get them to their daily 6 a.m. discovery of whether or not there was any work. But the 1947 Town and Country Planning Act can be regarded as one of the defining pieces of legislation which set a tone for the use of the environment in Britain. Much amended and consolidated in later years, its basic provisions are still the guiding principles for the nation's spatial layouts. Even when attacked by deregulatory-minded politicians, changes tend still to be at its margins rather than its core.

The Act made land use planning compulsory for the first time over the whole country. The county became the planning authority unit and was responsible for a 20-year plan that had to be updated every 5 years. There were financial provisions for compensation and betterment but these were unworkable in practice and some had to be repealed. Its drawbacks included the lack of a national scale of planning, and a concentration on the detail of development control such that no time was left for forward-looking and positive planning. The first has never been seriously attempted, but the second has become much more satisfactory in later years. The whole set a standard for developed countries which may now have been surpassed but which was at the time a world leader.

Sensibilities and attitudes to environment

The prevailing intellectual cast of the century is usually codified as **modernism**, which peaked internationally either side of World War I; in the case of Great Britain it was influential after *c*.1908. There were numerous movements within it, such as expressionism, surrealism and Dada. Together, these emphasised disjunct spatial relations, collage and non-representational forms at the expense of the realism that had

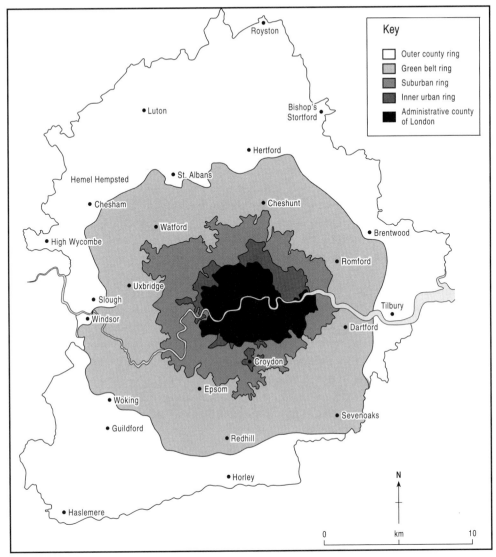

FIGURE 7.9 *The four rings proposed by Abercrombie in 1945 for London. The environmental consequences of this plan would have been to bring less-intensively used land nearer to the main urban populations. The attenuation of the green belt on the west has been very strong, not least because of the growth of Heathrow Airport.*
P. Abercrombie, Greater London Plan 1944. *London: HMSO, 1945, opposite p. 30, Drawing no. 4.*

preceded them. In writing, *vers libre* became dominant in poetry, and the stream of consciousness pioneered by Joyce liberated fiction writing. It is possible to see the construction of modernism in at least two ways. The first emerges as a withdrawal into private languages in order to protect the realism of the aesthetic from the social and political forces which appeared to threaten it; the second is held to be more reflective of the times, in which introversion, fragmentation and crisis were apparent in many ways. Neither is likely to be totally divorced from environmental events or thinking. Modernity, though, has also included the influence of science and technology, faith in which had probably never been stronger though even here dissenting voices can be heard.

Pylons to Skylons

Though anathematised by the country-loving elite, the line of pylons striding out from power station to town and village alike was acceptable to many people. It brought the benefits of electric lighting and the possibility of labour-saving devices and, more than that, symbolised the possibilities that the men (*sic*) in the white coats could open up. In agriculture, better crops would result from the chemical fertilisers that might in due course replace muck-spreading altogether; in the home, new pharmaceuticals might be efficacious in tackling old enemies like TB and pneumonia, just as x-rays had improved certain forms of diagnosis. Much of this indeed happened, though the benefits of antibiotics and synthesised pesticides like DDT came in the wake of World War II rather than any earlier.

Paradoxically, the inter-war period was the time when much of this confidence was undermined intellectually (though not in practical terms) for this was when physicists developed the quantum mechanics which demonstrated a radical uncertainty about the nature of the world's smallest particles. The British contribution to this largely continental field of learning was Lord Rutherford's demonstration of the possibility of splitting the nucleus of an atom and releasing very large quantities of energy. This, he thought, would have no practical application.

This heroic age of physics came to a kind of fruition in 1945 at Hiroshima and Nagasaki. The trust in big science has never been so absolute since then, although the predictive capacity of its theories at meso-scale is still quite good when treated as a set of probabilities. The capacity of science and technology to deliver destruction to British lives and environments and to have little part to play in clearing up, for which personal dedication and rough politics were still the means, has probably been behind the refusal ever since of the polity to deliver unlimited public funds to big science.

Only fit for picnics

The defining poem of this part of the twentieth century is without doubt T. S. Eliot's *The Waste Land* of 1922. However, its environmental resonances are difficult to explicate when compared with references to the countryside scattered through his oeuvre, where for instance a chorus in *The Rock* lists amongst other sournesses,

> In the pleasant countryside, there it seemed
> That the country now is only fit for picnics (Chorus I, ll. 33–4).

This theme of plangent loss is common in the creative arts of the period from 1915 to 1950. It need surprise no one, given the passage of two immense wars and the occurrence of a great deal of social, economic and environmental change. It is moulded from individual human tragedies as well as the loss of Britain's world dominance and finds comfort only in the familiar scenes of an imagined country rather than the unfolding drama of new ways of life. Its audible hallmark must without doubt be Edward Elgar's cello concerto (Op. 85) of 1919, which subdues the composer's earlier heroics into a restrained melancholy (Fig. 7.10). It has become theme music for TV plays and advertisements which are essentially nostalgic in their appeal, but we might detect something deeper and darker than that. In its reaction to his 'Pomp and Circumstance' type music, this concerto disavows the jingoistic and reaches for consolation into the countryside which was so influential for Elgar.[84] In some ways, this shift of vision echoes Edward Thomas's 'Fifty Faggots' of the hazel and ash underwood which was the habitat of robin, mouse, blackbird and wren, that must,

> Light several Winters' fires. Before they are done
> The war will have ended, many other things
> Have ended, maybe, that I can no more
> Foresee or more control than robin and wren.

FIGURE 7.10   *The distilled English nostalgia of the opening soloist's theme of Edward Elgar's Cello Concerto Op. 85 of 1919.*
Complete Edition of the Works of Edward Elgar series 4 London: Novello, 1988, vol. 32, p. 6.

But what it and others almost universally celebrate is rural England: not even rural Scotland or Wales let alone the cities.[85] Like other examinations of the fine arts of the time, this opinion extrapolates from an elite but of course an influential group which influenced the growth of, for example, the pro-access lobby and the oncoming of planning legislation. Its hardiness is demonstrated by Prime Minister Major's evocation in the 1990s of the kind of country he wanted to see, with warm beer, old ladies cycling to church in the early morning mist and, inevitably, the sound of willow upon leather.

The end of industrial dominance in the world and the foreseeable end of empire seem to have fuelled a renewed love affair with the English countryside and nature, so that it is the object of intense scrutiny and approbation. There is a flood of topographic writing about it in the inter-war period, a school of painters who established themselves most famously at St Ives in Cornwall, the better to be able to portray deep rural England, and even music. The work of Ralph Vaughan Williams (1872–1958), for instance, has been said to be full of cowpats.[86] This mood intensifies again with World War II, when again the countryside is seen as the core of what is worth fighting for. Examinations of the iconography of propaganda material, such as travelling exhibitions of paintings, have shown the centrality of the biscuit-tin image of the English village in attempts to evoke patriotism and sacrifice.

The increasingly fragmented nature of post-1945 life can be seen in the contrast between George Orwell's famous novel *1984* which is eloquent on the dangers of control by the state, and which in material terms describes 1948 rather accurately, and the sculpture of Barbara Hepworth (1903–75), one of the St Ives school. Her sculptures, which are best seen when set in the open air, move increasingly towards an identification with organic forms. They are often pierced, for example, to allow sight through them and bring the background into focus along with the sculpture itself. So here we can perhaps see a vision of the unity of human creations and environment, albeit knowing any spontaneity of that identification seems to have vanished into the planning regulations.

There is an acute commentary on much of this time (and a bit beyond) in the work of W. H. Auden (1907–73) whose early poetry has several references to the then newly derelict lead mining area of the North Pennines, especially around Alston and upper Weardale. A former lead-smelting mill's chimney might point 'The finger of all questions' which in time came to include living elements of the environment

> The trees encountered on a country stroll
> Reveal a lot about a country's soul
> . . .
> A culture is no better than its woods

which recognised one of the themes of this book, namely the dynamics of nature and of human affairs:

> This land is not the sweet home that it looks,
> Nor its peace the historical calm of a site
> Where something was settled once and for all:

Which came in 1954 to an aerial view (Auden liked flying) of Gaea

> and Earth, till the end, will be Herself
> . . .
> to Her, the real one, can our good landscapes be but lies,

all of which seems to add to an abiding confidence in the value of the natural world, an awareness of the folly of some human endeavour, but a pleasure in living with it all the same.[87] Nevertheless, any overall assessment of the time has to include the reworking of an identity which included an environment that had somehow failed to become a green and pleasant land. The Skylon at the Festival of Britain's (1951) South Bank site was if nothing else detached from the earth.

The 1950s: legacies and foreshadowings

History suggests that some periods may be thought of as turning points. So far, the later Mesolithic, the advent of agriculture in the Neolithic, enclosure, the shift to an inorganic economy, and the importance of overseas trade, may be cited as key factors in the development of the environment in Great Britain. The 1950s may also be significant, though in a less material way, as a transition between an older Britain with a greater degree of continuity and today's feeling for rapid and unforeseen change. If we look for something that stands above the rest for importance then the 1947 Town and Country Planning Act must occupy that position. Not only did it allow some control of development but it presaged the 1949 Act which established the Nature Conservancy and also exempted farming and forestry. Its consequences are everywhere we look and in probably every environmental system in the country.

In 1950 a great deal more of the past was visible than it is today. The redevelopment of towns and the intensification of agriculture since then has changed a great deal, especially in areas of great pecuniation like East Anglian agribusiness and their opposites like deserted colliery areas and abandoned steelworks. In 1950 we could look back to more traces of former environments than is possible today; paradoxically, there are more such environments today in museums, as if they had been captured from the wild and put behind bars for our safety. The greatest legacy, quite obviously, was that of the Victorian city and its industrial accompaniments. Quite a lot of them, including some of their less desirable emissions, were still present and culminated in the great London smog of 1952. Beyond the towns, the pattern of fields was largely

PLATE 7.5 *The agricultural scene in Shropshire in the mid-1990s. It is as though poised between the popular conception of the 'traditional' and the up-to-date. Fields have not been recently amalgamated, though the machinery used in cereal harvest is clearly modern. Rural electrification is a more recent happening than many visitors believe; the woodlands look 'natural' in the landscape though their uniformity suggests an industrial use for the product. The hills carry sheep at densities which owe most to the subsidy structure of the CAP.*

that produced by enclosure: so well rooted was this configuration that in the public eye it was seen as immutable and a return in the post-1950 period to the larger fields common in some parts of medieval Britain caused a great deal of outcry from bodies like the CPRE (Plate 7.5). Many smaller monuments have inevitably disappeared. The gutting of the remnants of medieval city buildings and plans had already taken place and the primacy of the road vehicle was well established even in the towns.

The oncoming adaptation of the nation to road transport was clearly the most important nascent event in 1950, with plans being laid for the first motorways,[88] for the phasing out of the tram, and for the shift to petroleum products which was entailed. In parallel the volume of support for conservation of various kinds was growing and becoming more vocal, especially since there was now a base in the government structure for wildlife. The Nature Conservancy had powers of compulsory purchase even though their use was, to say the least, sparing. Thus two trends already present were being gelled into more powerful interest groups and lobbies, each able to claim for itself patches of land and water: the process of fragmentation into ownership, never absent since the later Mesolithic, was being solidified.

CHAPTER EIGHT

A post-industrial world, 1950 to the present

INTRODUCTION: MILLENNIUM FEVER

End-of-millennium discussions of the environmental history of Great Britain in the last 50 years, and especially since 1970, are set within the context of global environmental concerns in which climatic change, pollution generally, the possible exhaustion of resources and the loss of biodiversity all occupy centre stage (that is, our television screens – powerful additions to our 'shafts of sunlight') from time to time. We have also to recall that in this indeterminate world, it is difficult to decide which of our present concerns are likely to remain thus in, say, 50 years from now.[1] Picking which of the flood of recent events, pieces of legislation and policy statements to include becomes very selective indeed, and some will turn out to be the wrong choices.

The passage from the pre-1950 period is not marked by any obvious 'events' but can be discerned in the shift from one major energy source to another. Between 1950 and 1970 oil became as great a supplier of power as coal. In 1950, coal supplied 200 mtce to the nation, and oil 20 mtce.* In 1970, they were equal at 140 mtce each. In 1950, the price of oil imports was twice the price of domestic coal, but in 1970 it was about three-quarters of it. Not only did the earlier cost invigorate a search for domestic sources of oil but also it linked Great Britain to major suppliers in the outside world. In addition, after 1960, nuclear plants began to provide electricity for the National Grid, supplying 10 per cent by 1970. By the mid-1980s, natural gas had also become a major fuel. Thus from a consumption of 4.5 tonnes of coal per capita in 1900, the average person was granted the opportunity to deplete 6.2 tonnes of coal equivalent in 1987.

Oil fuelled the shift in travel modes to the private car (Table 7.1 and Fig. 8.1), which has brought further environmental effects of many kinds including increased access to the countryside for leisure and residence, and the further accretion of suburbs. The number of cars grew from 5.5 million in 1960 to nearly 20 million in 1990. The car accounts, too, for about 80 per cent of all mileage on the roads. The road network itself absorbed another 15,000 ha of land between 1985 and 1990, reflecting the fact that we all travel more: over the last 25 years, the average distance travelled each day

---

*mtce = million tonnes of coal equivalent, a measure based on calorific value which enables comparisons to be made between different fuels. Also used in some sources is mtoe, which uses oil as the standard.

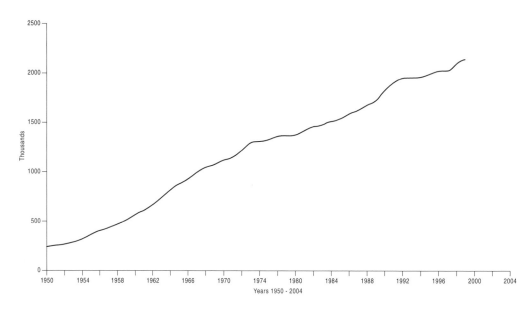

FIGURE 8.1 *The inexorable rise of the number of cars (in thousands) since 1950 into the 1990s. (Due to different counting methodologies some of this curve is basically an estimate but the trend is scarcely in doubt.)*
The Department of Environment, Transport and Regions, Transport Statistics Great Britain, 1998 Edition, *London: The Stationery Office, 1998.*

per person has increased by three-quarters to almost 30 km. Emissions include 90 per cent of all carbon monoxide and 47 per cent of all black smoke, as well as the synergistic creation of photochemical **smog**. If on a global scale the world's environmental problems come down to two factors: meat and motors, then Great Britain's number one is without doubt the motor.[2]

In environmental terms no age has perhaps been more conscious of 'the environment', yet no other has brought about such a strong impact. This may have helped to produce some of the unease which has been a major feature of life in the last 40–50 years, though not of course the only basis. The rapid changes in both society and space (as in the development of New Towns, for example) are well captured for the 1950s in the novels and short stories of Angus Wilson: 'One day the earth would tremble beneath your feet and nothing would happen, the next day would seem so calm and yet the social seismograph registered an earthquake.'[3] In part, these tectonics have been underlain by an increased individual prosperity, famously summed up at Bedford in 1957 by Harold Macmillan (then Prime Minister), in his 'never had it so good' speech, in which the British people enjoyed 'a state of prosperity such as we have never had in my lifetime – nor indeed ever in the history of this country.'[4]

*1 cusec = 1 cubic foot per second.

ATMOSPHERE AND CLIMATE

It is a safe bet that we talk about the weather more often than about any other aspect of our environment. That resonance is mostly local in outlook, with an occasional tut at the peculiarity of places that have floods and tornadoes. The atmosphere nevertheless connects us to the rest of the world at different scales: emissions from our chimneys may end up in raining out over continental Europe or via the upper atmosphere may contribute to global levels of 'greenhouse gases'. We can confidently reject the idea that our national life is in some way a product of the climate,[5] although it still impinges on us, for all our technology. For example, there has been considerable notice taken of the fact that the annual mean temperature of Great Britain has risen by 0.5°C since 1900 and that three out of the four warmest years of the century have been 1989, 1990 and 1995.[6]

Extreme events

The climate of the British Isles is usually classified as equable, but extreme events do occur. On 29 May 1920, for instance, the River Lud suddenly flooded Louth in Lincolnshire and in rendering 1250 people homeless also caused 23 deaths. On 15 August 1952 in north Devon, very heavy rainfall from a thunderstorm descended upon an already saturated Exmoor and caused the narrow-sided and steep courses of the East and West Lyn (whose confluence is in the village of Lynmouth) to flood to unheard-of levels. The amount of water in the combined stream over a few hours during the peak flood was estimated at 18,400 cusecs,* which is close to a record figure for the Thames at 19,500 cusecs. The flow of this immense amount of water was held up first by the piling of tree debris against bridges and then by the accumulation of boulders, of which the West Lyn moved more than 50,000 tons in the one day. The water rose in the village at the rate of 6 inches (15 cm) every 15 minutes for about 3 hours and then fell quickly, leaving 28 people dead, 17 bridges destroyed and damage over a wide area to walls and roads. To the south, the valley of the Exe was the scene of a number of landslides.[7]

Another storm which caused great damage was the hurricane which hit southern England on 16 October 1987. Over a 3–4 hour period, south-west winds up to 114 mph (182 kph) swept south-east England. The previous comparable storm in that region was that of 1703, though there were four other major storms in the 1950–99 period. East of a line from the Isle of Wight to Norwich, some 15 per cent of the timber volume was brought down. Some historic specimens, for example, in Kew Gardens, were lost. Overall, the damage seemed to be random and patchy; 90 per cent of woodlands of native species was unaffected except for the loss of branches, and salvage operations produced more bare ground than the storm.[8]

Increased attention to recording weather and climate phenomena has revealed that tornadoes are a common feature of the British scene: they may be expected on 20 days/year. In 1950, one of them left a trail of damage from Buckinghamshire to Norfolk and on 23 November 1981 some 105 separate tornadoes were observed. They mostly occur on autumn afternoons and the greatest density seems to be in Bedfordshire.[9]

Air pollution

Here we can point to a defining event: the great London smog of 4–10 December 1952. An anticyclone was stationary over London and the smoke output of the city was trapped under an inversion. A normal concentration of smoke particles of about 250 $\mu g/m^3$ became as high as 14,000 $\mu g/m^3$ in central London. As the days progressed, mortality mounted (Fig. 8.2): later calculations suggested there were 4000 extra deaths.

Government response was slow but there was eventually the Clean Air Act of 1956.[10] The Act was the first piece of legislation to adopt a classification of grades of

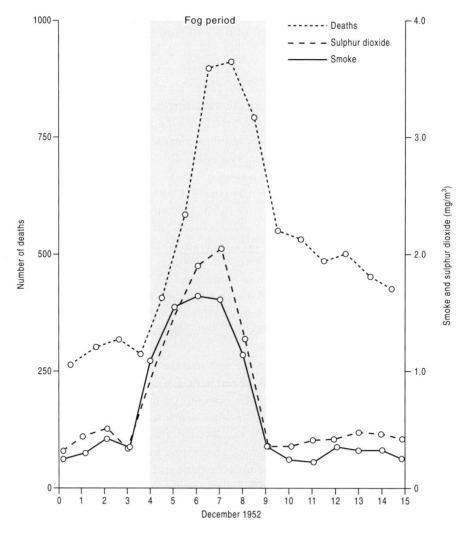

FIGURE 8.2   *Pollutant concentrations and deaths in the London smog of 1952. The official report cut the deaths off so that after the first few days they were attributed to 'influenza'.*
P. Brimblecombe, The Big Smoke: A History of Air Pollution in London since Medieval Times, *London and New York: Routledge, 1987, p. 168, Figure 8.3.*

smoke and to apply them to the control of domestic emissions. It coincided with a period of cheaper gas and electricity (and of many fewer servants to clean out fire-places) and so the reduction to smoke levels of 20 per cent of the normal 1950s values was not all that difficult to achieve.

The Act did not address the emission of sulphur. Levels in the atmosphere led to health problems and to the export of enough sulphur dioxide (SO_2) to cause a political row with Scandinavian countries whose fresh-water acidification could, they argued, be traced to high chimneys at British power stations. They were partly wrong, as it happened. (Interestingly, south-east England now gets about 6 per cent of its SO_2 from shipping in the Channel and North Sea.)[11] The UK eventually joined the '30 per cent club' of European nations determined to reduce their sulphur emissions by 30 per cent, and sulphur levels are indeed falling. Their place is taken to some extent, however, by nitrogen oxides, ammonia, and photochemical smog derived from vehi-cle exhausts which results in ozone, among other compounds (Fig. 8.3).[12] One result of the drop in SO_2 levels is a resurgence of lichen growth. Species thought to be extinct, such as *Lecanora pruinosa*, have recently been found in 30 sites; churchyards especially are zones of recolonisation. Great Britain as a whole has about 1700 species:[13] many of these were restricted if not extirpated in cities when sulphur fall-out was high but some are now recolonising their former habitats on walls, trees and asbestos sheeting (Fig. 8.4). One churchyard in the clean air of Pembrokeshire has 160 species. In the rural areas, it has been recognised that soils vary in their ability to buffer acidification from the sulphur which is deposited from the atmosphere. The concept of **critical loads** has been developed, which shows that the soils of the uplands of the North and West, together with some sandy podsols of lower areas, are most vulnerable to deposition rates which exceed the critical loads for sulphur from human-produced sources.[14] A variety of urban air pollutants still exists and their long-term effects on all forms of life, including humans, is poorly known, except perhaps that some of the organic compounds have proved carcinogenic in industrial contexts.[15]

Air pollution received a temporary seasonal boost in the 1980s when cereal farmers started to burn post-harvest straw rather than chopping it up and ploughing it in. In 1981, straw burning emitted about 27 per cent of British ammonia emissions, which came to 20 kt of nitrogen per year. On one day in August 1984, satellite photography recorded between 300 and 400 burning fields in Great Britain. The fires were liable to cause damage to trees, hedges and buildings; road accidents also resulted from the smoke.[16] The NFU introduced a code of practice in which burning was avoided at weekends and bank holidays, but were overtaken by legislation in which the practice was forbidden from 1993 onwards. A satellite survey in 1995 showed a maximum number of fires in mid-August to be only 22.[17]

Global linkages

The obvious contaminants of the air have dropped in concentration as public aware-ness has been reflected in legislation. A Great London Smog Mark II will presumably be needed to deal with the emissions from motor vehicles. Dealing with the effects of

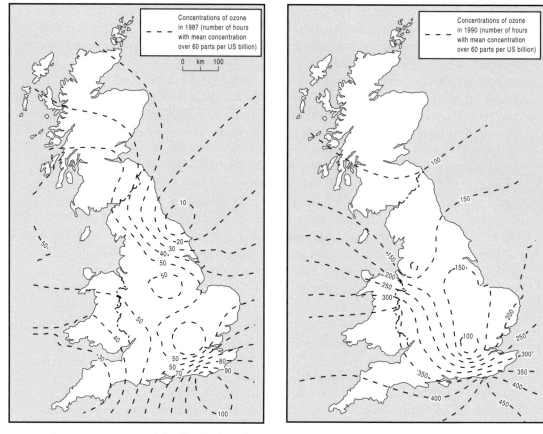

FIGURE 8.3　*Maps of the distribution of ozone at ground level in 1987 and 1990 as surrogates for the new smog in which vehicle emissions and photochemical reactions have replaced the soot and smoke of previous decades. In 1987 it was a dull sunless year and 1991 was a sunnier year. Note the gradient out from London as the ozone forms.*
A. Davison and J. Barnes, 'Patterns of air pollution: Critical loads and abatement strategies', in M. Newsom (ed.), Managing the Human Impact on the Natural Environment: Patterns and Processes, London and New York: Belhaven Press, 1992, pp. 118 and 119, Figures 6.6a and b.

carbon dioxide (CO_2) and other 'greenhouse gases' will be much more difficult. Not only are they either invisible or largely unrecognised (like methane, for example) by the public but also they are intimately tied up with material standards of living, especially that of per capita access to energy. The scientific predictions concerning global warming are full of entirely proper uncertainties and so to disbelieve them is not difficult. Further, some of the popularisations (palms at Portsmouth, outdoor café life in Holmfirth) overlook the disadvantages like increased storminess, coastal retreat, water shortages and perhaps colder winters.[18] Following the various conventions of the 1990s, the UK is on course to stabilise its CO_2 emissions by 2010; that much is surplus fat for us. Thereafter, real reductions in energy use per capita will be required and thus changes in lifestyle. Will any elected government bite that bullet?

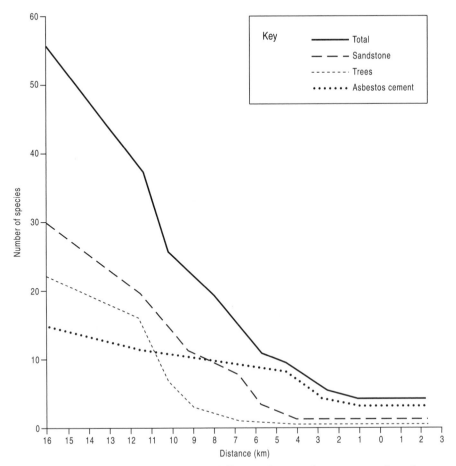

FIGURE 8.4 *Number of species of lichens on different substrates along a transect from the centre (at 0) of Newcastle upon Tyne in the 1980s. The chief agent in their demise has been held to be sulphur but this type of graph suggests that nitrogen compounds and Volatile Organic Compounds (VOCs) may also inhibit lichen growth.*
K. Mellanby, Waste and Pollution: The Problem for Britain, *London: HarperCollins, 1992, p. 74, Figure 9.*

LANDFORMS

The greater intensity of land use in rural areas has been reflected in rates of soil erosion. Experimental work has shown that the mean annual loss in Bedfordshire on fields under grass is 0.2–1.0 t/ha/yr but on bare ground it is ~11.0 t/ha/yr, normalised for slope. When storms occur, then losses equivalent to 150–200 t/ha/yr have been calculated.[19] Severe water erosion was documented for over 600 sites between 1967 and 1976, and one parish in east Shropshire exhibited such evidence for 27 per cent of its surface.[20] Water erosion happens on undulating arable fields every 3 to 10 years and so in some years about 70 per cent of lowland England is affected. Estimates

PLATE 8.1 *Intensive agriculture produces soils which are bare for much of the year and where organic materials are now little used (see the bags of chemical fertiliser in the background), the 'skeleton' of the land shows through, especially where the harrow has been. Sussex, 1960s.*

suggest that about one-quarter of England and Wales are at moderate or high risk of erosion (Plate 8.1).

A common consequence is flooding as the runoff spills out onto roads and settlements. Winter cereal cultivation on the South Downs in the 1980s increased the flooding of properties in wet autumn months: one house and garden was flooded on eight separate occasions between 1983 and 1993: response has been 'dilatory and unimaginative' says one authority.[21] In the uplands, soil erosion is usually less than 1.0 t/ha/yr but when firing occurs this can rise to 4.8 t/ha/yr. High sheep numbers have been associated with the onset of sheet erosion in the Peak District.

Slopes erode, landslides ensue, and rivers change their course at both gradual and rapid rates at the present[22] as well as in the past: floods and a landslide in the Lakes closed the main Ambleside–Keswick road for a number of days in 1997 showing that vulnerability to environmental hazards still exists. As Figure 8.5 shows, landslides are quite frequent in Britain, with some combinations of rock type clearly being the most conducive to slipping.[23] Happily none of the 300–400 earthquakes which affect Great Britain annually (the most susceptible zone is the Welsh Marches) are normally sufficiently strong to induce further landslips: the strongest in recent years was a Richter 5.4 tremor on the Lleyn Peninsula of North Wales in 1984.

A postscript to the study of the soil–landform complex comes with the suggestion that 'wet' soils in southern England are home to higher rates of infant mortality than 'dry' and 'intermediate' soils, with a difference of 31.9 per cent between the dry and

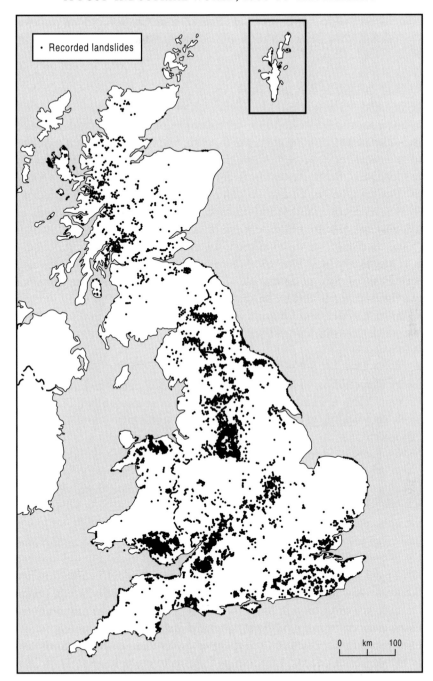

FIGURE 8.5 *The distribution of recorded landslides: a total of about 9000 are plotted in this map. The gaps in mid-Wales and the Southern Uplands may have a geological component but could also be due to under-recording. It is clear that certain geological/landform configurations have a very high incidence.*
Department of the Environment, Landslide Investigation and Management in Great Britain: A Guide for Planners and Developers, *London: HMSO, 1994, p. 1, Figure 1.1.*

the wet. The relationship holds after the effect of social class has been removed and, if true, is a fascinating demonstration of a continuing influence of environmental factors.[24]

Fresh water

Great Britain is normally labelled as a wet country. Total annual replenishment in the UK is now estimated at 120,000 gigalitres from surface waters and 9800 gigalitres from the main groundwater sources.[25] About 25 per cent of the surface water is actually abstracted for use but this average includes the Scottish Highlands, with lots of rain and snow, but few people. Within England and Wales, the regional variation in the balance between effective rainfall and licensed abstraction is remarkably high. London and south-east regions are effectively zones of deficit, with 50 per cent of normal rainfall being used to supply users directly. In a dry year, this produces deficits and in any case re-use is normal. Thames Water has sponsored a drought-tolerant garden in Dulwich Park by way of a symbol.[26] Less than 1 per cent of abstractions is used for spray irrigation in agriculture but most of that is in East Anglia and in the 8 or 12 weeks of summer, so demand forecasts of rises over 100 per cent in the face of climatic change are a little less innocuous than they look.

Diverting the hydrological cycle

The intensity of diversions of the natural flow is especially great in England and Wales and, as previous chapters have shown, has a history of some depth.[27] The last 50 years have scarcely shown any slackening of the pace of searching for more and better water supply, although the water wars for upland catchments typical of the nineteenth century have abated or at least involved different antagonists. In the face of increasing demand for urban and industrial supplies, the standard response has still been to construct reservoirs and to increase the quantity of abstraction from rivers. Few parts of Great Britain have been immune from the search for dammable valleys, though as always the uplands are evaluated first because of their higher rainfall and lower costs. But several large impoundments have been built in Rutland, Huntingdon and Essex in the last 30 years. Since southern England is an area of high demand and relatively low supply, rivers like the Thames contain much recycled water proceeding from one use to the next via a series of treatment plants. It is said that between Oxford and London a molecule of water will likely have passed through eight human guts.

Calls for a national water grid (first officially surfacing in the Water Act of 1945) always foundered on the costs of the pipelines involved, and with privatisation now complete the degree of cooperation required seems to make it unlikely. If there were one, then a key element would be the Kielder Reservoir in Northumberland, which replaced farmland and forest, and which was at the time of its construction the largest reservoir in Europe. It is connected via pipelines to the basins of the Wear and Tees to the south (Fig. 8.6) and was intended to supply part of the burgeoning demand for industrial water for chemical and refining plants on Teesside. Another important element was a large impoundment in Upper Teesdale, at Cow Green (Plate 8.2). This was

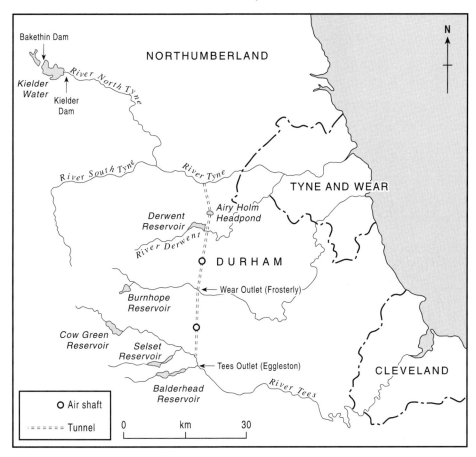

FIGURE 8.6 *Large-scale water transfers in north-east England. Kielder Water is the largest impoundment in northern Europe and can not only supply industrial Teesside with water but also regulate the level in the Tyne, Wear and Tees to the benefit of some wild species, including salmon.*

the scene of a major controversy in the 1960s since it meant the loss of a portion of the local flora in which there were a number of rare plants, most of them being arctic-alpine relicts.[28] In general, reservoir construction has been one of the more obvious areas of land use conflict in recent years.

The intersection of demand, supply and land use makes groundwater supplies look attractive. In the 1970s, groundwater supplied up to 44 per cent of all abstractions in Lincolnshire and up to 59 per cent in the Avon and Dorset regions. The national level was then about 25 per cent. To make such resources renewable, various recharge schemes for this and other examples have been proposed and a few deployed. In them, surplus water (usually in the winter) is put back down boreholes to await usage during the season of deficit. It is more complicated than it sounds and the costs of injecting water in places where it will not immediately be shed back into a river are sometimes prohibitive.[29] Use of the Chalk under London for at least 150 years resulted in a 90 metre drop in the water level by the 1960s. Nevertheless, the water-table is now

PLATE 8.2 *The flooding of the agricultural land in Upper Teesdale in the early 1970s was not controversial but the conflict with the arctic-alpine biota was pursued through all the possible legal channels. If it had been later then eco-warriors would almost certainly have been involved. The Cow Green Reservoir has, as forecast, raised temperatures in the near vicinity. This picture (1981) shows one area traversed by streams whose gravels held some of the rare flora and whose blanket bog had a few specimens of the dwarf birch Betula nana.*

rising as there are so few water-using industrial plants like breweries in the central zone: in places, the Underground is kept usable only by continuous pumping.

It is scarcely a revelation, therefore, that grand plans for estuarine barrages emerge. The reasons for damming off an estuary may include navigation and power generation but often centre on water collection. The Wash and Morecambe Bay have often featured in speculations but the Cardiff Barrage will actually be built. As always, the objections feature environmental aspects of the development: in this case, the loss of tidal wildlife habitat is considerable and the estuary seems likely to dam up radioactive effluent from the nuclear power plants in Gloucestershire as well as silt and untreated sewage from the Severn. But the appeal of large projects such as this is tremendous. Ostensibly this is because a major benefit is the creation of a lot of employment as well as the more direct economic returns. At a deeper level, we might suspect that the appeal is due to the Promethean aspects of a large project: an attack of Concorde syndrome, so to speak.

Impacts of abstraction and use

In many senses, water is one of the environmental components most affected by human activity for the longest period. Increased demand and greater technological capability in the last 50 years have continued that set of processes. Almost every aspect of surface flow has been altered: urbanisation affects the rates of runoff as well as the contaminants in the water; channelisation of rivers affects their erosive power and their fauna; damming changes silt loads and water temperatures downstream. Some 98 per cent of all runoff passes over or through land so there is enormous scope for interaction: new synergisms are always emerging and there are probably a number of which we know nothing. A number of the immediate impacts of water abstraction, use and purification are seen in Table 8.1, but there are many more, especially those resulting from capital projects like the construction of new reservoirs, treatment works and piping. As with so many processes now, energy is consumed in supplying the resource and dealing with effluent. Pumps are the obvious application and, for example, the Thames Water Company has a 'bubbler boat' which oxygenates the Thames tideway with 30 t/day of oxygen during periods of low flow and high temperatures. In compensation, the same company generates about 200 Gwh of electricity from the fermentation of organic wastes, about 20 per cent of its electricity consumption in the 1990s.

The most visible of the environmental effects of water use comes in the virtual disappearance of some rivers in the 1980s. So much surface water was abstracted, or taken out of their groundwater sources, that they dried up completely, or very nearly so, in summer. In Chalk and other limestone terrains this is not new: many streams were called 'Winterbourne' for that very reason. The National Rivers Authority (as it then was) had quite a long list to tackle: for each problem a local management scheme had to be formulated and agreed with all parties. In the cases of, for example, the Misbourne in South Buckinghamshire and the Ver near St Albans a good measure of success has been achieved. The upper 10 km of the River Ver had dried up in the face of an increase in abstraction from 10,000 m^3/day to 45,000 m^3/day between 1950 and 1985.[30] The National Trust has re-engineered a stretch of the River Cole in Oxfordshire so as to produce its sixteenth-century appearance, putting back the bends and shallow reaches so as altogether to make the system (along with some water meadows) retain water rather than shed it quickly into the Thames. A total watershed plan along similar lines is being formulated for Upper Wharfedale in Yorkshire. Nevertheless, English Nature reported that 18 **SSSI**s were at risk from water abstraction in 1996. Such sites are part of the overall series of catchment management plans formulated in the 1990s by the NRA and the Environment Agency.

The same sort of progress towards cleaner rivers has been made, with the legislation providing for ever higher standards. The public awareness of water quality issues has also risen, so that the number of pollution incidents reported has risen in concert with the quality of water. The major sources of pollution are gradually being tackled: for some the rising water bills provide for the cost of treatment, for others the problem is being shifted to another environmental sector rather than being 'solved'.[31] The major sources of freshwater pollution now are: sewage, industrial effluent, agriculture, extractive industry and cooling water. This is an informal ranking, since the different characteristics of, for example, the **BOD** of sewage and the heavy metals from industry, create different problems, which cannot be directly compared.[32] Silo

TABLE 8.1 Water–environment linkages

Operation	Actual or potential environmental impacts		
Abstraction from boreholes	Damage to habitats from reduced levels or flows in rivers, lakes and wetlands		
Abstraction from rivers	Damage to habitats from reduced levels or flows		
Reservoirs	Water level changes affect wildlife		
Water treatment	Energy consumption from pumping	Disposal of wastes from treatment processes	
Water supply	Leakage from pipes	Energy consumption from pumping	Traffic disruption and waste production during repairs and maintenance
Sewerage systems	Surface water pollution from storm overflows	Energy consumption from pumping	Traffic disruption and waste production during repairs and maintenance
Sewage treatment	Landfill for non-biodegradeable wastes	Emission of greenhouse gases	Smells
Sludge processing and recycling	Potential contamination of land	Water pollution from spreading of sludge	Emissions from incinerators; impact on marine habitats
Effluent discharge to rivers	Impact on water quality from permitted discharges		

These are mostly routine processes. Failure at a sewage works would produce a much greater pollution risk.

leakage has a BOD about 200 times that of domestic sewage, for example. Domestic output of the latter is about 135 litres per person per day. In the face of a tradition of simply throwing everything in the rivers, the falling representation of poor quality river water is something of a national triumph, though improvements are still there to be made. One of the strongest NGO influences in water quality improvement is now the coarse angling fraternity (*sic*, pretty well) who are the most numerous outdoor recreationists in England and Wales. Their up-market brethren the salmon and trout fishermen are also enthusiasts for legislation like the EU Freshwater Fish Directive (78/659/EEC) which sets water quality objectives for designated stretches of water which enable the fish to live continuously or breed there. England and Wales have about 19,500 km of such waters and compliance with the Directive has been improving. It is, however, difficult to assess observance since fish populations vary with management just as much as with water quality. Fish have been suffering in other ways: some are getting partly feminised or hermaphroditic. This seems to be due to **oestrogens** in the water. Their source seems to be bacterially unlocked oestrogen-mimic compounds from industrial detergents and **PCB**s, especially the alkylphenols.[33] Other endocrine disruptors may be present in the environment and may not in fact be limited to fish.

Water quality even in large lakes has been affected. Lake Windermere develops surface scums of blue-green algae in the summer, due to some continuation of direct sewage discharge to the water. In the south basin of the Lake, increased phosphorus levels (a 20-fold increment between 1945 and 1986) resulted in markedly lower levels of oxygen in the bottom waters. This in turn affected the population of Arctic char (*Salvelinus alpinus*), a fish which does not tolerate the warmer layers of water which develop in the summer. Treatment of sewage (from a resident population which doubled this century as well as the influx of tourists) has stripped out enough of the phosphorus to allow the char population to start to increase again.[34]

The outstanding issues in population–water–environment relations in Great Britain appear to be three-fold at present. The first is the extent to which more environmental manipulation will be tolerated in order to supply deficits, especially where domestic consumers are kept short. Although the south-east looks like the region with most potential difficulties, West Yorkshire has seen more standpipes in recent summers. Clearly, there is more to it than the number of reservoir gigalitres per 10,000 head of population: the age and condition of the pipework is an important variable, as is the regulatory environment. A final context is the vulnerability of the whole population–water–environment system to climatic change. Winter water runs off earlier in a warmer world and summer runoff is reduced. Anything which varies with flow volumes (such as nitrogen concentration) would undergo summer peaks.[35] If more summer rainfall became convectional, then flood frequencies might increase markedly. Hence, a warming trend will pose problems for water management.

FORESTRY AND WOODLANDS

In the year 2000, the area of woodlands and forests in the UK was about twice that in 1900. Even so, a cover of 9 per cent is one of the lowest in Europe, where 20 to 40 per cent is more common. In 1924, about 65 per cent of the tree cover was broadleaved; now the same proportion is coniferous. In spite of the economic rationale for the conifers, self-sufficiency for wood products is only about 8 per cent. Regional shifts in recent decades have emphasised the importance of the uplands in afforestation schemes.[36]

The types of woodland now present can be classified[37] as:

- 'Ancient' woodland where there has been continuous woodland cover (either deciduous or of Scots Pine in the far north of England and of Scotland) since

1600. If the woodland pre-dates the initial fragmentation of woodland in a particular place then it is said to be 'primary'. A number of ancient woodlands have been placed under conservation management by English Nature or charities such as County Wildlife Trusts and the Woodland Trust. The social context of woodlands has meant that there has been a great reappraisal of the virtues of 'old' woodland, analogous perhaps to old buildings as cultural phenomena.[38]

- 'Recent' woodland grown up since 1600. If it has grown on previously unwooded land since the fragmentation, then it is also called 'secondary'. Small areas include coniferous shelter belts, deciduous copses, pheasant coverts and other amenity and sporting plantations, and experimental energy plantations of, for example, willow. There also exist a few woodlands under old systems of management: Chalkney Wood in Essex, for example, has been under continuous coppicing since the sixteenth century.[39]

The area under trees is being extended by the designation and planting of new swathes of countryside, including some existing forests, some lower grades of agricultural land and some derelict land, called 'Community Forests' (Fig. 8.7). The aim is for a 15 per cent cover of England by woods in 2050 and a key element is a 'National Forest' in the Midlands with 30 million trees being planted from 1990 onwards. Among other functions this is to be productive (its implementation is in the care of a Limited Company), and to be ecologically redemptive since it is to contain a proportion of native species and will reclaim some derelict land. A totally different project aims to restore the 'wildwood' in the Carrifran valley of the Southern Uplands, using pollen analysis as an indicator of late Mesolithic vegetation.[40] Less radical but more widespread are attempts at near-to-nature forestry, which includes the retention of old and dead trees, control of grazing and browsing by deer, sheep and cattle, and the enlargement of small woods.[41]

The commission estate

This period has seen many of the earlier plantings by the Commission come to maturity accompanied by the visual changes associated with clear-felling and replanting. It now has an estate of 1.089 million ha of which 861,000 ha are forest and woodland, and its annual cut is about 76 per cent of the increment. To replace the felled areas, some 8344 ha were replanted in 1996–7, for example, along with 250 ha of new planting.[42] All this has happened in an atmosphere of heightened public awareness of conservation and amenity issues. One result has been both a greater degree of public use and appreciation of the Forestry Commission's lands together with a stronger condemnation of any of their plantings seen as unsympathetic to their local landscapes. In fact, the largest areas of planting in Britain are in Scotland, with the Cheviots, Breckland and South Wales as the only really large areas elsewhere. Kielder, though, has one afforested block of over 1000 km^2. In the period 1940–80, there was a 462 per cent increase in coniferous plantation in Scotland.[43] After the 1981 Forestry Act, about 75 per cent of all plantings in the uplands were by private enterprise with 90 per cent of all private planting being in Scotland. There at least, afforesting more than 30 per cent of the surface has negative effects upon agriculture, including a reduction in the number of full-time occupiers of farms.[44]

The ecology of coniferous plantings is well known.[45] The previous environment (often moorland) is gradually replaced with a tree monoculture low in species variety but with a deep *mor* humus layer. In the early phases the bird populations are not reduced but once at the pole stage and thereafter they lack density and variety of species. Nevertheless, the rides may support populations of nightjars and provide good terrain for sparrowhawks. If there are enough prey species overall, then raptors such as the goshawk, merlin, tawny owl and peregrine falcon may increase in numbers, though the golden eagle (*Aquila chrysaëtos*) and the hen harrier (*Circus cyaneus*) are adversely affected.[46] Mammals tend to benefit: increased numbers of roe, red and sika deer, pine marten (*Mustela martes*), fox, polecat and wild cat have been observed. A summary of the ecologically negative aspects of afforestation would include:

- Moorland may survive only in reserves in heavily planted areas such as South Wales and Galloway. The association between open terrain and the visibility of archaeological remains such as field systems and settlements is lost as well as regional landscape character and well-loved views.
- Some near-natural habitats have been destroyed: examples include undamaged blanket bog, sand dunes and raised bogs.
- At least 66 higher plant species have been substantially reduced in Scotland, and over 50 species which are locally common only have declined nationally. Two species of heaths and moors have become extinct: *Tricophorum alpinum* and *Pinguicula alpina*;[47] among bird species, 22 have been displaced or reduced.
- Rates of acidification and podsolisation have risen on poor soils. Ploughing has increased runoff rates but reduced total water yield from the mature plantations. Salmon and trout populations have suffered from lower invertebrate productivity.
- All invertebrates have been affected by the spraying of trees with organophosphate insecticides to combat pine beauty moth and other pests. However, the cumulative area of the forests sprayed in the last 50 years is only 2 per cent of the plantation area.
- Foxes and crows use the forests as a base to act as predators on moorland birds beyond the forest margin. Many such species will not nest near large blocks of afforested terrain.

Much of the continuing vitriol directed at the forests has, however, arisen from amenity and recreation groups worried as in earlier years at the intrusion into open terrain and in particular at the diversion of footpaths and interruption to walking routes. Beyond that, the plantations alter the visual scale in the uplands and much of the feel of wildness is diminished by the fences, roads, planting lines and the other obvious evidence of human presence. The Commission has responded with detailed design criteria involving shape, character, visual force, scale, landscape diversity and especially the nature of the forest edge.[48] They aim to replace the ruler-straight with the diffuse and have often planted deciduous belts around large areas of conifers in

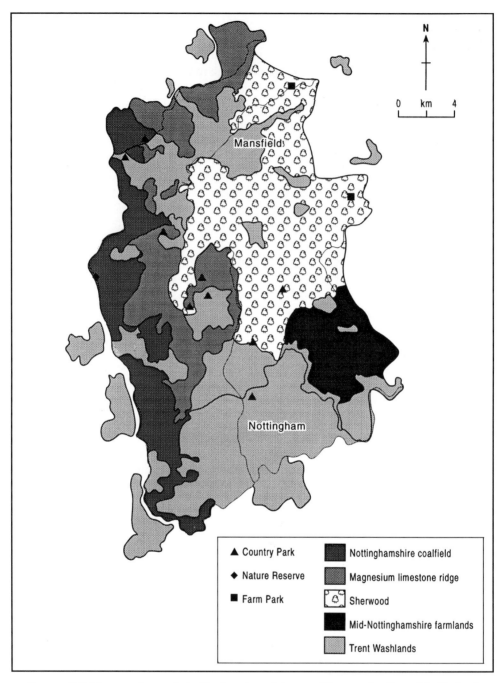

FIGURE 8.7 (above and opposite) *The location of the new 'Community Forests' in England which are now being developed. (The National Forest in the Midlands is not shown.) For one example, the Greenwood Forest, the overall land-use pattern within the unit is shown in greater detail. The information on the map has been drawn from the general website for all the community forests* www.communityforest.org.uk/ar1999.

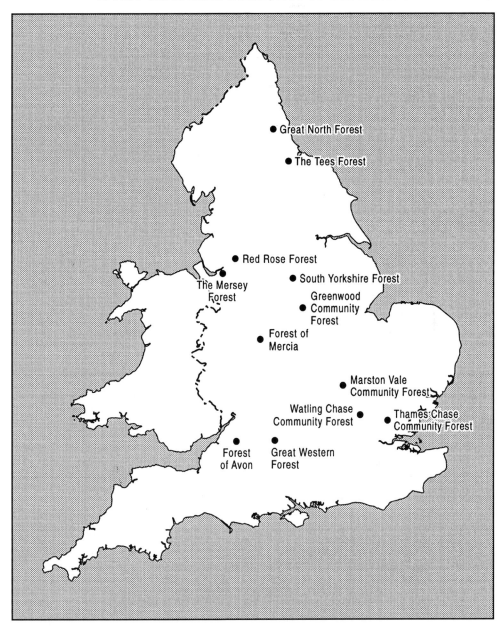

lowland areas: driving through the Forest of Dean, for example, is often to be in a corridor of deciduous trees. Beyond such simple measures, it is possible to develop a co-evolutionary policy for the conifer forests which increases their value to wildlife and small-scale scenic diversity. The re-establishment of areas of native species' woodlands is one course of action.[49] Yet the popularity of planted forests for outdoor recreation is quite high and they hide lots of cars which would otherwise be visible over wide areas. In the 10 years to 1997, the Center Parcs company has developed three very popular resorts in 400 acres of such forests, with cabins and central facilities linked only by walkways and cycle tracks. Moreover, they are the only places where competitive car rallies can be held.

The continuing implacable objection to the upland forests appears to come from one or two recreation and amenity groups and the writers of letters to *The Times*, but they have been reinforced by two trends. The first of these was the use of forestry as a tax haven for the very rich, especially when conservation values were threatened as in the Flow area of Caithness and Sutherland. The planting of an area of peatland of some 400,000 ha with 80,000 ha of conifers represented to the Nature Conservancy Council the most massive loss of important wildlife habitat since World War II.[50] The second is where privatisation threatens to remove the compensating worth of recreational access to upland forests: the Commission was forced by the government during the 1980s to formulate a sell-off programme and by 1996 had disposed of 125,000 ha. Only 1610 ha of that land is subject to access agreements that safeguard recreational use under the new owners. New forestry is likely to move downhill into land spatially marginal to ordinary lowland farming (the rules for farmland acquisition were relaxed in 1986): enclosed areas on the margins of uplands allow the survival of the chough (*Pyrrhocorax pyrrhocorax*) and corncrake (*Crex crex*), so that care about their possible loss of habitat will be needed.[51]

Forestry in northern Scotland

In the twentieth century the afforested area of Scotland trebled: 30 per cent of the blanket peat on the Kintyre peninsula, for example, has been afforested since 1945. Another area particularly affected was the Flow Country of Caithness and Sutherland. ('Flow' is a northern term for any flat deep peat bog.) The Flow Country comprises the largest expanse of blanket bog in Europe and originally covered about 400,000 ha. Bog is the natural climax vegetation of the area and the Flow has never been used intensively because of its poor nutritional levels. Successive grant and tax relief policies designed to encourage forest planting, especially of quick-growing conifers, produced an investment bonanza from the mid-1970s to mid-1980s. Land was bought up by national forestry companies whose interest in the land was confined entirely to its economic viability. By 1986, about 67,000 ha of the Flow Country had been forested or designated for planting. The effects of tree planting, of Sitka Spruce and other introduced species, has been to disrupt drainage patterns, water tables, evaporation rates and soil structure within and far beyond the areas actually planted. Sediment loads, increased water run off, erosion and inquination by chemical fertilisers have altered the water catchment and hydrology. Plant, animal and bird species have been lost or severely reduced and water loss has caused permanent shrinkage of the peats. Because of the widespread effects of tree planting only 8 of the 41 hydrological systems in the Flow Country are so far unaffected.

Reference: D. A. Stroud, T. M. Reed, M. W. Pienkowski and R. A. Lindsay, *Birds, Bogs and Forestry* (Peterborough: Nature Conservancy Council, 1987).

Deciduous woodlands

The values of these wooded areas, small in area compared with the upland conifer forests, lie in features such as their contribution to biodiversity via species and habitat variety, their presence in the landscape as a 'typical' feature, their productive value as sporting estates and (though less common) their output of underwood products. Unmanaged examples can be 'restored' to a more purposeful set of conditions. They are mostly in private ownership so that the institutional and financial conditions have to be exactly right to encourage woods which, for instance, provide baseline conditions for scientific work as well as conservation areas and scenic ingredients.[52] Only since 1984 has the Commission recognised that it has a role to play in conserving and indeed restoring broadleaved woods of 'ancient' character, though Geoffrey Dimbleby pointed out the virtues of birch in ameliorating soils in his Broxa Moor work of the 1950s.[53] This recent involvement may to some extent reverse the trend in which 46 per cent of such woods were lost in the 50 years after 1933 so that a mere 2.6 per cent of England has been continuously wooded since AD 1600. It is thus highly fragmentary and indeed has been so for centuries but some modern developments have accelerated the process. The construction of the M25, for instance, increased by three times the number of very small (<6 ha) ancient woods along its route. Many plants and invertebrates find it difficult to colonise across the gaps caused by such fragmentations; by contrast, the great spotted woodpecker (*Dendrocopos major*) needs 10 ha for breeding, but it can utilise several discrete patches and only needs a piece 0.26 ha surrounding the actual nest site.[54] Most ancient woodland is outside an area of statutory protection and only patchily in sympathetic ownership, so that more comprehensive ways of ensuring its survival (which inevitably means particular forms of management) are needed.[55] Where coppicing is re-adopted, then damage from both fallow and roe deer is likely, with roe likely to do the most harm by its browsing.[56]

The conversion of small woods to pheasant covert is more appealing to many landowners. A mixed wood with an understorey of shrubs and a herb layer is structured so as to send the flushed birds out over a valley or over tall trees. Hand-reared birds (one gamekeeper can now raise and release up to 6000 birds on his own) are now dominant over their wild equivalents and they add little to biodiversity of an estate compared with their wild counterparts.[57] There is little compatibility with informal recreation by the public, especially at nesting and shooting times, and estates run for profit are especially anxious since customers must have a good quantity of birds, well presented, to aim at. Buckinghamshire and Nottinghamshire recorded the highest shooting incomes in the 1990s.[58]

The tree as such is an important part of many people's daily environment. The sycamore can probably claim to be one of the most important species in this respect. Disliked by purist conservationists for being imported (sometime in the fifteenth–sixteenth century; it was still scarce in the late seventeenth century) and by some city-dwellers for the excretion ('honey-dew') shed by resident aphids, it occupies a key role in much-visited scenes like, for example, Haworth (Yorkshire) and the High Street in Oxford, as well as providing shelter around myriad farms, especially in the uplands.

If a 'green' project is sought and imagination or money are in modest supply, then planting trees is usually the outcome, though probably not the sycamore.

Whose woods are these?

It is remarkable that with the maturation of the Forestry Commission's estate, the public had for the first time right of access to woods for their general enjoyment, not simply as part of a delimited common rights agreement or on a public footpath. Many woods remain in private ownership, though subsidised by public funds without any reciprocal rights of access. Global linkages are explicit in the case of woods being useful for soaking up carbon. The actual capacity of even fast-growing conifers to absorb carbon is not huge, however, so that the current accumulation into plantations is about 2.5 million tonnes of carbon per year. This is about 1.5 per cent of the UK's annual emissions of carbon and is about the same as the absorption levels of rivers and wetlands.[59] To maintain this level would need the further planting of 25–30,000 ha/yr of conifers or 10,000 ha/yr of poplars.

AGRICULTURE

The whole national and international perception of rural Britain is enmeshed with its rural landscape, whose ecology and appearance is in turn tied up with changes in agriculture. Changes in husbandry since the 1960s include a phase of intensification in which the main pursuit was increased production. We shall examine first the main trends in this intensified agriculture, then the effect of Britain's participation in the **Common Agricultural Policy** (**CAP**) of the European Union (EU), and finally the reactions to the post-1960 era in the environmental sphere.[60]

Farming, agricultural production and land use

Since the inter-war period, there has been a political presumption in favour of a prosperous agriculture. Increasingly this has meant a bias towards larger units of production ('agribusiness') and has led to the many accusations that the pertinent government ministry (called in 1999 the Ministry for Agriculture, Fisheries and Food, or MAFF) has been so allied to food producers that it fails to protect the interests of human consumers, let alone the environment. In the wake of **BSE** ('mad cow disease') and *E. coli* outbreaks confidence fell so low that in 1997 a semi-independent agency to pursue food safety policies was set up by the government. But its starting line is a rural Britain with the biggest sheep numbers in the EU, the third-biggest cattle herd and the third-biggest cereal crop together with the biggest share of large farms. Almost one-third of British farms are larger than 50 ha, compared with the EU average of 7.5 per cent.[61]

The outworking of science, technology and policy has resulted in a number of practices which have changed the nature of farming and indeed often the face of the land. The main elements of such actions are summarised in Table 8.2. All these processes took place within the context of recurrent episodes of concern about the loss of agricultural land to other uses. In fact the high point of conversion of agricultural land to urban-industrial uses was in the 1930s, at a rate of about 25,000 ha/yr. In the 1970s, this was down to 16,000 ha/yr and fell as low as 8,000 ha in the recession of 1978. So the urban area of England and Wales is now about 11 per cent (for Great Britain, 8 per cent), with 77 per cent still remaining in agriculture. Any synoptic map of land cover will confirm that Great Britain is a much-farmed land; it will not tell us how uncertain an occupation farming has become, so that neither production nor protection measures are likely to convince the farmer to plan for a very long term.[62]

TABLE 8.2 1960s agricultural intensification

Process	Environmental linkages	Other characteristics
Consolidation of holdings	Increase in size of fields, so fewer hedges, walls and ditches	More production from fewer units; East Anglia becomes one of the major productive areas of the EU
Machinery	Weight of machinery on soils queried; effect on wildlife at, for example, harvest-time	Continuation of long trend since 1800 but much accelerated after 1960
Use of chemical fertilisers	Much more P and N_2 in runoff, especially N_2 after 1960s	Easier to handle and with less labour than manure
Use of biocides	Large quantities miss target organisms and enter runoff and food chains	Phasing out of persistent organochlorines quite early in Britain in comparison with other countries
Underdrainage	Shrinkage of organic soils, loss of wet-tolerant plant species in, for example, meadows	Like many other processes, has been heavily subsidised
Industrialisation, for example, indoor production of pigs and poultry	Heavy energy use (usually mains electricity) and intensive waste production, for example, pig slurry and dead chickens	Rapid increase of battery hens from few but large units post-1960

The conflict over the passage of land into urban use has been mainly located in the urban fringe. The high price of housing means that no agricultural worker can afford to buy it, so farming has to be machine-intensive and large loans are common. In south-east England in 1990, good quality agricultural land was worth £1500–2000 per acre (£600–810 per ha) whereas land zoned for housing would bring £0.5–1.0 million per acre (£202,000–405,000 per ha). The mediating factor was planning legislation, especially where there was a presumption against more housing. Thus 'busting the Green Belt' became an important aim of developers.[63] Aided by some government ministers in the 1980s who were not sympathetic to any retardation of money-making, attrition was often allowed on appeal. Notably, however, one minister objected to housing near his own country retreat and was thus responsible for the greatly increased use of the acronym NIMBY ('**N**ot **I**n **M**y **B**ack **Y**ard') in this as other planning contexts, notably that of nuclear power and waste disposal.

Within agriculture, there has not been a great deal of shift in the proportions of cropland and grassland, for it was the ploughing programmes of World War II that restored Victorian levels of tillage after the 1870–1930 decline. But the mixed farm is an endangered species since specialisation is now much more common: more cattle in the west for example and more arable crops in the east, where livestock numbers are lower than in 1938. The overall defining trend of the period after 1960 has without doubt been the intensification of production, mostly of the categories suggested in Table 8.2. The first two are to a great degree interdependent, for the use of machinery has enforced the consolidation of holdings, most noticeably in the amalgamation of fields, with the consequent disappearance of hedges, ditches and walls as landscape elements and as habitats (Plate 8.3). The massive application of machinery came after 1945, with the advent of machines like the combine harvester, which needs a field of 100 acres (40 ha) in order to be economic. Since the number of workers has also fallen, it is no shock to find that agricultural power per worker which was at 1 hp/cap in 1910 should be 50 hp/cap in 1980, with on-farm consumption of petroleum fuels quadrupling between 1938 and 1974. The direct descendant of the engineman in *Tess* (p. 188) is R. S. Thomas's Cyndyllan of the 1950s:

> Ah, you should see Cyndyllan on a tractor.
> Gone the old look that yoked him to the soil;
> He's a new man now, part of the machine,
> His nerves of metal and his blood oil.[64]

The nerves of metal have meant that the post-war length of field boundaries of all types fell by one fifth and that of hedges by a quarter though there had been considerable loss (perhaps 96,000 km) before 1940, due to urban expansion.[65] Perhaps 25 per cent of the 1945 length of hedges has been lost, at a 1945–70 average rate of 8000 km/yr, with the peak rates of disappearance in the 1960s. Regionally, the rate of loss in eastern England was about ten times the national average, yet in the Midlands only about half the national rate.[66] The net loss in 1984–93 was 158,000 km. As well as deliberate clearing, hedgerows were lost to burning and excessive trimming. The removal of 2.5 km of hedges of one farm in Devon was calculated to save 15 per cent in work time which involved machinery; the cultivable area rises 0.02 to 0.1 per cent when hedges are cleared. If the area is susceptible to erosion, then potential soil loss increases, possibly by as much as 20 per cent.

Another matter for concern in the 1960s was the direct effect of the machines themselves upon soils. A Ministry enquiry which reported in 1970 allowed that the passage of heavy machinery on some soils was producing problems of soil structure but thought that the main problem was the lack of drainage of such soils: if arterial

PLATE 8.3   *The arable agriculture of eastern England in particular produced large fields which were dubbed 'new prairies' and which were to some extent like pre-enclosure systems, though the latter would have held much more wildlife in odd scraps of field, meadow and marsh. A lone tree seems to have survived intensification near Essendine in west Lincolnshire in 1981.*

drainage were improved and under-draining extended, then only marginal attention to machine-loading of soils need be given.[67] Fertility of the soils in terms of mineral nutrition has not been in question since the price of oil has been low enough for the production of cheap nitrogenous chemical fertiliser. One tonne of it has the same nitrogen content as 25 tonnes of farmyard manure and its application requires less labour. Hence, in the 1945–60 period, the nitrogen, phosphorus and potassium levels of most soils were kept high with bagged fertiliser; after 1960 there was a greater concentration on nitrogen. These levels are in part a response to knowledge that an output of 5–6 t/ha of wheat or barley extracts 100 kg/ha of nitrogen and 20 kg/ha of phosphorus from the soil.[68] There have been dramatic effects too of plant breeding on cereals since strains responsive to nitrogen have been brought in. Shifting the biomass from the stalk to the head has improved yield at often remarkable rates. Introduced in 1972, the wheat variety Maris Huntsman occupied 34 per cent of the acreage in three years.

Another source of nitrogen concentration results from the way in which pig and poultry production has become aggregated into large units. By 1960, about 90 per cent of all broilers came from 1000 holdings and in 1970, 5 per cent of holdings were responsible for >50 per cent of the output. Pigs are not dissimilar: in the 1970s, two-thirds of pig holdings went out of production and four-fifths of all sows came to be kept in herds of >50 pigs. More of these were outdoors, however, than in the case of broiler birds, which are totally housed and hence suffered very badly during the time of power cuts in the 1970s. Both types of production result in concentrations of excreta (called 'slurry' in the case of the pigs) and a fair proportion of dead but unsaleable chickens. Intensification was extending even to orchard planting densities in the late 1980s, bringing British apple and pear orchards towards French, Spanish and Dutch tree densities.[69]

The use of a variety of **biocides** has probably featured highest in the public consciousness of industrial agriculture. Sprays were introduced in the late nineteenth century when extracts of plant poisons such as nicotine, derris and pyrethrum were applied to fruit, hops and vegetables. Although by the 1930s less than one million acres/yr were being sprayed, there was huge expansion after 1950 when the new synthetic chemicals became available to farmers. By 1975 at least 65 per cent of the cereal crop was treated with herbicides, which meant that no hoeing either by hand or by tractor was necessary. The later 1970s saw biocides being applied to two-thirds of the potato crop and virtually all sugar beet. In terms of active ingredients, the nation's cereal crop in one year received 6 tonnes of organochlorines, 107 tonnes of organophosphorous compounds, 588 tonnes of fungicides and 8026 tonnes of herbicides.[70] It seems probable, too, that agriculture will be the site of most use of Genetically Modified (GM) plants, which are now under trial by large companies, designed to lower biocide and fertiliser use, but with the capacity to spread their modified genes into their own weedy relatives with unknown consequences. In 1998 there were 325 test fields in Britain, mostly in a belt between Lincolnshire and Suffolk. In 1999, two companies were fined for failing to maintain the barrier which was to stop the drift of GM pollen on to unmodified oilseed (bees can carry modified pollen up to 5 km) and the question of contamination of organic crops has been raised. So much damage has been done to trials by direct action groups that in mid-1999 the government announced that it was considering permissions to keep trial sites secret.[71] There have also been five field trials of GM trees, all of which were discontinued by 1999, those of poplars falling to eco-activists.

Energy, materials and environment

An energy audit of a mixed farm in Worcestershire carried out in 1989–90 gives us a clear picture of the energy and matter flows on a modern farm,[72] summarised in Figure 8.8. The fossil fuel input for the year was 13,318 GJ (giga joules) and the organic matter output was equivalent to 21,191 GJ, so that the capture of solar energy was 7873 GJ. That is, the solar energy retained in the system was just over half the value of the fossil fuel input, and was also about 0.07 per cent of the incident solar energy. The gross energy efficiency of the farm was 0.20 per cent. In terms of atmospheric linkages, the crops absorbed about 1043 tonnes of CO_2 but the holding emitted 2768 tonnes of CO_2, 24 tonnes of methane, and 2.8 tonnes of nitrous oxide. So farms like this are net contributors to enhanced levels of the 'greenhouse gases', here emitting 2.7 times as much carbon as it absorbed.[73]

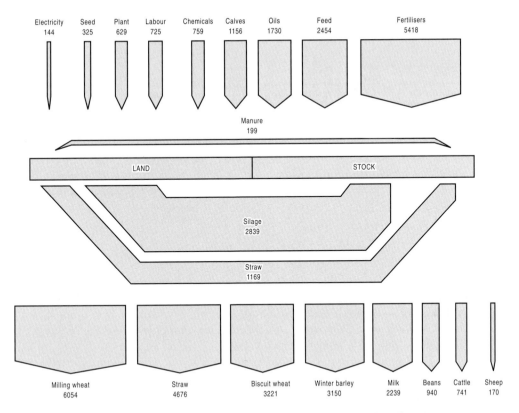

FIGURE 8.8 *The annual flows of energy (excluding those from the sun) for a farm in Worcester-shire in the early 1990s. (Values are GJ of energy transferred.) The dominance of industrial inputs is easily seen and compared with the amount of energy that comes from people in the form of labour.*
A. M. Mannion, Agriculture and Environmental Change: Temporal and Spatial Dimensions, Chichester: Wiley, 1995, p. 197, Figure 7.1.

Environmental impacts of intensified agriculture

Next to urbanisation, the intensification of agriculture has received more public attention than most other recent environmental change. Apart from the visual elements of transformation, such as the larger fields and the greater number of buildings in the countryside,[74] the chemical residuals and overall habitat change have evoked the most concern. To some extent agriculture has become a pollution production process (Fig. 8.9) which has not been combatted by any coherent framework of policy and institutional change. Only the application of EU standards, albeit with a liberal inter-pretation, has made any difference.[75] The element causing most long-term concern is nitrogen, large amounts of which have entered the groundwater and rivers, often increasing markedly after *c.*1965 (Fig. 8.10). Its recent sources include fertilisers not taken up by crops, slurry escapes and silage effluent. One result has been the

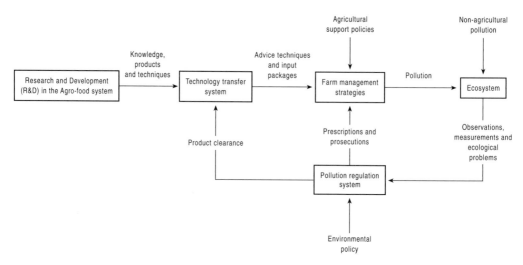

FIGURE 8.9   *Agriculture can produce pollution as well as food and fibre. This flow diagram shows how many inputs into the system can affect the amount and types of pollution and the importance of the single feedback loop in their amelioration.*
P. Lowe, T. Marsden and S. Whatmore (eds), Technological Change and Rural Environment, *London: Fulton, 1990, p. 62, Figure 3.3.*

accumulation of groundwater with a nitrate level above the EU/WHO recommended level of 50 mg/litre for drinking water. The regional concentration shows that levels are low in the uplands but much higher in the lowlands, where intensive farming is combined with lower rates of runoff to increase the concentrations found. South-east England has relatively low values because of the action of the Chalk in absorbing nitrates in solution, but eventually these will reach the water-table and appear in rivers.[76] The reform of the CAP now under way means that it may become possible to control nitrogen through land use change, including buffer zones; in the past only water treatment would have been a possible, and very expensive, approach.[77]

The environmental potentials of residuals from biocides focused upon the accumulation of long-lived compounds such as DDT, which were not target-specific and which became biologically magnified via predator–prey chains. Seed dressings such as aldrin and dieldrin (used against the wheat bulbfly) were eaten by gramnivorous birds and passed on to buzzards (*Buteo buteo*), sparrow hawks and peregrine falcons. These birds were also at risk from DDE, a breakdown product of DDT, which was in fact more toxic than its precursor substance. Action on these issues was taken in 1961 (before the appearance of Rachel Carson's *Silent Spring*) and few organochlorines are now licensed for use. At all events, predatory bird populations which suffered from high levels of toxic residues which caused eggshell thinning and hence virtually stopped reproduction, have largely recovered. Pigeon fanciers are agitating for falcon culls since, they claim, their racing birds are being taken by peregrine falcons *en route*

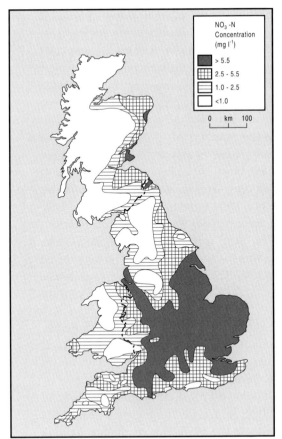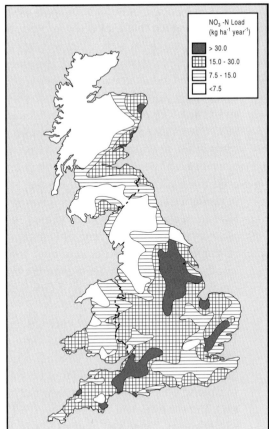

FIGURE 8.10   *Maps of existing nitrate concentration ground water (mg per litre) and of loading*
in terms of additions to that concentration in kg per ha per year.
Betton, C., Webb, B. W. and Walling, D. E. 1991: Recent trends in NO–N concentration and
loads in British rivers. IAHS Publication 203, 169–80.

and suburban bird tables are robbed of greenfinches by the swoop of a sparrowhawk.
Organochlorines are rarely toxic to humans unless ingested directly; the organophos-
phates used in, for example, sheep dip are directly poisonous and so farmers are urged
to apply them (and their use on sheep has been compulsory) with appropriate caution;
nevertheless a number of farmers claim that symptoms of CNS malfunction are
probably due to contact with organophosphate insecticides.

Of all the habitat changes, the removal of hedgerows has had the highest profile.
The ecological consequences hinge on the serious loss of habitat for about 20 species
of plant and perhaps a marked effect on a further 30 or 40 species. Impact on animals
is especially marked on birds, which lose food supplies, nesting sites and cover. Of
some 90 lowland terrestrial species of birds, 65 of them breed only in hedges and 23
species breed commonly in hedges. Out of 28 species of lowland mammals, 21 breed
only in hedges and 14 commonly do so. This all suggests that hedgerow removal is

likely to be more significant in areas where there are plenty of small woodlands and it was also argued that they acted as corridors between larger wooded areas. After a great deal of research, however, key workers concluded[78] that, so far as bird life was concerned, the conversion of woodland and scrub was more significant; the preservation and indeed reinstatement of hedges was likely to be aided most by the knowledge that hedges were essential for the successful breeding of the grey partridge which is a sport bird. The exact role of pesticides is still controversial. A report by the JNCC in 1997[79] implicated them in the diminution of the numbers of 11 common species of birds in the previous 20–30 years (tree sparrow [*Passer montanus*]: –89 per cent; grey partridge: –82 per cent; skylark: –58 per cent; swallow [*Hirundo rustica*]: –43 per cent; even the starling [*Sturnus vulgaris*] at –23 per cent), but the farmers' representative organisations thought that other farming practices were more important factors. There have been species which have gained in the same period, too. The cuckoo (*Cuculus canorus*) is up 38 per cent, the magpie (*Pica pica*) 138 per cent, the stock dove (*Columbus oenas*) 246 per cent and, topping this list of examples, the great spotted woodpecker at 303 per cent, gaining perhaps from the fragmentation of woodlands. An overall index of wild birds of farmland that had 1970 = 100 would end the century at about 65.

Birds are without doubt (1) often identified as agricultural pests and (2) adaptable. They probably have no impact on farm output at the national level but can do so at the scale of the individual holding. The damage is from the loss of yield and from the costs of repairing damage that they cause. Cereals, for example, may be grazed by swans and geese, which also puddle the soils. Rooks will dig up the seeds of barley and oats, and starlings will descend on cereal seedlings, especially those grown near to the winter roosts. They have increased this habit since the 1970s due to shallow drilling of cereal seeds. Only sparrows and greenfinches are able to eat standing crops of cereals though rooks will apparently jump off a fence onto an ear of corn and bring it down to the ground to eat. Orchards have always been vulnerable to bullfinches, which eat out flowering buds, as will starlings; some corvids will enter livestock housing. A number of adaptations are caused by loss of other food sources: geese deprived of salt-marshes by reclamation will look for nearby croplands; bullfinches in contrast will turn from orchards to ash tree seed.[80]

It looks as if agriculture is currently the major cause of changes in wildlife populations, though this must be small compared with some periods in the past, like times of rapid enclosure or industrial growth. In 1978–82, as an instance, some 204 ha of wildlife habitat in Shropshire were affected by agriculture and none by development or forestry. One process which parallels those in other fields is fragmentation, when a formerly contiguous area of habitat becomes split into many parcels. Elements of diversity like old orchards have become fewer. A further, if less tangible, fragmentation is that since c.1970, farmers and conservationists have been antagonists in the debates over land use and production practices[81] (Plate 8.4). Even a farmer-supported research station can allow its director to say that 'Pesticides and inorganic fertilisers have got us into a situation where farming looks like a 19th-century smokestack industry . . .'[82]

PLATE 8.4 *One attitude to environmental modification is seen in the Greenpeace protest in the autumn of 2000 about the imports of GM soya as chicken feed. There was a widespread public reaction to the idea of GM foods, which contrasted with the government's then relaxed view of the process, which it had to modify in short order, albeit with a few squawks.*
Photograph: Greenpeace UK/Cobbing

Reactions

Environmentally related reaction has been important. One response is based on the idea of the 'typical' rural landscape, especially in England. The loss of the hedgerow has been seen as the removal of an immortal element of the scene. Some hedges are almost certainly the lineal descendants of very ancient features: a Saxon origin is not at all unlikely for them. Many more, however, are creations of the period 1700–1850, when Parliamentary Enclosure required owners to hedge or wall their fields and some 160,000 km of hedges were laid. Thus in some places the 'new prairies' of the 1960s and 1970s were a reversion to a seventeenth-century or even medieval state, a condition perhaps no less authentic than the familiar patchwork of field, farm and hedge that visitors to the countryside expect to see. MAFF's eventual reaction was to set a target of restoring some 900 km of hedgerow, and in the years from 1972 to 1988 the length of those features increased by 15 per cent. A second response is underlain by concerns about the loss of wildlife, and has produced MAFF schemes in the mid-1980s for the voluntary retention or restoration of untilled areas and woodlots on farms, as well as ponds. This is called 'set-aside'. Its wildlife conservation value is limited since the areas involved can be rotated round the farm and so are basically ephemeral in their impact. Places with diverse floras, such as the marsh pastures of Suffolk and Norfolk or the hay meadows of Upper Teesdale[83] become eligible for designation as Environmentally Sensitive Areas (ESAs) where both protection of the status quo or some measure of post-intensification management are undergirded by subsidies. An initial list of 150 candidates produced by the Nature Conservancy Council and the Countryside Commission became 21 in 1988 (Fig. 8.11); 6 more were announced in 1994. In the case of the Pennine Dales ESA, two tiers of management are allowed. Both require that the farmers do not plough, level or re-seed the land and carefully control the fertilising of the grasslands. The Tier 2 scheme removes stock by 15 May and forbids inorganic fertilisers altogether.[84] A similar scheme for landowners prepared to submit 10-year management plans for conservation and recreation in a range of rural landscape types has been designated 'Countryside Stewardship' and has paid, for example, £200/ha/yr for the management of historic water-meadows. Regrettably perhaps, all these schemes mean a fragmentation into productive and protected areas at the farm level.[85] A pioneering venture from *Tir Cymen*, the Countryside Commission for Wales, involved whole farms being managed for non-intensive farming and for the benefit of wildlife and amenity; both annual and capital payments can be made.[86] Thirdly, the excess nitrate problem was officially tackled in 1990 with the designation of (1) Nitrate Sensitive Areas (NSAs) where voluntary restrictions on the use of chemical fertilisers and slurry were encouraged by the availability of financial compensation, and (2) Nitrate Advisory Areas (NAAs) where there is free advice but no money. These actions followed the pressure exerted by EU Directives on drinking water (COM 80/788) and nitrates (91/676). The UK government decided that land use control was the best way to conform with these directives: in the affected areas there is a code of good practice which, for instance, avoids the application of chemical fertilisers, slurry or manure in autumn, and refrains from ploughing grassland in the autumn as well. 'Catch' crops to avoid bare soils in winter are also recommended.[87] There is lastly the revival of interest in 'organic' farming. This practice, found in Britain as a conscious aim since the 1920s, tries to reflect a holistic integration of farm biota, production and the overall environment (Table 8.3). In the 1990s, the area of fully organic farming has been estimated at 25,000 ha, with a further 24,000 ha undergoing conversion. Of this total, some 18 per cent is intensive horticulture, with the rest being cereals, field-scale vegetables and livestock. Nutrient cycles are kept up with clover leys, so that a farm will have at least 50 per cent of its land in clover or some other nitrogen-fixer; clover is sometimes planted as an undercrop to cereals. This also reduces the bare area and hence soil loss.

Figure 8.11 (opposite) *Environmentally Sensitive Areas announced in 1991: one of the first of many tentative moves by MAFF towards slowing down the trends to intensification produced by post-1960s agricultural policies. There have been subsequent additions.*
W. M. Adams, 'Places for nature: Protected areas in British nature conservation', in F. B. Goldsmith and A. Warren (eds), Conservation in Progress, *Chichester: Wiley, 1993, p. 204, Figure 11.6.*

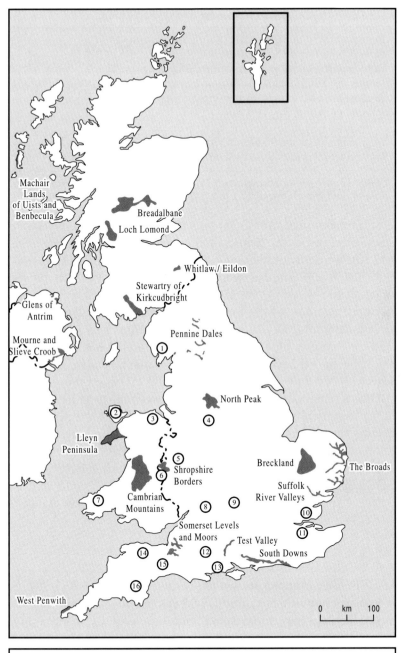

Machair Lands of Uists and Benbecula

Breadalbane

Loch Lomond

Whitlaw / Eildon

Stewartry of Kirkcudbright

Glens of Antrim

Mourne and Slieve Croob

Pennine Dales

North Peak

Lleyn Peninsula

Shropshire Borders

Breckland

The Broads

Suffolk River Valleys

Cambrian Mountains

Somerset Levels and Moors

Test Valley

South Downs

West Penwith

0   km   100

Key				
	1	Lake District	9	Upper Thames Headwaters
	2	Anglesey	10	Essex Coast
	3	Clwydian Range	11	North West Kent Coast
	4	South West Peak	12	North Dorset / South Wiltshire Downs
	5	Shropshire Hills	13	Avon Valley
	6	Radnor	14	Exmoor
	7	Preseli	15	Blackdown Hills
	8	Cotswold Hills	16	Dartmoor

Table 8.3 Principles of organic farming

To produce food of high quality in sufficient quantity
To work with rather than against natural systems
To encourage biological cycles within the farming system
To operate closed systems for organic matter and nutrients
To allow livestock to express their innate behaviour
To avoid pollution
To maintain genetic diversity of the farm and its surroundings
To provide workers with safe, satisfying work with an adequate return
To consider the wider impacts of farming systems

Source: Adapted and rephrased from A. M. Mannion, *Agriculture and Global Change: Temporal and Spatial Dimensions*, London: John Wiley and Sons, 1995, p. 200, Table 7.2.

In spite of all these quite radical changes in farming, the countryside remains a place of high affection in the British (and especially English) mind.[88] This can be symbolised in the longevity of the radio soap opera *The Archers*. After a trial in 1950, this was launched in 1951 and has been in production ever since. It was intended to be an accurate picture of country life but directed at the urban community and hence a vehicle for understanding of the former by the latter. Its engagement with up-to-date and occasionally controversial issues is complimented by a strong attachment to tradition and conservative values, with deep roots in English history in a rural setting.

It is fashionable to cast agriculture as an environmental bad guy rather than a set of ecosystems with development potential for wildlife.[89] Looked at over the long term, though, it shows a remarkable series of swings between the production of necessities and an 'alternative agriculture' in which there is some butter for the bread and indeed even some jam. These have been chronicled by Thirsk[90] who notes that the present diversification of the rural scene is a post-1980s 'alternative' phase which is one of a series that can be traced back as far as the Black Death (Table 8.4). It is possible, then, that agriculture is about to become more environmentally friendly. But its history since, say, 9000 BC is largely one of confrontation with the processes of nature and we ought not to expect too much.

Sacred space

This category is little changed from the previously demarcated period except for continued secularisation of the traditionally religious sites. The labelling of so many monuments as 'heritage' has transformed them into a kind of commodity, a trend which has been reinforced by the introduction of admission charges at some cathedrals.[91] The growth of 'New Age' spirituality and the persistence of organisations of 'Druids' has meant that places like Glastonbury have acquired a numinous significance for adherents. The Druids are taken seriously enough that a small number are sometimes allowed into the stone circles at Stonehenge for the midsummer sunrise, though to the dismay of the tabloid press no virgins are sacrificed. The potential for an interesting display of the monument and its contemporary environments is immense.[92]

TABLE 8.4 Phases of 'alternative' crops, after Thirsk

Mainstream: Medieval period before the Black Death	Cereals and meat production imperative but cereals in excess after the Black Death.
Alternative experience 1350–1500	Land to grass: sheep for wool and cheese, for cattle, rabbits, deer parks. Dovecotes, fishponds (carp), industrial crops, for example, saffron.
Mainstream	Demand for mainstream products rises again: grain sought: rising prices, Civil War. So more ploughing and use of intensive rotations and of fertilisers.
Alternative experience 1650–1750	Key date is 1656 but signs up 100 years before. Luxury foods: development of gardens, orchards, fishponds. Better horses. Horticulture for lettuce, cress, asparagus, artichokes. Industrial crops: madder, flax, woad, hemp. Extension of rabbits, potatoes, bees, hops. (Vine unsuccessful.) Woodland planting and management.
Mainstream	Grain and meat after 1750: population growth – cereals above all else.
Alternative experience 1879–1939	Survival after 1879: large areas to grass, especially leys, plus more intensive production of grain with N_2-fixing rotations (clover, sainfoin, lucerne) plus lime. Covered cattle yards. Industrial and cash crops: sugar beet, tobacco, hops. Dairying, horticulture (including grass), mushrooms, poultry, game, warrens, pigs, woodlands, orchards.
Mainstream	Intensive production under wartime conditions and then with extensive UK and CAP subsidies.
Alternative experience 1980s–	Organic farming, coppicing, meadows, rapeseed (as fuel), flax, hemp (paper, fuel), willow, herbs and flowers, orchards, vineyards, goats, venison, ostrich, game, boar, geese, trout.

Source: J. Thirsk, *Alternative Agriculture: A History from the Black Death to the Present Day*, Oxford: Oxford University Press, 1997.

There are still some relics of folk belief centred on wells, where votive objects may still be seen and where in Derbyshire the practice of 'well-dressing' has domesticated earlier beliefs. The whole suggests that there is a still a current of pagan or 'folk' religion in Britain, though its adherents are seldom highly visible. This veiled quality is still not quite true of the mainstream churches, especially when their clergy behave like everybody else. A few sites have become national sacred sites, however, such as the rebuilt church on the island of Iona, west of Mull in Scotland, and to a lesser extent the pilgrimage centre at Walsingham in Norfolk. There is a distinct concentration on rural areas: most of the urbanised population believes that it is the countryside which contains permanent spiritual characteristics.[93] Some, though, insist that the 'sacred turf' of a major football ground is in some way hallowed and when the ground is sold for development possession of a square from the pitch is like acquiring a religious relic in pre-Reformation days.

There seems to be a revival of the maze. Not exactly sacred space, but having a kind of nod to the relation of life to the thread of time, mazes from the Middle Ages have survived as at Wing in Buckinghamshire, which is a turf maze, unlike the hedge forms which are most familiar; brick, water and maize have also been recent contributions to the repertoire. East Anglia seems to have the highest density, from which no conclusions can be drawn.

FRESH-WATER WETLANDS

One type of habitat that increased during the century was the drainage channel. These are defined as artificial channels in lowland areas which carry water throughout the year, at least in the main drains.[94] The lengths of channel present (c.32,000 km of main channels and 96,500 km of subsidiary channels) ensure that this type of water-course is a more prevalent feature of the environment than main rivers (30,500 km) or canals at about 3000 km. They are often below the level of the highest tides and play a key role in keeping the land available for agriculture. Pumping may often be needed to discharge the water to the sea; most of the areas of greatest concentration lie close to the coast. Agricultural intensification has usually meant underdrainage and thence the loss of year-round water in the drains. One consequence has been the loss of habitat for the common frog (*Rana temporaria*). Coupled with less insect food for amphibians since biocides are so commonly used, this species is now an uncommon sight. Among those that remain, an increased incidence of albinism has been reported, mostly in the south-east and south-west of England since c.1990; a connection with climatic change is suspected but no mechanism has so far been elucidated.

Some 5.8 per cent of Great Britain is underlain by peat, though much less now carries wetland ecosystems. Scotland is actually seventh in the proportional world rankings of peatland area (10.4 per cent), Wales is eleventh (7.7 per cent) and England twenty-first with 2.8 per cent. (The world champion is Finland with 33.5 per cent.) Though peat is found in both lowland and upland areas, many of the lowland areas have been transformed (that is, 'reclaimed') and the most extensive peaty ecosystems are the blanket bogs of the uplands.[95] A general decline in the quantity and biodiversity of extant wetlands is typified in the Norfolk Broads. Created as peat diggings in medieval times (see p. 114), the natural processes of infill have reduced the area of open water. Recreational boating has eroded the banks and so added to the silting. A reduced volume of water has had to bear higher levels of phosphorus and nitrogen from urban sewage outfalls and agricultural runoff. Much marginal vegetation has declined and the water is generally turbid. In an attempt to reconcile the various interests, a Broads Authority was set up in 1988, with the kind of powers possessed by

National Park authorities, though without the name. Active management to restore some of the ecology is in progress.[96] No such unifying approach has been emplaced in the Somerset Levels, where peat digging for garden mulches and agricultural intensification has changed the older landscapes and biotic patterns. Equally, relict fragments of wetland like the meres and mosses of Cheshire–Shropshire, pieces of fen in Cambridgeshire and Huntingdon and in regions like Dorset, have survived because of designation and management as reserves under one type of jurisdiction or another. Few have been isolated entirely from processes such as eutrophication, drying-out and pesticide runoff and it has been increasingly realised that the fauna and flora so prized by scientists and naturalists is often a product of a recent phase of human manipulation rather than simply a natural phenomenon.[97] The type of 'mining' of lowland peats like Thorne and Hatfield Moors in Yorkshire does nothing for wildlife and hydrology even if the sites are handed back when worked out to English Nature, who have installed 2500 dams to keep up water levels. Much to its credit, the B&Q chain of garden and DIY centres has refused to sell peat compost dug from areas of scientific interest.

Drainage and reclamation have combined to reduce the area of reed-bed in Britain. Once an important roofing and flooring material, its area has fallen to 922 sites totalling 6524 ha with one-third of those in East Anglia. There are only 71 large reed-beds of at least 20 ha, though 90 per cent of the beds over 5 ha are under some form of protection. Overall, diminution of their area is occurring (1978–93 saw a loss of 13 per cent), which helps to account for the rarity of the bittern and the reed bunting (*Emberiza schoeniculus*).[98] Cutting and use of the reeds, as happened at Wicken Fen, is no barrier to conservation.

HEATHS, MOORS AND OTHER COMMONS

These semi-natural systems have shared in the national taste for phases of intensification. There have been times when reclamation for agriculture has seemed highly profitable. Between 1949 and 1970 some 11,000 acres (4450 ha) of heath between the Alde and Orwell were thus taken in, aided no doubt by the summer irrigation which was first applied in Britain in this very region. There are pressures from increased numbers of recreational users, and the structure of agricultural subsidies has encouraged farmers to increase the density of grazing stock. Many such areas form the remaining common land of the nation.

The commons – a tragedy?

Common land has declined in total area and become more fragmentary as many owners have enclosed portions for reclamation, and bodies like the armed services have exercised domain over large areas. Yet the bundle of ancient legal conditions that accompanied common rights has persisted. Common land is subject to use by commoners who may have all or some of the rights to graze animals, fish, shoot, take wood or peat, or cut bracken. The land and its mineral rights is, however, held by the lord of the manor. Only in areas which before 1974 were designated Urban Districts did the public have *de jure* access away from rights of way; such areas accounted for about 10 per cent of commons in the 1950s.

To try and tidy up the status of commons and to ensure their future rational management, there was a Royal Commission on Common Land in England and Wales, which sat from 1955 and reported in 1958.[99] The report took stock of the nature and distribution of common land and recommended legislation for the compilation of a national register of common land and rights, and for the setting up of management plans to oversee their use; 'village' greens were to be included as well. The stocktaking revealed that the occurrence of commons was strongly bound up with the distribution of heath and moorland, so that two-thirds of the English commons were in the northern counties, whereas half the lowland commons were within 80 km (50 miles) of London (Fig. 8.12). Approximately 70 per cent of designated common was in England and the rest in Wales; the total was 1.5 million acres or 607,000 ha.[100] In total, this category comprised about 4 per cent of the land surface, which was roughly the same as the proportion of forest and woodland in 1945 or of houses and gardens in the 1930s.

The primary use was grazing, which took place on 80 per cent of the land, with some 47 per cent of it uncontrolled by anybody ('unstinted') and the other 33 per cent under some form of annual leasing of rights ('stinted'). The next largest use was amenity and recreation especially in south-east England. The greater effectiveness of use which the Royal Commission desired to bring about was anticipated in a sample survey of management, both actual and potential, carried out in the 1960s by a Cambridge-led team financed by the Nuffield Trust.[101] This added considerably to the knowledge of village greens, since some of them were open to the public not by virtue of legislation but by customary rights, for which the minimum period for recognition was 20 years. Subsequently, Parliament has never legislated for the active management of the commons by representative bodies (not even setting up a Quango) and so only their whereabouts is known and pertinent rights are registered. The modification of those rights to meet changed conditions has never occurred, however, with the result that the common lands are not as fitted as they might be to play a role in a country-side dedicated to pleasure as well as production.[102]

Mountains and moorlands

Table 8.5 gives a picture of the current interaction of upland land use and terrain in the three countries. About 30 per cent of the total land surface is high ground and, within this total, between 14,000 and 17,000 km^2 is dominated by heather (*Calluna vulgaris*) which is widely perceived as the characteristic landscape element in, for example, Scotland and the North York Moors. The rest is grass moor and sedge-dominated vegetation, with patches of rushes of the genus *Juncus* and wet flushes with a variety of species of bog-moss, *Sphagnum*. The boundary between the lower categories in the Table fluctuates since economic conditions may make it profitable to improve rough grazings but they may subsequently prompt the reversion of enclosed land to moor with the resultant breakdown of walls. Thus the 'moorland edge' may move up and down on a scale of decades; 'island' farms may be sharply delineated in

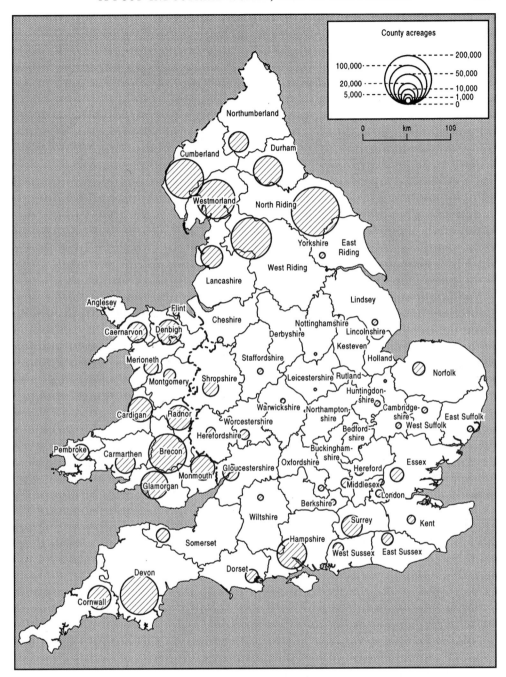

FIGURE 8.12 *The distribution of common land in England and Wales in 1957. The over-whelming importance of upland common (mostly moorland and heath) is clear, though large areas of heath in the south-east are vital outdoor recreation resources.*
L. D. Stamp and W. G. Hoskins, The Common Lands of England and Wales, *Collins New Naturalist, London: Collins, 1963, p. 104, Figure 8.*

the landscape by the improved grassland around them or may melt slowly back into the moorland if the fields are no longer fertilised and sown, and the dwelling allowed to decay or be sold as a second home.

TABLE 8.5 Land use in mountain and moorland regions

Land use	Main locations	Notes
Sheep pasture	All areas	The main use in terms of production; can be combined with nearly all other uses: forestry the clear exception. Stocking density lower when deer or grouse preserved.
Deer forest	Confined to Scottish Highlands	A post-eighteenth century development; now seem to be too many deer.
Grouse moor	Drier moors in Scotland, N. England and Wales: all on eastern side of upland zones	A mid-nineteenth-century development, now often commercially run for profit. Involves vegetation management and predator control.
Forestry and woodland management	Ubiquitous	Woods often managed for recreation as well as production by Forestry Commission; less so by private owners. Deciduous woods often managed for wild life.
Outdoor recreation	Ubiquitous, though proximity to urban populations may influence intensity of use	'Traditional' low-intensity uses such as walking and climbing now only part of a hi-tech spectrum with, for example, mountain bikes, four wheel drive vehicles, hang gliding and winter sports.
Scenic protection	In England and Wales, there are National Parks and AONBs, Scotland has different system	Development is restricted to some extent; grants available for approved purposes.
Mines and quarries	Ubiquitous, but size varies greatly	Popular for very large developments since away from settlement concentrations; best of all for 'super-quarries'.

These landscapes are mostly bare of trees, with two classes of exception. The first comprises the relatively recent plantations of conifers made by the Forestry Commission and private landowners. The second are patches of ancient woodland. Examples are the remnant Scots Pine woodlands of the Scottish Highlands, the oak copses of Dartmoor such as Wistman's Wood, and some of the ash woodlands of the

Pennine limestones. The treeless condition, it will be remembered, is largely a product of human activity from prehistoric times onwards: it is not a 'natural' landscape. While the moors and mountains are perceived today as wild areas, many of them were quite heavily industrialised: the Highlands supported iron smelting in the eighteenth century, the Pennines long exported lead on a large scale, and Welsh slates roofed much of Victorian Britain.

The uses made of the mountains and moorlands are still manyfold and the main range is discussed below. Most of them are not particularly intensive (with the exception of quarrying, obviously) and so do not produce high densities of people or machines in any one place; they still look wild in many respects. An exception to this is military training: moorland areas such as the Otterburn region of Northumberland, Dartmoor and the Brecon Beacons experience varying degrees of live firing and survival training. Any walker on a hilltop has the experience of seeing jet aircraft screaming past at a lower altitude. But there is still quiet to be found, although many upland areas seem to be on commercial flight paths, so that intrusion of the industrial world is easily perceived.

Table 8.6 summarises the main land uses in the mountain and moorland regions, with a few non-systematic notes. It confirms that the main agricultural use of moorland is for sheep grazing. A distinction is usually drawn between 'upland' farms which run both cattle and sheep, and 'hill' farms where only sheep are kept.[103] Both utilise heather, grass and sedge moors as pastures. The hill land is not systematically managed like the lower enclosed land: it may be burned from time to time and the stocking density may be controlled (increasingly through the use of wire fencing) but nutrient input and species composition is left to nature, the sheep and any supplementary feeding.[104] On 80 per cent of moors, pregnant ewes are given supplementary feed in winter, which may concentrate the animals to the point where they create bare ground and so accelerate soil erosion. Firing is carried out to burn off inedible plant material such as old 'leggy' heather and the dry litter of the deciduous purple moor grass (*Molinia caerulea*). The practice is perhaps 200 years old in Scotland but of much greater antiquity further south.[105] Sheep numbers doubled after 1947, back to the level of the 1860s, and the EU support for marginal farms and for sheep has added to those high densities. Whereas in 1977 only 29 per cent of the moorlands of England and Wales had more than 2 ewes/ha, in 1987 it was 71 per cent. In recent decades, 26 to 30 per cent of the gross output of agriculture in Scotland has been from sheep. However, the sales of sheep represent only a small part of the nutrient flow (about 4 per cent of the heather may end up as sheep tissue)[106] and this is replaced by minerals in the high rainfall. Fire and the animals may well cause the vegetation to alter: as stocking increases, the composition of the sward changes, with heather losing ground to fescue grassland and to mat-grass. Burning combined with heavy grazing may also diminish the ability of heather to compete against bracken (*Pteridium aquilinum*).[107] The cloaking of hill land with bracken fern is one of the main visual elements of upland Britain today: in Scotland there was a 79 per cent increase in the area of bracken in the years 1940–80.[108] The fronds cast a heavy shade so that only a few grasses grow beneath, and the plant itself is not eaten by sheep for it is toxic to them. (When more cattle were kept on the hill, they trampled the young fronds. Also, the fern was cut for bedding and packing, soap making and even for silage, but none of these practices is now common.) Bracken's rate of increase was estimated to be as fast as 3.77 per cent per year in central Wales in the late 1970s; elsewhere in Great Britain, rates of spread of 1 to 2 per cent per annum are common. In the 1980s the area of land covered by bracken in Wales was about 1240 km², which is equivalent to the area under tillage, or the whole county of Gwent; for Scotland, 4720 km², which is about 6 per cent of the land area; and in England, 404 km², the same as the Isle of Wight (Plate 8.5). The total for Great Britain was reckoned by Taylor as 6361 km², the area of Lincolnshire.[109] Amenity is also involved since where there are roe deer, bracken may carry the ticks which are the vector of Lyme disease. Further, people do not like to walk through large areas of it since it is very tiring, it may also harbour adders, and its spores contain a known carcinogen, ptaquiloside.[110] The only vertebrate beneficiary seems to have been the whinchat, *Saxicola rubetra*.[111] Control measures (mechanical and herbicidal methods are possible) are expensive in relation to the productive value of land released;[112] any warming at higher altitudes seems likely to reduce the number of the frosts which control its upper limit.

TABLE 8.6   Upland farming management zones

Type	Percentage of upland	Approximate area (million ha)	Use	Management
Inbye land	28	2.1	The fertile core of sheep and cattle pasture	Dunging, fertilisers, reseeding, drainage
Outbye land	35	2.7	Unimproved grassland to heather moor: sheep only	Burning. Fencing beginning to replace shepherds; fox control
Grouse moor	6	0.5	Sporting shooting; some sheep grazing as well	Burning, predator control, disease control, grit for birds
Deer forest	17	1.3	Sporting shooting, some sheep grazing as well	Some culling of hinds; illegal culling of raptors

About 1 million ha (14 per cent) cannot be put into these categories: much of it is in military use or so wet that it supports no grazing.
Source: D. Baldock and G. Beaufoy, *Nature Conservation and New Directions in the EC Agricultural Policy*, London and Arnhem: Institute for European Environmental Policy, 1993, Annex 1.

In the Isle of Man, the mountains, with their heaths and moors, occupy about one-third of the land area and until the mid-nineteenth century provided common pasture. Rough pasture for sheep is now rented out in blocks divided by stone walls and lies generally above 200 m. Some 12,000 ewes graze the hills but few blocks carry

PLATE 8.5 *The Long Mynd in Shropshire in 1985 exhibits a sea of bracken and not much other vegetation. Except in very dry years, it is not very productive pasture for sheep and is poisonous to domestic stock. Similar scenes are typical of most other hill land areas.*

the maximum for which the hill sheep subsidy is paid, and many farmers pasture sheep at much lower densities.[113]

The low productivity and income levels of the hill farms provoke landowners and tenants to consider intensification. Reclamation is popular when prices are high and seem to be stable. The most popular practice is to fence the land and convert the 'wild' vegetation to managed grassland (a three-year process), though occasionally during the 1970s crops of barley were planted on intake land in the hills. Between 1946 and 1981, some 17 per cent of moorland in Great Britain was 'improved', either by afforestation or reclamation for arable crops. On the North York Moors between 1950 and 1979, for example, 6050 ha of reclamation was carried out, of which 3588 ha was 'primary' reclamation of moor hitherto not in agricultural use.[114] The rest was from moor which had itself reverted from improved pasture, mostly in the 1930s. In the Brecon Beacons, the years 1948–75 witnessed the conversion of 4542 ha, of which 2738 ha were secondary reclamation. The places mentioned are National Parks and the conflict between access to land, visual amenity and agricultural change was most acute in those regions. Exmoor was the focus of a series of controversies until an indicative plan managed to reconcile, at least partially, the warring blocs.[115] Moorland reclamation was nowhere subject to planning control and in total perhaps 800,000 ha of rough grazing were converted between 1946 and 1981, much of it temporarily as far as crops were concerned. These regions of Great Britain are likely to be the next target for HNV (High Natural Value)[116] farming extensification plans drawn up in EU headquarters. Such plans are likely to de-emphasise moor drainage, a subsidised practice which has done little to produce economic benefits and a great deal to bring about scarring of the landscape, downstream flooding after rainstorms, sedimentation of streams, the invasion of rushes and possibly the spread of liver fluke in sheep as well.[117]

The dominance of heather on the drier moors of northern England, on the eastern flank of Wales and on the eastern side of Scotland derives in part from the rotational burning of the vegetation in order to encourage the growth of young shoots of heather. This is an important part of environmental management to encourage the breeding of high densities of red grouse. In Great Britain as a whole, there are about 460 estates on which grouse shooting is important: 297 in Scotland, 153 in England and 10 in Wales. They manage approximately 16,800 km^2 of moor for this purpose, of which 14,000 km^2 is in Scotland; the annual bag of birds shot is about 450,000, with 250,000 of these from Scotland. Sportsmen in 1989 paid about £60 per brace shot, with a net addition of £15.3 million to the Scottish economy.[118] Most Scottish moors yielded less than 25 birds/km^2, but one-third of English estates had sufficient grouse to allow bags of over 100 birds/km^2. Grouse numbers are apt to fluctuate cyclically since on wet moors and in wet years threadworm infection is rife and reproductive performance is poor. Medicated grit and even direct dosing is advocated, as may be some measure of predator control, especially of crows and foxes. Other keepers try to kill birds like merlins and hen harriers, and there is an ambience of conflict between nature conservation and moor keeping, just as there is between walkers and keepers since the latter want no disturbance to birds in the nesting season or soon before a shoot.

Estimates from satellite photography suggest that the area of heather moor is declining: the extent over which managed *Calluna* moor is dominant in England and Wales was about 510,000 ha in 1980, compared with 631,000 ha in 1947. About two-thirds of the difference lies in conversion to grassland, mostly as a result of high stocking densities.[119] The uncertainty of grouse numbers, compared with the hitherto more reliable returns from sheep, mean that a landowner with a choice is likely to favour sheep rather than grouse. In 1995, MAFF introduced a Moorland Scheme designed to regenerate heather: moors with over 25 per cent heather are eligible and farmers agree to reduce stocking rates to agreed levels in return for a payment of £30 per ewe removed.[120] Here is an example of an upland equivalent to the ESA, since the primary motivation is the protection of a particular kind of scenery and ecology. To provide an historical context, it can be noted that the decline in heather moorland seems to be at least 200 years old and is sometimes about 400 years in train. In each instance, the decline in *Calluna* cover is attributed to increases in the numbers of sheep.[121] No doubt high sheep densities support the populations of escaped or released lynxes, panthers and pumas which fuel the 'beast of Bont/Bodmin' and other big cat stories which appear from time to time. A population of perhaps 100–200 of these predators exists in the wild but it is not likely that they are breeding. MAFF does not think they exist.

In Scotland, a major sporting use of moorland and mountain alike is the deer 'forest', which like its English precursor, the Royal Forest, does not necessarily have trees. Created largely in the nineteenth century after the first flush of profits from sheep had declined, some 1.2 million ha remain as sporting estate and there are estimated to be around 300,000 red deer in the Highlands, the number having doubled in the 1965–95 period (Fig. 8.13).[122] Since 1950, they have colonised many forestry plantations. The Red Deer Commission would like to see the total number reduced by one-third but the means to achieve such a planned reduction are absent in the present milieu of neo-liberalism. Though evidence of 'overgrazing' by deer is hard to acquire, there is no doubt that they prevent the regeneration of semi-natural woodlands, degrade heather moorland, damage plantations and spill into lower areas (including farmland) in search of winter food and shelter.[123] Not enough females are shot to keep the population lower and the popularity of deer farming means that the effort of killing hinds on the hill in winter is not compensated in the price of venison. Some estates provide winter supplementary food, which in effect adds to the problem.[124] The problem is complicated by the conflicts between the income from deer shooting, arguments about 'sustainability' and the image of the Highlands being needed for tourism and, increasingly, a distinct Scottish identity.[125]

Until the nineteenth century, the wild lands of England and Wales were environments to be shunned by genteel folk. The Romantic movement changed that: the influence of Wordsworth was clearly critical in articulating some wider change of outlook. Thereafter, the nineteenth century and the early part of the twentieth century saw a great number of novels set in mountain and moorland areas, some of which have

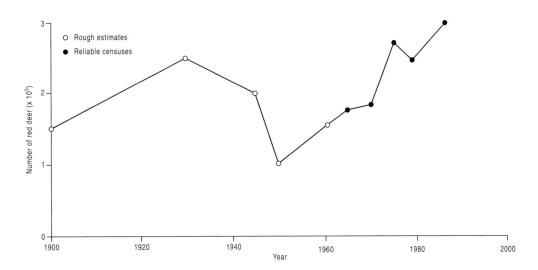

FIGURE 8.13 *The 'official' figures for the total numbers of red deer in Scotland 1900–86. The more reliable figures are depicted as solid circles; the rest are estimates.*
C. Sydes and G. R. Miller, 'Range management and nature conservation in the British uplands', in M. B. Usher and D. B. A. Thompson (eds) Ecological Change in the Uplands, BES Special Publication no. 7, Oxford: Blackwell Scientific Publications, 1988, p. 328, Figure 3.

become canonical. Those of the Brontë sisters (especially Emily's *Wuthering Heights* of 1847), R. D. Blackmore's *Lorna Doone* (1869) and Conan Doyle's *The Hound of the Baskervilles* (1902) all come to mind. Their equivalent for the post-1950 period is difficult to find: there is so much fiction that some of it is bound to be set in the Lake District or Devonshire, but none of it seems destined for the status of its nineteenth-century foreunners. Their place has maybe been taken by today's poets: the early verse of R. S. Thomas, situated in mid-Wales near Welshpool and near Machynlleth in west Powys, celebrates 'The Moor':

> It was like a church to me.
> I entered it on soft foot,
> Breath held like a cap in the hand.[126]

But Thomas also worried that

> There is no present in Wales
> And no future;
> There is only the past,
> Brittle with relics.

In south Yorkshire, Ted Hughes found an upland landscape whose valleys are full of a decayed industry, yet has no hesitation in praising the role of the 'Moors', which,

> Are a stage for the performance of heaven.
> Any audience is incidental.[127]

But the storytellers of popular entertainment love moors in the sense that successful television series have been set with them as the chief element of the scenery. The performances may not be heavenly but the audiences are numbered in tens of millions: the soap opera *Emmerdale* has a clear location in one of the Yorkshire Dales, as had the adaptation of the veterinary tales of James Herriot, as *All Creatures Great and Small*. The country bobby at the centre of *Heartbeat* rides his motorbike through the distinctive landscape of the North York Moors and the village of Goathland has been transformed (or deluged, depending on who is talking about it) by the subsequent visitors. There must be an element of appeal in this landscape for the audience: it cannot simply be the relative ease and low cost of filming in open landscapes without visible landowners to be placated. One of the most popular of all response to moorlands, however, is to drive the car to a viewpoint overlooking a long stretch of wild country of that type and to read the Sunday newspapers.

The appraisal of the uplands of all parts of Britain is complex and likely to remain so: interest in their resource productivity vies with that of their aesthetic and wildlife interest: they veer between being essential resources and essential environments. The marginality of some of the productive uses (sheep farming, winter sports) produces overuse in some locations (high stock densities leading to soil erosion; low winter profits leading to funicular railways up mountains like the Cairn Gorm for the summer trade), and so the uplands are tied into the sway of forces well beyond those of their immediate time and place. Mountains in general, and upland Britain in particular, feel the effects of human and natural changes more acutely than the lowlands: their resilience is lower and so irreversible change becomes more likely.[128]

MILITARY LANDS

The continuation of the Cold War, which included the need to host airbases belonging to the USA meant that post-World War II release was generally slow, with 77,300 ha let go between 1961 and 1971. In the late 1990s there were about 400 disused wartime airfields still recognisable in England. In total, the services retained 268,000 ha in 1972, of which five-sixths belonged to the army, used mainly for training purposes. Some 243,000 ha was in rural areas and the rest in 'urban' uses, like dockyards. Controversy often surrounds military land use, but the Nugent Report of 1973 pointed out that 40 per cent of the army's holdings were grazed, that there were 8000 ha of woodland and that the public had access to one-third of the estate.[129] On Salisbury Plain, the army held 37,000 ha but 24,000 ha were in agricultural use. The end of the Cold War has, however, intensified the military demand for training areas, as troops are withdrawn from Germany. Pressure for increased live firing in the Northumberland National Park, on the Otterburn ranges, has led to conflict with other interested parties. The continued presence of the army in areas which they took over for wartime purposes but have never released, such as the former village of Lyneham in Dorset, also causes resentment. At the same time, some parts of Salisbury Plain have never been ploughed, and the services have committed themselves to the conservation of the Chalk grassland flora and fauna.[130] The replacement of the three well-known 'golf balls' at Fylingdales by a single truncated pyramid has in fact been regretted by more than one landscape commentator. Less acceptable to everybody are the piles of rubbish at the bottom of the Holy Loch submarine base near Dunoon, deposited by the US Navy in its years of tenure. A plan in 1997 to clean up the seabed was controversial since some local people felt that the materials were better left to get sedimented down than be disturbed, especially since there were some unlabelled drums with totally unknown contents. The biggest symbolic event of the 1990s must without doubt be the return of Greenham Common in Berkshire, site of the United States' base for cruise missiles, to open space. The 900 acres (364 ha) of military landscape is to be restored mostly to heathland. The missile silos have to remain since they have to be open to inspection until 2006 under a treaty with Russia. One apparent land use paradox is the missile tracking base on St Kilda, the islands of which are designated as a World Heritage Site.

The military legacy has persisted on Gruinard Island, off the coast of Wester Ross. In 1942–3, the Chemical Defence Establishment used it to conduct experiments with anthrax spores. Though primarily a disease of domesticated mammals, anthrax can be contracted by humans and death may result. The experiments contaminated all 200 ha of the island as well as killing the sheep on which the bacillus was tested. The island was purchased from its owners in 1945 for £500 with the proviso that the family might buy it back for the same sum should it ever be declared safe. Eventually the island's most heavily contaminated areas (that is, with the highest density of dormant spores, about 45,000 per gram of soil) were disinfected with formaldehyde and the island was declared 'safe' in 1988. Even so, the family did not want to have it back, fearing lawsuits from any visitor who contracted the disease from the remaining 3 spores per gram of soil which constituted the 'safe' limit.[131] The possibility of persistence and accumulation is always present with nuclear materials, which are fabricated and transported within Britain for the armed forces. From time to time, leukemia clusters in the Harwell area of Berkshire are blamed on nuclear materials but conventional statistical analysis usually melts them away.

CONSERVATION AND LANDSCAPE PROTECTION

The post-war years have been a stage for tension in this section of the national environment. On the one hand there is an increased public awareness of 'nature', born of school curricula and of television offerings. On the other, there is the continued move towards greater intensity of production, in both the sense of land conversion for urban-industrial use and in greater production from the food industry. Since the 1980s, too, there has been a greater presumption on the part of central government that decisions about land use were to be left more in private hands, with less collective say. So the role of public bodies in conservation matters has been diminished.

Wildlife protection

The 1949 National Parks Act also established the Nature Conservancy, which successfully carried out research into conservation and managed a system of protected areas. Its history in the 1970s and after has been unhappy, with governments splitting off the research function to an Institute of Terrestrial Ecology and then breaking the Conservancy into national divisions. The reasons for this dismemberment have been traced to pressure from landowning interests and from government departments that feared restriction to their programmes.[132] The same period has also seen a growth in the involvement of the voluntary sector and in particular the activity of county Wildlife Trusts which have leased and bought land in echo of the national picture. Local authorities' interest was often quite minimal until the **Agenda 21** provisions of the Rio Accords brought them more firmly into the biodiversity arena in the mid-1990s. Non-statutory biosphere and biogenetic reserves have now been created; the government's biodiversity programme embraces species recovery, species action plans, habitat creation and habitat restoration. One hundred and sixteen key species and 14 types of habitat are identified. All are based on 'natural areas' which seems to imply a lack of historical understanding somewhere. In species terms, ideas about restoring the beaver to Scotland seem to be at the fringe of practicality, both ecological and political.[133]

The legislation to protect wildlife, though never perfect, has been quite thorough but was never enforced strongly: lack of personnel, local pressures, *force majeure*, and bad PR by conservation bodies all have meant a steady break-up of the areas of land and water devoted to wildlife in Britain. The major piece of law has been the Wildlife and Countryside Act of 1981. It required landowners to be re-notified of all areas of natural interest on their holdings (SSSIs), yet in 1978 MAFF improvement grants (which tended to destroy wild life) amounted to £540 million whereas the Nature Conservancy Council (now minus its research functions) had £7 million.[134] SSSIs were damaged at a rate of about 10 per cent of the designated areas per annum and the antagonisms between farmers and conservationists rose to new levels. The EU is now involved, with its Habitats Directive of 1992 becoming law in the UK in 1994; special protection areas (SPAs) for birds are among its first concerns; by 1998, there were 173 of these, covering 700,000 ha. A renewed sense of urgency about SSSIs has allowed English Nature to announce that 90 per cent of them are managed in a satisfactory way and that in 1996–7 only 0.34 per cent of English SSSIs (3261 ha) were being degraded, mostly by heavy sheep grazing in the uplands.[135] (The biodiversity action plan requires £27.4 million/year to be spent by 2000. A 10-km stretch of by-pass road costs about £101 million.)

On the ground, the 1990s saw the consolidation of a set of protected areas, imperfect though their management might be. There were 333 National Nature Reserves (199,000 ha) and 6178 SSSIs (2,041,000 ha) which altogether covered about 8 per cent of Great Britain. Scenic protection (see below) was far more extensive. In addition, the 1981 Act had extended protection to many species of plants and animals wherever they were found (Fig. 8.14). About 400 mammals, amphibians, invertebrates and plants have been given some degree of priority for conservation since the Rio Accords of 1992 but 1972–98 attrition rates of, for example, 62 per cent for the bullfinch and 42 per cent for the linnet are difficult to reverse. There is only one known roost of the Barbastelle bat and only 340 recorded sightings of it in the last 150 years. In most of these biodiversity loss cases, loss of habitat and the use of pesticides are seen as central. Great Britain is home to three species of bird threatened with global extinction: the red kite, the white-tailed sea eagle and the corncrake. It houses over 30 per cent of the world population of the bluebell (*Endymion non-scriptus*), which is endemic to Europe, 50 per cent of the grey seal's (*Phoca vitulina*) world numbers. Sixty per cent of the world's population of gannets (*Sula bassana*) nest on the British shoreline and 80 per cent of the world's pink footed geese (*Anser brachyrhynchus*) overwinter in Scotland and England.[136] In contrast, deliberate attempts at regional or total extinction have recently included badgers, coypu (*Myocastor coypus*) and the ruddy duck (*Oxyura jamaicensis*). Unwanted but small-sized species like introduced flatworms and crayfish will be hard if not impossible to eliminate.

The failures of nature conservation in the last 30 years are all too well known. Damage to SSSIs and NNRs, the continued persecution of predators, illegal badger digging and the persistent spill-over from intensive agriculture have all made the nation needlessly depauperate of wild life compared with, for example, France. However, a number of birds have been re-introduced to countrysides which they formerly inhabited: the osprey (zero to 130 nesting pairs 1954–99) and the red kite are good examples. Urban reserves and city farms have spread knowledge and interest, and

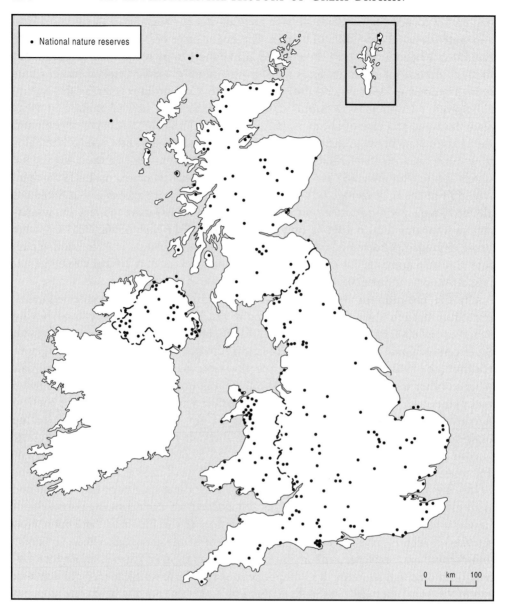

FIGURE 8.14 *The UK: National Nature Reserves in 1995. One perceived problem is that these are static fragments in a dynamic environment, particularly that of climatic change.*
Department of the Environment, Digest of Environmental Statistics 1996, *London: HMSO, 1996, p. 150, Figure 8.2.*

habitat creation (of which the RSPB's continuing management of open water and reed-bed at Minsmere in Suffolk is the outstanding example) has worked in favour of some species. On Teesside, ICI recreates inter-tidal mudflats in the course of pumping out brine for the manufacture of soda and chlorine. The county councils and a commercial firm have converted an area of birch, conifers and bracken near Leighton

Buzzard (Bedfordshire–Buckinghamshire) to what is described as 'its former glory as a flourishing heath' by mechanical clearance and sowing heather seed, and the RSPB wish to go further at Minsmere and change 158 ha of arable land back into a Sandlings heath.[137] The value of unsuspected habitats is being recognised, as with churchyards, for example. In Norfolk, for instance, six species of meadow plant have at least half of their county population within churchyards; the cowslip (*Primula veris*) and the meadow saxifrage (*Saxifraga granulata*) are examples. Similar stories in other places can be told for mosses and butterflies and for lichens; the glow-worm (*Phophaenus hemipterus*) was for many years only seen in one churchyard in Sussex. Some species of plant flourish at the foot of headstones and walls where mechanical mowers cannot reach: calamint (*Clinopodium calamintha*) and wall pellitory (*Parietaria judaica*) are examples from Essex.[138] In the case of animals the wildlife picture is continually being complicated by escapes: colonies of porcupines or wallabies have been chronicled, for example. In the 1990s 'wild' boar began to get out from farms and to breed: a feral population of from 200 to 400 with concentrations in Dorset and on the Kent/Sussex border was estimated. Animal rights activists who released mink from farms contributed to the demise of other species both in the wild and in captivity.

If there is one lesson to come out of this history, it is that sequestration of nature in designated areas is not enough: it adds to the fragmentation which sees each piece of environment as a specialised concern, usually of somebody else. Pieces of 'environment' are now held by all kinds of voluntary and statutory bodies and by individuals and under different pieces of UK and EU legislation and voluntary agreements. This diversity is doubtless better than a state monolith at the mercy of the prejudices of the current minister but may not effectively tackle the priorities of the day, preferring perhaps the showy and the atypical to the more mundane key species of ecosystems, for example. Something more will sometime be needed[139] since a static nature reserve may not cope with a warming climate for instance.

Landscape and cultural heritage protection

The 1949 Act eventually resulted in the establishment of 10 national parks in England and Wales; the Norfolk Broads were added in 1988 as an analogous administrative area though not an actual park; the New Forest and South Downs were designated in 1999. The confusions of the 1949 Act have been mirrored in the parks' management ever since its implementation. The parks were supposed to protect valued scenery, especially where this was open moorland and mountain. But they were also mandated to keep a healthy rural economy (so that forestry and farming were never subject to development control) and to facilitate outdoor recreation. Juggling with these preoccupations and on limited budgets, some of which was from local rather than national sources, has always been a difficult affair. More fundamental matters, such as the ecologically degraded nature of much upland habitat have never been addressed seriously, since the existence of moorland has always been taken as the starting point for both scenic protection and nature conservation, not the preceding forests.[140] The Environment Act of 1988 addressed itself to the purpose and running of the parks for the first time since 1949.[141] The same concepts of what the parks are was at its heart[142] for their local nature was affirmed with the appointment of parish councillors to the planning authorities, and there was still no control of agriculture and forestry. Scotland was still excluded, though the new Parliament will likely change that quite quickly.[143] (The Areas of Outstanding Natural Beauty [AONBs] of England and Wales are paralleled in Scotland by National Scenic Areas; all are shown in Fig. 8.15.)

The successes of the National Parks and the AONBs are probably to be seen in what is not there: intrusive blocks of conifer forests, many more large quarries, more military live firing, villages allowed to ribbon out along roads rather than fill in vacant sites, and even more slices taken off the parks by roads like the Okehampton by-pass on the fringes of Dartmoor. Television masts have never met serious opposition but mobile phone transmitters are calling forth the Lake District solution where one of them is disguised as a tree. New farm buildings are subject to control by persuasion rather than regulation but the farming community of the uplands is sufficiently precariously poised to view any controls or even any opposing views with suspicion. A continuing sore is the way in which the price of housing rises in National Parks so that local young people cannot afford it. Attempts by park planning authorities to restrict housing to local people have not been upheld by the courts.

Elsewhere, Ancient Monuments are at great risk: possibly 22,000 have been destroyed since 1945 (about one a day) and only 76 per cent of Ancient Monuments are regarded as being in a complete state. Agriculture is regarded as the cause of most piecemeal destruction, though overall loss is greatest due to development and thereafter to gravel and mineral extraction. Provided excavation and recording has taken place, this may not be thought to be significant but if we think of a historical continuity undergoing fragmentation then the dislocation of our culture becomes more apparent. The one feature that seems secure is the country house, protection of which seems a classless phenomenon. It is hard to imagine them being demolished at the rate of about one per week in the 1945–55 period.

More leisure, new technology

Rural Britain in the post-1950 years has experienced a leisure explosion. In part, there have simply been more people who wish to spend part of their free time in rural surroundings; some prefer wild places, others the more domestic landscapes of field and village. The great difference from previous eras has been the accompaniment of technology. The main carrier is the private car and most public bodies have had to make provision for this: the National Parks, for example, have built many car parks in popular places.[144] The internal combustion engine has also powered the speedboat, which is not easily compatible with quiet lakes (Windermere in the Lake District has had a long-running controversy) or estuaries with high densities of sailboats. The mountain bike has added its weight to footpaths already eroded by many pairs of boots, necessitating a great deal of footpath remedial work which is very expensive in upland areas. In most areas, there is no legislation which enables the appropriate authorities to control these new technology-intensive recreations, whose supporters claim that every use of the land was at one time new anyway: where is the rationale for stopping something just because it is an innovation? One outcome has been a

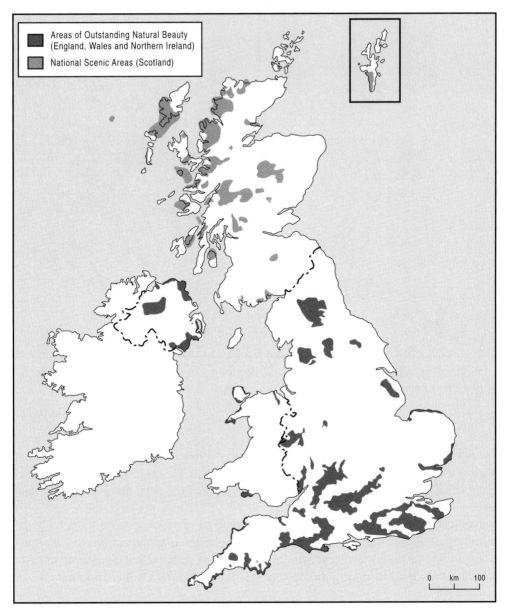

FIGURE 8.15 *The second tier in England and Wales and the first in Scotland at the time (1995). In both categories the priority is development control for scenic protection.*
Department of the Environment, Digest of Environmental Statistics 1996, *London: HMSO, 1996, p. 150, Figure 8.2.*

sharp diminution of perceived tranquillity in rural areas: maps are published by the CPRE which compare areas remote from main roads and motorways, power stations and busy railways (Fig. 8.16). The remaining reservoirs of peace are North Devon, Shropshire and the North Pennines. (Aircraft noise is presumably not included.)

Attempts to develop winter sports in the Scottish Highlands have provided a sharp

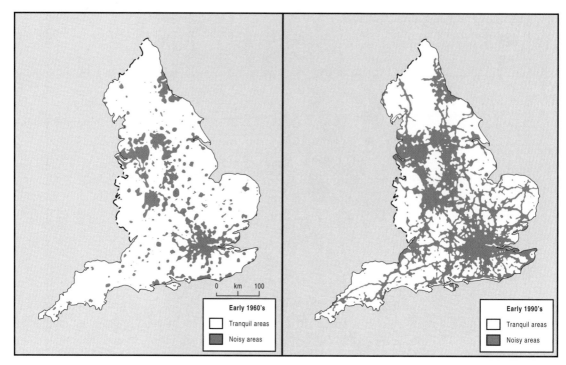

FIGURE 8.16 *Noisy England: a comparison by the CPRE of the early 1960s with the early 1990s. The clear advantages of living in County Durham can be seen, with Exeter as a second best. CPRE and Countryside Commission,* Tranquil Area Maps, *October 1995.*

focus for such antagonisms, especially with the post-1960s development of the Aviemore region of the Highlands, with ski developments in the Cairngorm Mountains. Apart from the obvious landscape and ecological effects of the lifts, the proposal (finally approved by Scottish National Heritage in 1997) for a funicular railway up Cairn Gorm itself which would decant people onto the fragile *Rhacomitrium* heath on the summit has summed up the state of the balance between conservation and development in late twentieth century Britain. Concessions such as fencing people in once they have reached the mountain top seem unlikely to protect this ecosystem.

Conflict or cooperation

In the 1980s and 1990s, Great Britain became a much less happy place in which to live. Environmental relationships were not the cause, nor were they a primary element in its expression. But they shared in its public face, for any communal good was to be served by private enterprise, and this was not to be hindered if at all possible. So if many disadvantaged groups suffered a loss of voice, it is not surprising that nature (whose cries were never that loud anyway) was scarcely heard at all in such an adversarial world. There was a presumption in favour of production over protection and the EU (plus some of the more vocal NGOs) was left to modify policy at the margins.

It is perhaps miraculous that so much of the protective legislation has in fact survived and it is probably thanks to teenage burglars and car thieves that so much Parliamentary time was taken up which could otherwise have been devoted to downgrading the protective acts from 1949 onwards. Lowe's assertion in 1983 that conservation had become 'a major cultural force' looks a bit overstated.[145] The national mood of sourness exposed other nerves as well. The underfunding of bodies like the NCC meant that they were often remote from the communities in which their objects of protection were found. In extreme cases, such as the Flow controversy in northern Scotland, the developers' case could be strengthened by presenting conservation as English colonialism,[146] and smaller-scale repetitions of name-calling (on both sides) have been frequent. These manifestations of hostility reinforce Bill Adams's call[147] for nature conservation to be a felt element of everyday life, rather than a specialised fragment of a particular 'product' which is fenced off, like an eighteenth-century country house. There may be wider applications of this thought. For a start, ignorance of the environmental history of many nature conservation areas means that policies for conservation are basically the preservation of a particular historical 'horizon', often divided off from its present physical surroundings.

RURAL SETTLEMENT: THE RURAL ECONOMY

The colon is deliberately placed, for in these years rural settlement and the rural economy are not necessarily intimately connected. The growth of commuting has meant that many people who live in rural areas make their living in the cities. Their environmental effects are not great beyond the changes inherent in new housing but incomers are likely to be conservative of instances of apparent permanence of, for example, trees and buildings. The level of migration to rural areas is, however, high enough for the label 'counterurbanisation' to be applied.[148] Only the house journal of the 'old' countryside, *Country Life*, remained virtually unchanged until the 1980s.[149]

Such a search for continuity is less important to those who live by the countryside. Central agricultural policy has tended to make the rich farmers richer and leave the poorer ones static; the message then becomes, 'If possible farm a large unit.' Landowners who have land with planning permission for housing can often retire to Spain in short order. In part, their riches are created by the planning system itself and we all pay for this to the tune of about £13 per household per year. Taxpayers have also contributed about £3 million to the costs of the 51,000 ha of Heritage Landscapes whose owners are freed from inheritance taxes. Proposals to de-intensify farming are crucial to any radical change: it is claimed that a more environmentally friendly set of farm policies (with reform of the CAP a central plank) would result in the regeneration of rural communities as well as bring back some diversity of wildlife.[150]

Planning will normally be the most sharp where the middle classes have settled; these will not include the areas of greatest agricultural intensification like the Fens or remote areas like mid-Wales. Thus the balance of types of countryside will vary regionally. Allanson and Whitby[151] differentiate four types of rural regions:

- The *preserved countryside* of, for example, the English lowlands, with anti-development and preservationist attitudes impelled by newer, middle-class settlers. Such groups can be said to be interested in pulling the ladder up behind them.
- The *contested countryside*, where farmers and development interests are in opposition to some newcomers who adopt the standpoint which is commoner in the first category, detailed above.
- The *paternalistic countryside*, where large estates and big farms are dominant and where these groups still shape the land effectively. Falling income will bring demands for diversification whose introduction will depend on the landowners' attitudes.
- The *clientilist countryside* dependent upon direct agricultural support (as in remote upland areas) and local politics focus on state aid and concerns about employment.

Diversity is not in itself to be deplored, of course, but it is easily conceivable that there will be a rural equivalent to the kind of segregation that occurs in towns, where the planning system reinforces the position of the dominant social group rather than, for example, putting environmental concerns at the head of the agenda. In spite of the neo-liberalism of the 1980s and after, the planning system is still an important mediator between different groups of actors in the countryside,[152] but because of its long-standing limits it has little to say about environment except in the narrow, protectionist sense of inhibiting land-use change.[153] As the countryside becomes less a place for farming[154] and more a place for a diversity of rural activities, including residence and work, it seems that only ideology will prevent the planning system extending its reach into forestry and farming. Biotechnology at present provides a very uncertain extra set of unpredictable factors. All these will have to fit into a set of regions where the national and the international will come up against very local cultural attitudes.[155]

THE URBAN-INDUSTRIAL ENVIRONMENT

Some injustice is probably done to these topics by pushing them together, not least because there is industry in rural areas. Nevertheless, they have sufficient in common in their environmental and cultural relations that they can be treated together.

The expansion of urban-industrial land uses

The expansion of factories, roads and housing is rarely uncontested nowadays, for they are seen generally as precursors of the type of rapid change which brings a NIMBY reaction from those directly involved. As a very broad generalisation indeed, such changes are less disputed when they take place in areas of high unemployment; thus the conversion of large areas of green fields on the outskirts of Sunderland for a Japanese-owned car factory attracted far less adverse attention than new road schemes in south and south-west England. The use of 'green field' sites for out-of-town shopping centres reached such proportions in the 1980s that the then Minister announced that fewer permissions would be given, especially since the traders left in the town were being badly squeezed. The opposition to even small supermarkets can be very strong, especially in, for example, small towns with a coherent sense of community: the argument over the building of a store on one field on the edge of Hay-on-Wye (Powys) will serve as talisman.

Sharp controversy is inevitable where designated Green Belts are concerned. Since their inception after the 1947 Act, they have often held a special place in the geography of towns, with their possession and maintenance having a particular moral significance, not least in preventing one urban centre from being physically joined to the next. They have, of course, been the site of some industrial development, some housing and road building, and have attracted people to recreational areas, so that they are neither wild nor lonely places. Nevertheless their visual difference has won them many friends. Development-minded local authorities and companies found sympathy from the national government of the 1980s and attempts to break up the nation's 1.65 million ha of Green Belt were sustained. Bodies like the CPRE fought on many fronts with, somewhat ironically, their chief allies being the established residents of the Green Belts who did not want their property values diluted by the building of more executive houses.

Coasts have been another area of controversy since 'reclamation' of estuarine mudflats and salt-marshes seems a cheap way of providing land for industries which need access to the sea or large quantities of cooling water. The losers are usually the seabird populations: the reclamation of large parts of the Tees estuary for chemical plants and oil refineries is an example. The long-term problems of such areas are discussed in the material on coasts, below.

Emissions

Cities and industries are great concentrations of energy and matter. They take in both, store and process them and then want to be rid of waste products, in ways which are analogous to the functioning of an organism. In an unregulated society, anywhere will do for a quick dump. Britain currently recycles about 8 per cent of household waste, placing it near the bottom of a league headed by the Netherlands at 60 per cent.[156]

The impacts on air quality have been mainly discussed in the section on the atmosphere. One recent trend linking both town and country has been the interest in burning municipal wastes so as (1) to get rid of them and (2) to use the energy in all that paper, wood and plastic material that they contain. One strategy suggests another 130 incinerators by 2015. The worry comes with fallout from the exhaust stack and in particular the possibility of trace quantities of substances like TCDD (dioxin) being present. Some unexplained concentrations of cattle diseases and malformations in the vicinity of industrial high-temperature toxic waste disposal plants have aroused concerns which have in general not been allayed in spite of high levels of research. But worries about wastes as fuel have been behind protest movements in places as different as Upper Weardale and the north Tyneside suburb of Wallsend.

In the case of water, the long-standing problem of sewage is still important. Untreated sewage is notably intrusive at times of low flow, when it may take up most of the oxygen in the water and so kill the fish. If extraction levels are high, then this may be heightened. One summer, the Drax power station in North Yorkshire showered the neighbourhood with powdered sewage after its cooling system took in the untreated material from a nearby town at a time of low flow. Even treated sewage retains its concentrations of nitrogen and phosphorus and so eutrophication is possible, especially in warm weather. Surface water may also receive the unwanted effluents of past industry: we have all seen a small stream issuing from an old adit or mine drain that contributes a spongy orange carpet of flocculated iron compounds. In the aftermath of the privatisation of the coal industry nobody wanted responsibility for the continued pumping-out of accumulated water in the derelict workings of the West Durham coalfield. It was feared that if key pumping stations near Bishop Auckland were not maintained, then the whole River Wear could become a kind of orange porridge, with consequent effects upon runs of salmon and sea trout, sewage disposal and the tourist industry around Durham City. Eventually, a residual caretaker solution was found but not before the intervention of the Minister of the Environment.

The time-hallowed disposal site of municipal and industrial solid wastes has been a hole in the ground. Not much has changed since 1950, except that the scale of the operations has increased and the composition of the waste has moved away from dominance by ashes and cinders to paper products and plastics, as well as much organic, putrescible waste from fruit and vegetable preparation. A rising trend now (1999) results in about 500 kg/cap/yr of household waste. A national landfill tax introduced in the mid-1990s has spurred some interest in recycling schemes but it is still the expectation that 60 per cent of solid wastes will enter landfill sites. These sites generate noise, gull populations, rats, and dust when in operation.[157] They may leak solutes into the ground water, so lined pits or impermeable clays are needed. Landfills account for about one-third of all water pollution incidents. Once filled up and finally covered, then methane will probably be generated for about 30 years from anaerobic fermentation of organic materials. This can be tapped as an energy source, though its use can only be local; if unburnt it contributes to the 'greenhouse' emissions total. Toxic waste disposal sites need special licences and are supposed to be carefully inspected but illegal fly-tipping of such wastes by small firms and individuals is still common. Former tip areas can be built upon, though special care is needed if there is any suspicion of toxic materials or of methane production. As with hilltops, investment in holes has been a good way of making money in the last 30 years.

Undeveloped land in cities has been likened to an 'urban common' and the vegetation of embankments, walls, waste tips and similar substrates has sometimes given rise to very characteristic biogeographies.[158] Bristol, for example, has areas dominated by buddleia (*Buddleja davidi*) which forms a scrub with birch and willow as well as a dense strip along railways. Brought from China in the 1890s, its attraction for butterflies is well known. Glasgow, by contrast, has many elm seedlings since Dutch elm disease spared many trees in that city. Manchester is home to great spreads of knotweed (*Fallopia japonica*) and giant knotweed (*Fallopia sachalinensis*), usually near water. Sheffield is characterised by a number of fig trees (*Ficus caria*) along the Don and other watercourses: they may well have been spread by warm water during steel-mill days. Teesside is home to calcicolous species which spread on the spoil heaps of limestone flux from steelworks: the carline thistle (*Carlina vulgaris*) and centaury (*Centaurium erythraea*) are plants of short, dry grassland on calcareous soils. A more or less ubiquitous urban plant is fireweed or rosebay willowherb (*Chamerion angustifolium*), a native species which expanded somewhat in the late nineteenth century but exploded during World War I where woodlands were felled and then again in World War II onto bombsites, with its apparent apogee in the summer of 1941 in London.

Derelict land and its re-use

One of the eyesores of Great Britain is derelict land. It expresses a reproach to its former users in a country which is so densely populated and intensively used. (To children and lovers, of course, it may well be a very valuable resource.) Surveys in England in 1974, 1982, 1988 and 1994 (there have been fewer such appraisals in Scotland and Wales) show at the latest count some 39,600 ha of derelict land in England, of which some 87 per cent was thought to justify reclamation. Some of this land is also regarded as contaminated, with the potential to cause damage to any future user. A proposal for a national register of contaminated land was dropped in 1995 and voluntary registers introduced instead. In Wales, a 1988 survey logged 745 sites which might have been regarded as contaminated by the legislation then in preparation. A 1990 survey in Scotland recorded 8297 ha of derelict land, mostly in urban areas, though 14 per cent of the area was in Green Belts and other protected areas.[159] The uses which give rise to dereliction form a long list headed by spoil heaps from collieries and metal working (29 per cent of the derelict area of England), excavations and pits (15 per cent) and railway land (16 per cent); the abandonment of military sites added another 6 per cent and there is a residual category of 'general industrial dereliction' (21 per cent). The overall quantity of derelict land fell by 45,700 ha between 1982 and 1988 (11 per cent) and 14,000 ha of land had been reclaimed, aided mostly by central government grant; a further 95,000 ha were reclaimed in 1988–93. So its creation is still continuing.[160] Figure 8.17 shows the characteristics of derelict land types and their environmental relations as well as their role in putting humans at risk. Practically every likely pathway to humans and environment can be placed at hazard on derelict sites and it is a relief that not all of them exhibit every possibility.

Aided by government grant since 1949 and especially since 1963, reclamation has brought many benefits, not least in the supply of land for housing, afforestation and recreation. The face of regions like the West Durham coalfield and the valleys of South Wales has been largely transformed in those years, as collieries closed and as the backlog of tips was transformed, a process subjected to great scrutiny after the Aberfan tragedy of 1966. The treatment of contaminated land is not so thorough because of the great costs involved in the separation of the contaminants from the matrix which they pollute: covering is most often used but is not exactly fail-safe. One often overlooked end-use for land at least partially reclaimed is nature conservation: several small areas of ponds, scrub and grass on the sites of former steelworks in West

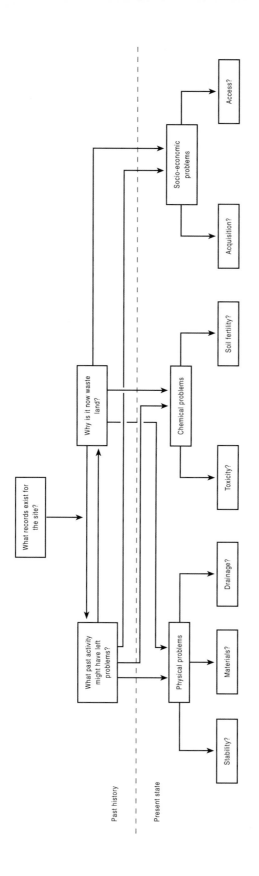

FIGURE 8.17 *The reclamation of waste and contaminated land is not simply a matter of levelling it off and planting trees or executive homes, as this diagram shows. The list of investigations which have to be carried out is daunting to any local authority or developer since, even if the land is cheaper, there may well be extra costs in finding out about these features, let alone correcting them. Far easier to build on farmland. Department of the Environment,* Transforming Our Waste Land: The Way Forward, *London: HMSO, 1986, p. 52, Table 5.1.*

Yorkshire, for example, are now managed for their wildlife and educational value. Small areas of scruffy land in all stages of succession, such as are found near railways, are important habitats for birds and small mammals.

Awareness

Since towns and cities are concentrators of materials, they are the obvious catalytic sites for the changes in attitudes needed if re-use and recycling are to achieve higher levels.[161] At present, there is a difference between levels of recycling in industry and that by households, with the latter needing to improve (Table 8.7). Some 83 per cent of municipal solid waste goes into landfill, 9 per cent is incinerated and recycling only accounts for 7–8 per cent. A government target exists for 25 per cent by 2000. Starting with skips for paper, glass and metal cans, some local authorities have moved to collecting household wastes which have only been minimally separated and then machine-sorting the non-putrescible wastes which can be sold off, sometimes at a profit. Such ideas have spread to large institutions, such as Universities, though those on compact sites have a distinct advantage. Urban farms have also been a post-1960s venture (shades of the city cowsheds of the pre-railway era), with an emphasis not simply on the cuddly but on expanding the experiential horizons of inner-city children beyond the television set. Similarly, a more tender attitude to urban greenery and open space is a recent development: the opposition to the conversion of a playing field in Oxford to a Business School would not have been successful had the would-be benefactor put up his money in the 1960s. Urban public transport may undergo a revival in the next decade if the plans of the late 1990s come to fruition.

TABLE 8.7   Percentage of selected materials recycled in the UK 1985–96

	1985	1990	1992	1993	1994	1995	1996
Ferrous	39	38	45	42	42	40	44
Copper	43	50	35	35	32	34	36
Lead	69	67	64	67	74	71	73
Aluminium	40	39	39	29	39	53	44
Paper and Board	28	32	34	32	34	37	n/a

Note: In 1996/7 total municipal waste in England and Wales was 23.3 million tonnes, of which 8 per cent was recycled. The government has set a target of 40 per cent for 2005.
Source: Department of Environment Transport and Regions, *The Environment in Your Pocket*, London: HMSO, 1997.

LAND USE: THE OVERALL PICTURE

Dudley Stamp's Land Use Survey of the 1930s has been followed by another attempt at a comprehensive picture in the 1950s and then a more modern approach based on sampling in 1996.[162] This gives a land use picture (Fig. 8.18) which is replete with environmental implications, showing that the urban proportion for the whole UK is now about 10 per cent and that a third of England is covered by arable land, yet in the rest of the UK the proportion is less than one-tenth. The percentage of open space in built-up areas (at 9.6 per cent) is actually three times that devoted to industry. In the period since the first LUS, the area of forest and woodland has doubled, to about 12 per cent. As Figure 8.19 shows, permanent grassland has been the major loss, from a high in the 1960s. New elements in the landscape of Great Britain were measured by counting the number of sampled 1 km squares in which particular features were found: set-aside farmland in 25 per cent, for example, new housing (under 5 years old) in 11 per cent, large supermarkets in 0.9 per cent and vertical structures like pylons and communications towers in 21 per cent, a figure almost certainly out of date by now. In all these counts, the proportions were higher for England than for the other regions. (The young people who did the fieldwork identified traffic and its related pollution as their major cause of environmental concern.)

One type of land use not identified in its more subtle forms by this survey is that of 'restored land'. The restoration of derelict land from heavy industry is relatively easily demarcated but recent years have seen some more examples of restoring an old order: the course of the River Cole and heath in Suffolk have already been mentioned. 'Ancient' woodland is being mimicked by planting native broadleaves and coppice

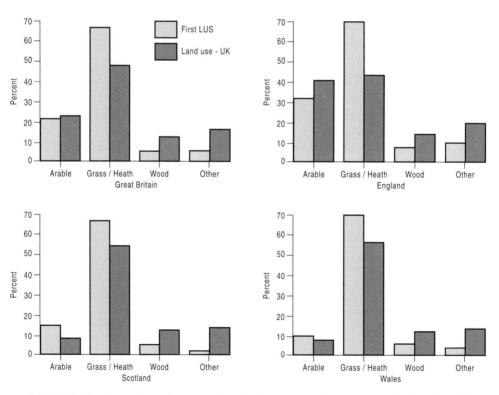

FIGURE 8.18 *Comparisons in non-urban land use proportions between the first Land Use Survey (1930s) and the 1996 survey. The first is derived from a total survey, the second a sample of rural squares. The data convey no great surprises except perhaps in the magnitude of the increase in wood, which is mostly coniferous plantation. Reclamation has diminished the area of heath and moor, though the categories are not easily comparable.*

R. Walford (ed.), Land Use – UK, Sheffield: The Geographical Association, 1997, p. 40, Figure 5.3.

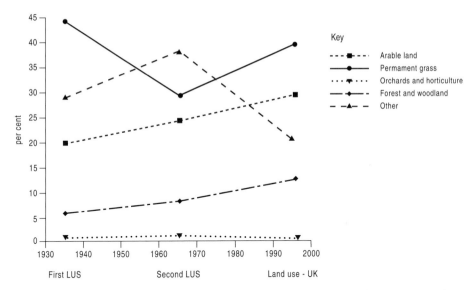

FIGURE 8.19 *Data from three land use surveys for England and Wales, the areas most subject to change. The proportions are of non-urban land and show a steady increase in the amount of arable land (which is of course being more intensively used) and a rise in the proportion of permanent grass, some of which is reclaimed from the 'other' of heath and moorland.*
R. Walford (ed.), Land Use – UK, *Sheffield: The Geographical Association, 1998, p. 41, Figure 5.4.*

management, and acceleration of trees into a decayed state is undertaken. The drainage of blanket bog ('gripping') now qualifies for grants to reverse the process and lowland bogs which had been worked over (as at Thorne Waste in east Yorkshire) can be managed for rich fen species which otherwise have locally disappeared: *Cladium mariscus*, the saw-sedge, is one example.[163] Encouraging bird species deemed to be valuable has a long history but some especially are being targeted: the barn owl is a good example. Given also that a number of country houses and archaeological monuments are kept in a particular condition by strict management, the question of authenticity becomes interesting: what exactly is acceptable in terms of landscape and nature? Are there limits to unauthenticity? What attitude, if any, should education take to such issues? Julian Barnes's novel *England, England* (1998) constructs a fantasy in which all the visitors to Britain see a simulacrum of everything they have come to see, on the Isle of Wight. It becomes the State. The rest of England (but not Wales, nor Scotland) collapses into a pre-industrial condition in which the Church of England is central and which refers to the Isle as the Mainland. It is an amusing but thoughtful commentary on the Portmeirion syndrome and (I suppose) post-modernism's lack of rootedness.[164]

LEISURE AND ENVIRONMENT

A great deal of leisure time is spent in the home and its surroundings, without discernible environmental links, if we except the energy demands of the disco and the vandalism of trees by the disaffected young. But the rural and semi-rural areas of the country have always held a strong attraction for groups who need more space or particular features. The advent of more technology, allied to more leisure time and greater disposable incomes, has heightened this attraction. Above all other possessions, the motor car has opened up rural outdoor recreation to many more people. It has also had the knock-on effect of being the means of transporting other pieces of technology and equipment. The trailer with the boat or trail bikes and the roof-rack with hang-gliding or caving gear are simple examples. The four-wheel drive leisure vehicle itself extends the range of the motor vehicle into off-road areas or green roads but it is nowhere near as influential as simply the number of cars carrying people. Occasionally, main roads into the Lake District have to be closed on a summer Sunday because the region is 'full', that is, its road system is jammed solid: 90 per cent of visitors to the National Park arrive by car.

Some of the most obvious effects of the 'leisure explosion' of the post-1950 period are the conflicts which have arisen between recreationists and others in rural environments. Some of these are other recreation users. A walker on a green way like the Ridgeway in Berkshire is unhappy to have to share the route with Land Rovers and Shōguns, for example; and equestrians make most tracks unusable for walkers and cyclists, as well as frightening some people with the unpredictability of their mounts. Canoeists often annoy anglers (coarse fishing is still the most popular outdoor recreation) and the speedboat is inimical to most other uses, including nature conservation. When planning procedures are applied by landowners such as the National Trust or by local authorities, then they tend to separate out all these activities spatially: another kind of fragmentation. Confrontation with production interests is a feature of most forms of access in rural areas. Farmers complain that walkers leave open gates and trample crops and even that the sheer numbers of people on public footpaths far exceed some notion of carrying capacity. Some landowners, in turn, go to great lengths to try and discourage walkers by ploughing and planting across paths, pulling down signs and illegally depasturing bulls in fields crossed by public pathways and bridleways.[165] Some species of animals become noticeable scavengers at popular picnic sites: sheep in North Yorks Moors villages, for example, and, after the visitors have gone, the rats.

Fox and stag hunting, as well as hare coursing, evoke strong emotions, though their environmental linkages are restricted in material terms to the conservation of certain kinds of landscape and the maintenance of populations of the hunted beast.[166] Great Britain currently has about 200 fox hunts and 4 deer hunts, employing 15–20,000 hounds. About 10,000 hounds are shot every year as too old or unsuitable for their job and 20,000 foxes plus 200 deer are culled. Hares are also hunted with hounds both by beagle packs and informally. In 1999, the government announced its intention to ban hunting with hounds, pointing out that drag hunting was increasing and could replace the unacceptable elements of killing by dogs. The poet Ted Hughes argued, though, that the red deer of Exmoor occupy a totemic position in local society and that if hunting is banned then the local lads will poach them to extinction.[167]

The major environmental link of outdoor recreation is, however, seen as ecological damage. At its simplest, this is seen in footpath erosion, where popular trails along the Pennine Way or the Lyke Wake Walk, for example, are erosion channels which are constantly being widened and which may act as sites of inception of further soil erosion. Bodies like the National Park authorities, the National Trust and county councils spend a great deal of money stabilising these paths, to be told by some users that they are the landscape. Nevertheless, the Dartmoor authority had to spend £400,000 in 1997–2000 in combatting recreation-caused erosion, under the slogan 'Moor Care, Less Wear'. More complex relationships may also exist, as when some species of birds but not others are scared by sailors during their breeding season to the point of deserting nests. In one study, a 'quiet' lake held higher populations of waterbirds on Sunday, suggesting that some species underwent a weekend effect of deserting lakes used for recreations which disturbed them. Sailing boats appear to be very frightening to ducks.[168] In all these cases, various forms of recreation management can be used to minimise impact and to attempt restoration. It is, too, important not to exaggerate the problem: the London green belt has only a few sites where serious damage is endemic, for the soils and vegetation are reasonably robust.[169]

At one stage in the 1970s, it was fashionable to suggest that leisure would be a virtually unmanageable environmental problem. That has not come about, partly because the professional groups with the money have had to work harder and those with the time, like the jobless and the part-time workers, have had no money. We might wonder whether all the use of the countryside has had any effect upon the European and even global environmental awareness of the participants: a subject for a simple, if tedious, social science PhD perhaps.

COASTS

Though Great Britain has a long coastline (about 15,000 km) in relation to its size, pressures for human-directed change have not decelerated in the last 50 years. Pressures to 'reclaim' were especially strong in the 1960s and 1970s when the expansion of oil refining and heavy industry was still taking place; work for the North Sea oil industry maintained some of the pressure even when other enterprises shrank. Other transformations included the building of many marinas, often seen locally as sources of income in the context of the decline of fishing, for instance. Marinas were less likely to displace seabirds since amateur sailors are best kept away from tidal mudflats. Pressures to protect have also been strong. The protection of the best scenery and wildlife sites has been aided by the central government designation of certain 'Heritage Coasts' and by the action of the National Trust in purchasing others (Fig. 8.20). By 1995, about one-third of the coastline of England and Wales (1539 km) had been put into the first category and 725 km into the second, with some £10 million of voluntary funds having been spent. In Scotland, 74 per cent has been designated a conservation zone. Other forms of protection have been sought for eroding sites, where the apparently natural processes of coastal landform change are deemed to be inimical to other interests. A major landslide at Holbeck Hall in Scarborough (June 1993) resulted in damage in excess of £3 million, and further south the soft cliffs of Holderness have lost 800 ha in the last 100 years. The eroded material tends to be carried into low-energy estuaries and embayments and doubtless it is some of the Holderness material that necessitates the resurveying of the shipping channel into Kings Lynn every two weeks.

Coastal waters, too, were scrutinised as never before during these years. The principal agent of change has been the EU, whose directives on water quality for bathing and shellfish found a great many parts of the British coastline wanting. So many were found to be polluted by raw sewage that the government was forced into the subterfuge of declaring that only 27 beaches were actually used for bathing; after the derision had died down, a further 350 beaches were admitted to have seen cold, pink animals hiding behind windbreaks.[170] EU infringement proceedings have followed the deduction that Great Britain is unlikely to comply with the directives within their stated period. Nevertheless, 18 beaches in Great Britain gained the EU's Blue Flag in 1995. Some small attempt has been made to bring marine conservation up to the standards of its terrestrial equivalent, but the knowledge base is poor and the will decisively lacking; in 1999 Greenpeace successfully won a court action claiming that the government had ignored its EU obligations in the EEZ. Marine pollution in general has been a problem, especially in the North Sea. Most industrial materials (especially metals, which come from sewage sludge as well as in runoff) have found their way into its waters in one form or another (Table 8.8) but the effects are always difficult to tie with certainty to a particular substance. Episodes of unexpected die-off of marine mammals or of finding malformed fish arouse concern but disappear into the limbo of 'more research' or 'suspected but not proven' and, in spite of the numerous international agreements about the need to improve the North Sea, little has been done. The effect of commercial fishing of sand eels by Danish vessels upon seabird populations is one of the few linkages upon which there has been agreement, to the point of reining it in by a decision of the Bergen Conference in 1997. Oil rigs have left 7.0×10^6 m^3 of cuttings beneath 1500 rigs and these appear to be leaching arsenic and nickel, which are toxic heavy metals; no solution was in sight in 1998.

TABLE 8.8   Direct inputs of metals to the sea from rivers in the UK 1985–96 (tonnes)

Metals	1985	1990	1994	1995	1996
Cadmium	79.9	63.6	36.7	30.7	26.6
Mercury	27.0	11.8	7.3	6.1	4.6
Copper	1,275.0	850.0	720.0	645.0	469.0
Lead	1,660.0	667.0	558.0	419.0	330.0
Zinc	3,630.0	3,920.0	3,462.0	2,805.0	2,110.0

Note: Year-on-year variations are affected by, for example, rainfall: 1996 was a very dry year. The general levels of decline are apparent and reflect changes in the status of manufacturing industry as well as attempts at regulation.
Source: Department of Environment Transport and Regions, *The Environment in Your Pocket*, London: HMSO, 1998.

Another contaminant difficult to trace without special instruments is radioactivity and the effluent from the Sellafield plant on the Cumbrian coast has been an

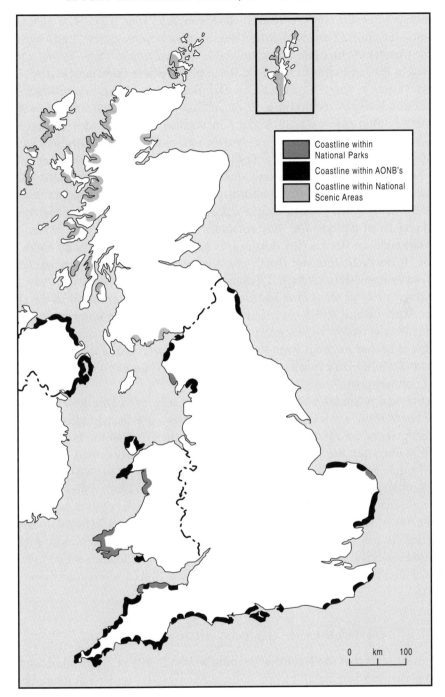

Coastline within National Parks

Coastline within AONB's

Coastline within National Scenic Areas

0   km   100

FIGURE 8.20 *Designated coastlines within the UK in the early 1990s. 'National Scenic Area' is a designation which applies only in Scotland. The additional category of 'Heritage Coasts' is found in England but most of these are contained within the AONB category.*
Department of the Environment, Coastal Planning and Management: A Review, *London: HMSO, 1993, p. 68, Figure 5.2.*

intermittent cause for concern since it started in 1952 (Fig. 8.21). The quantities of **radionuclides** emitted are well within the internationally agreed limits when they leave the plants but their fate is less amenable to monitoring from Vienna. Some are deposited in the sediments of the Irish Sea in places where they may be stirred up by currents. (There is about a quarter tonne of plutonium-241 and its products in the Irish Sea and levels of **technetium**-99 in shellfish were at 17,000 Bq/kg in 1996; the EU limit for foodstuffs is 1250 Bq/kg.) In shallow water off Morecambe Bay an increased frequency of onshore storms might well mobilise sediments which contain long-lived radionuclides (Fig. 8.22). The possibility that cancer clusters near Sellafield might be caused by aerosol transmission of radionuclides from seawater to beach and then be blown into houses caused distinct falls in property values at Seascale in the 1980s, irrespective of reassurance from the scientific authorities that the local concentrations of cancer were not enhanced in any significant way by emissions. Some radionuclides are carried northwards around the Scottish coast, across to the coast of Norway and into the North Sea. Various areas near to coasts as well as in deeper water have been used for dumping unused munitions: only in 1995 (when a new submarine cable was being laid) was an unmarked dump found in the channel between Scotland and Northern Ireland.[171] The sites are usually known, however, and so incidents are relatively few, except that fishing and dredging inevitably recover unexploded bombs or shells from time to time. One wrecked ammunition ship with perhaps 5000 tonnes of explosive materials lies in the Thames only 5 km from the Isle of Grain oil terminals.[172]

Most coastal processes and concerns have been subject to a fresh look in the light of predictions about sea-level rise. Most of the worries come from predictions that stem from the 'greenhouse effect' set of symptoms but there is also the fact that England south of a line from the Tees to the Mersey is downwarping, as part of adjustments that follow the retreat of Pleistocene ice (Fig. 8.23). A map of areas at risk (Fig. 8.24) reveals few surprises but the loss of high value agricultural land in the Fens, as well as densely populated industrial zones along the Tees, would affect the national economy considerably.[173] Should the predicted events materialise, then a national strategy will be needed; it will probably have to combine 'hard' defences such as strengthened and raised sea-walls along with managed retreat where it is too expensive to keep the sea out. Nobody, neither wildlife protectionists nor tourist town guardians, will be pleased.

FISHERIES IN THE POST-INDUSTRIAL WORLD

It is a curious fact that the Neolithic Revolution is still only very partial in the fishing industry, for most marine and fresh-water species are still hunted or gathered. True, the hunting may involve a modern diesel-powered trawler and sonar, but it still means going into a wild environment and catching a proportion of an animal population whose existence is to a large extent autonomous. It may also still involve people with horse-drawn carts hand-sifting through sand and mud at low tide for cockles. It does not now involve Great Britain in whaling at all.

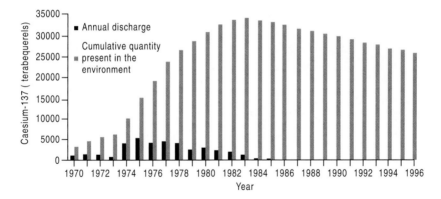

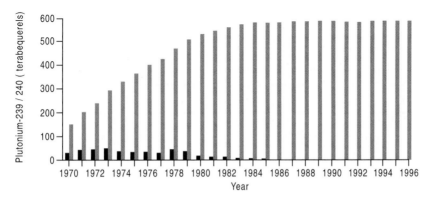

FIGURE 8.21 *Pollution from nuclear power plants. The upper histogram reflects Figure 8.22 perfectly, for the total quantity present in the environment falls at an interval after the lowering of discharge rates due to the decay of radioactive emissions. For plutonium, which has a much longer half-life (24,360 years) than caesium (30 years), there is no significant decline in the quantity present since radioactive decay is so slow.*
UK Environment Agency State of the Environment webpage, accessed on 2 August 1999, www.environment-agency.gov.uk/s-enviro/stresses/5waste-risings/.

Fisheries and the European Union

As with agriculture, Great Britain is locked into compliance with a set of regulations and financial instruments which relate to the whole Community. In this case, the EU tries to evolve management for the whole of the 200-mile **Exclusive Economic Zone (EEZ)** areas of its member countries; indeed it claimed in 1997 that it alone could 'agree' protection measures and the ministers of member states could only 'recommend'. Thus management tools, such as Total Allowable Catch (TAC) and the strategic measure for reducing fishing effort, the Multi-Annual Guidance Programme (MAGP) are set by the European Commission, though enforced by member states. Most European countries fish in non-European waters as well and all import fish, so that the international linkages are strong.

The basic data for 1992 in Table 8.9 show that Britain is a major user of the

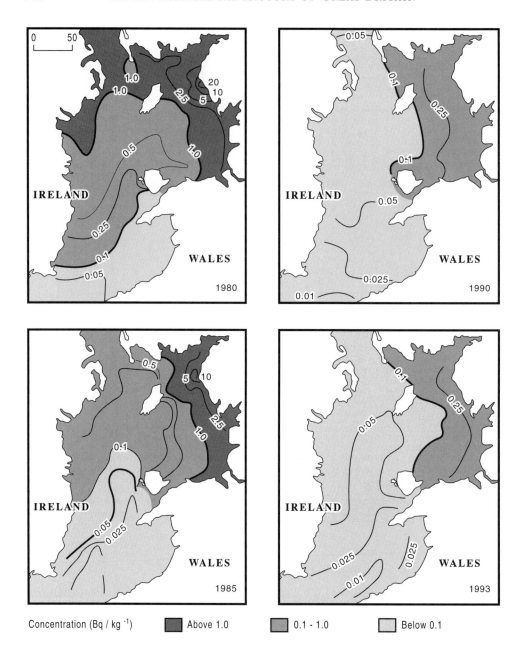

FIGURE 8.22 *Pollution from nuclear power plants. As international regulations take hold, radioactivity decays and technology changes, the quantity of radioactive caesium-137 in the Irish Sea diminishes. One problem with the lower amounts left, however, is their concentration in shallow-water sediments where they are vulnerable to mobilisation by storms; the effects of rising sea levels can only be surmised.*

Department of the Environment, Digest of Environmental Statistics 1996, *London: HMSO, 1996, p. 109, Figure 5.5.*

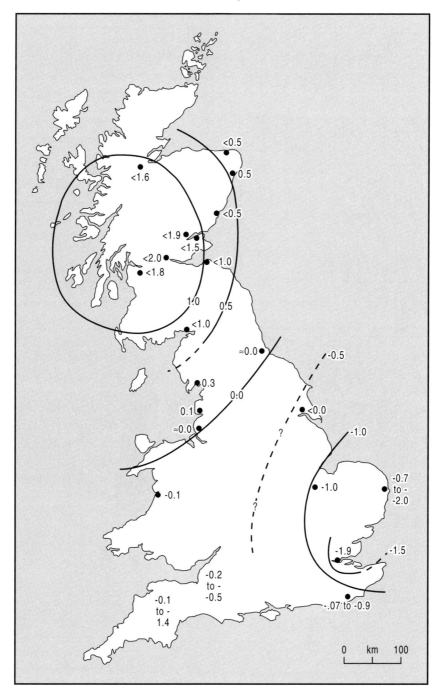

FIGURE 8.23 *Crustal movement in Great Britain at the present. Note the basic trends: that Scotland is undergoing uplift, while the southern half of England is downwarping, with neutrality from the Tees to the Mersey.*
I. Shennan, 'Holocene crustal movements and sea-level changes in Great Britain', Journal of Quaternary Science 4, 1989, p. 87, Figure 9.

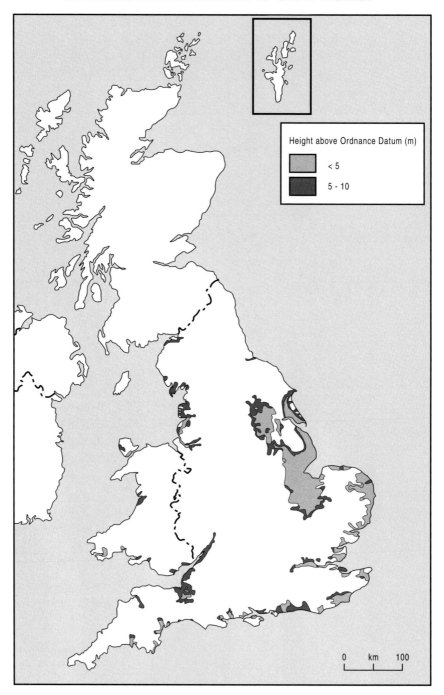

FIGURE 8.24 *Land areas potentially at risk from a sea level rise of 65 cm by* AD *2100. Tidal surges, storms and downwarping might all combine to threaten the Fens, industrial installations and a great deal of London.*
A. *Goudie and D. Brunsden,* The Environment of the British Isles, *Oxford University Press, 1994, p. 57, Figure 38.*

resources of the EU seas, contributing, from 10 to 15 per cent in most of the measures. (Spain and Italy dominate a majority of the numbers but Italy's effort is mostly Mediterranean.) In a temporal context, UK landings of fish from British vessels which ran at over 1 million tonnes/annum in both 1938 and 1948 were 759,071 tonnes in 1980 and 902,000 tonnes in 1991. The consumption of white fish has fallen by about 30 per cent since 1960 and for herring by 80 per cent. Employment of fishermen is steady at about 8,500 men (*sic*, more or less), compared with a 1938 figure of 26,000. Catching capacity increased from 35 to 40 per cent in the decade after 1970, aided by grants from bodies like the White Fish Authority and the Herring Board as well as regional bodies like the **HIDB**. Yet herring landings in 1970 were 146,000 tonnes and in 1980 only 5500 tonnes:[174] the pleasures of the kipper and the bloater now reach a select rather than a mass public.

TABLE 8.9   Fishing 1992: UK and EU

	UK	EU	UK as percentage of EU
Number of vessels	10,984	99,220	11
Power (Kw)	1,203,120	8,305,859	14
Tonnage	210,862	2,015,566	10
Catches (tonnes)	839,910	7,063,330	12
Aquaculture 1991: fish	55,500	280,933	20
molluscs and crustacea	12,150	701,770	2

Source: EU, *The Common Fisheries Policy. Fact Sheet 1–2*, 1994.
NB: These are all salt-water data: aquaculture of fish does not include trout farms.

Impact of fisheries

The biology of fish populations is complicated, with variations in reproductive capacity and survival (some cyclical, others less regular) being common. Scientific knowledge of the migration, spawning, growth and predation of the EU stocks is good but still subject to uncertainties in the existing data; there are also limitations on the accuracy of predictions. In addition, many fishing grounds were lost when 200-mile EEZs were announced: Iceland's declaration in 1975 was a particularly significant event since large quantities of haddock, cod and plaice came from those waters.

The designation of EEZs was supposed to make rational management of fisheries possible. As is well known, however, the European situation in both the North Sea and the Atlantic sectors has been one of over-catching to the point of reducing some species to very low levels. As the more expensive edible species have been reduced in number, shifts to industrial fish for the fish meal plants have increased: as cod, saithe, herring, whiting, mackerel and plaice became scarce, Norway pout, sand eels, horse mackerel and sprat were heavily fished and in turn have become over-exploited.[175] In 1998, it seemed as if the last remaining cod fishery, that of the Barents Sea, was about to collapse. The environmental task of the EU has been to reduce the danger that over-exploited species will become virtually extinct, as well as to reconcile the interests of food-catching nations like the UK with the industrial fisheries of Denmark and Norway. Success so far has been limited since the science is almost always diluted with the politics of appeasing the fishermen.

The uncertainties of deep sea fisheries have encouraged the growth of aquaculture in salt water. The sea lochs of Scotland are home to the EU's largest fish farming enterprise and the UK led the EU in farmed fish production in the 1990s when its freshwater trout farms were counted as well. The environmental relations of Scottish salmon farms (of which there were 430 in 1989 and 340 in 1998, though production increased from 30 kt to 115 kt in that period) are not a settled topic. Fish farms give off large quantities of ammonia into the water and these may result in blooms which are toxic to other organisms: scallop fishing was banned in the waters of the Hebrides in the summer of 1999. The ammonia loading by salmon farms is equivalent to a human population of 2–3 million and in summer of about 7 million. Before it was phased out, the antifouling agent Tributyl tin (TBT) was a source of mortality in fish farms. Intensive production in the sea therefore carries problems analogous to those on land for, for example, cattle and poultry, and, like them, new agents of unknown biological potential are constantly being introduced. In 1997, for example, two new chemicals were put before the licensing authority since dichlorvos was encountering resistant sea lice. Both have question marks over them in terms of toxicity to other organisms.[176] Many attempts at genetic modification are being tried, against a backgound of concern that fish will swim off and interbreed with wild populations. None, however, seem to relate to the fact that it takes 10 tonnes of wild fish to produce one tonne of farmed salmon. (The only farmed carnivores in Britain are salmon, trout and mink, with the last due to end in 2002.)

The inability of the EU and its member states to enforce rational conservation policies stems from a number of sources. There is, as evidence to the House of Lords in 1995 showed,[177] a considerable disconnection between the involved parties, but their Lordships were in no doubt about the seriousness of the problem and the need for drastic measures.[178] (Time for the exercise of real power, perhaps: hand the problem over to the supermarkets.)

Energy

The years after 1960 have been host to a number of changes in energy consumption in Great Britain. The first is the decline in the importance of coal, as it takes just one place in a mix of energy sources. The second is the passage of energy supply from state control to the private sector, with its implications for the regulation of environmental relations. Lastly, there is the nascent position of the 'alternative' sector, producing 'renewables', whose role is as yet unclear on the national scale.

The fossil fuels

The overall decline of coal in the consumption mix (Table 8.10) from about 90 per cent in 1950 to 27 per cent in 1990 (and 23 per cent in 1995)[179] has been accompanied by a shift in its sources. In the last decade alone, 48 collieries have closed, including 1 in Leicestershire that was only 10 years old. The major change in environmental terms is the move towards the cheaper open-cast production, with 31 per cent from this method in 1995, compared with 3.8 per cent in 1960. The restoration of exhausted mines is subject to strict control and it is often claimed that the drainage, access and layout of the restored land (most of which goes back into farming) is an improvement on the original surface. The rather bare nature of such restored land is obvious for some years, but the effects on drainage water are more long lasting. Runoff has to percolate through a very different set of strata from the original condition and this often produces a much more acidic water than before and which is inimical to most wild plants and animals.

TABLE 8.10 Energy consumption: the cross-over years

(as mt oil equivalent)	Coal	Petroleum	Natural gas	Nuclear electricity	Hydro-electricity	Total
1956	217.4	39.1	0.0	0.0	1.3	257.8
1960	198.6	68.1	0.1	0.9	1.7	269.4
1966	176.8	114.8	1.2	7.9	2.4	303.1
1970	156.9	150.0	17.9	9.5	2.4	336.7
1971	139.3	151.2	28.8	9.9	1.8	331.0
1976	122.0	134.2	58.8	12.9	1.9	329.8
1980	120.8	121.4	71.1	13.4	2.0	328.7
1990	63.7	72.1	49.0	14.2	1.5	203.3

From 1986, electricity was imported: in 1990 to the level of 2.9 mtoe

Source: British Geological Survey, *United Kingdom Minerals Yearbook*, Keyworth: BGS, 1988 and 1991.

The rise of oil and gas to a 65 per cent share has carried certain environmental consequences. Oil burned in power stations has kept up sulphur levels in the air, whereas gas (9 per cent in 1979, 32 per cent in 1995) adds less to them. Oil needs a great deal of processing and the refineries add emission burdens to air and water. On-shore spills contaminate coasts though refineries and road tankers occasionally lose refined products to their local surroundings. The actual extraction of oil offshore has been a source of oil pollution but improvements have reduced this loss; nevertheless, reported spills from rigs still number 100–200 in any year. Extension of the industry into the water off north-west Scotland has caused some concern, not least because exploration involves seismic testing, which is thought to affect the behaviour of whales and dolphins. Explorations will come within 65 km of the island of St Kilda, which is Britain's only 'natural' World Heritage site and which has the world's largest gannetry. Tanker accidents are the most spectacular impacts on wildlife, and their effects have been widely monitored in time as well as space. The immediate effects are known from the publicity: oiled seabirds and coated shorelines are the most obvious. Many species are affected, in rough proportion to their ability to escape the oil. Much better to be a salmon than a limpet, though fish taint was alleged in the case of a major spill in the Shetlands. Oiled birds usually die, even if rescued and cleansed. Not all tanker accidents are from vessels bound for Great Britain; some are using these waters as short cuts when they lose power. The *Sea Empress* (Milford Haven, 1996, 70,000 tonnes lost) was entering port but the *Braer* (Shetlands, 1993, 84,000 tonnes lost) was on passage. Illegal tank cleaning offshore adds to the pollution load caused by oil usage.

Nuclear power

In the aftermath of World War II, it was promised that the sword of Hiroshima would be beaten to the ploughshare of electricity that was too cheap to meter. It first came into the Grid in 1957 and after the mid-1960s it contributed about 3 per cent of the consumed energy, rising towards 10 per cent in the 1990s. The environmental connections have centred on the emissions of radioactivity from the fuel cycle and in particular from the reactors, though the earlier stages of fuel concentration and rod construction also emit some radionuclides. The plant at Sellafield on the coast of Cumbria has a reprocessing function and so is the most closely watched of all the installations; sometimes it seems as if it were the only one. Meeting the international regulations has meant that the quantity of planned releases has dropped through the years, though nobody yet knows about the long-term effects of very low but persistent levels of radiation on human health or on seabirds. Accidents are frequent but most are confined to the plant; the exception was a partial meltdown at Sellafield (then called Windscale) in 1957. This contaminated a downwind strip of land 80 by 15 km. All the milk produced in that area had to be washed away: it was hoped that the radionuclides would be dispersed in the Irish Sea. The other noteworthy emission from nuclear plants is hot water: per kilowatt of energy generated, this method uses a lot of cooling water. The older generation of reactors is coming to the end of its life and the decommissioning is another challenge for the industry, especially the end-use of the sites, some of which are in conspicuous places like Trawsfynydd in the uplands of North Wales (Plate 8.6). This plant closed in 1993 after generating power for 28 years and the operating company has proposed leaving the radioactive core on-site for 135 years.

Since the Sellafield nuclear particles are carried in currents round the north of Scotland, Norway as well as Ireland has been pressing for a zero-discharge policy, which is scarcely possible without closing the plant entirely. The Oslo/Paris Commission heard in 1988, for example, that the levels of the radioactive Technetium-99 have increased eight-fold in Norwegian waters in recent years. One special problem for Britain is the Dounreay fast-breeder reactor in Caithness where management was so poor that in 1977 a 75-metre-deep shaft containing 1000 tonnes of wastes, including over 50 kg of uranium, **plutonium** and sodium, exploded and threw low- and intermediate-level radioactive debris up to 65 m outside the boundary fence. The existence of a second shaft containing 700 tonnes of radioactive material was disclosed only in 1998. Clean-up is urgent, not least because coastal recession may reach the shafts and erode away their rock matrix.

PLATE 8.6   *A nuclear power station, emblematic of the new energy source that was to transform the price of electricity but which somehow has fallen short of its promise on many fronts. The Trawsfynydd plant is one of the very few not on the sea or an estuary but in the hill lands of North Wales; it is now decommissioned and will eventually be largely dismantled. Photographed in 1999.*

The military connections of the nuclear industry have produced a pathological addiction to secrecy. In 1996, damages of £6 million were awarded against the Ministry of Defence because it failed to notify a company which had bought land near Aldermaston (Berkshire: the site of nuclear weapons manufacture) that a flood had contaminated the site with plutonium in 1989. Campaigning groups and many ordinary citizens suspect that there are many more such cases, knowledge of which is withheld by government and large companies.

Atmospheric circulations mean that other nations' nuclear sites can impinge on Great Britain, as when a meltdown occurred at Chernobyl in the Ukraine in 1986. The cloud of radioactivity shed some of its nuclides onto the UK, notably southern Scotland, Northern Ireland, North Wales and Cumbria. Figure 8.25 shows the peak of **caesium**-137 in milk which resulted in Britain. In North Wales, the initial fallout affected 2.5 million sheep on 5000 farms.[180] In 1997, there were still 11 holdings in Cumbria containing 14,000 sheep which were under 'restricted movement orders': that is, the farmers were not allowed to sell off either the ewes or the lambs.

'Alternatives'

Official hostility to 'alternative' sources of energy in Great Britain has been strong: evidence of the malfeasance at high levels designed to derail power generation from wave energy is very convincing. Government only started to collect statistics in 1990. But with privatisation, the government had to create a 'non-fossil fuel' subsidy (the Non-Fossil Fuel Obligation) to ensure the survival of the nuclear industry.[181] Figure 8.26 shows the contributions made by different 'renewables' in the early 1990s to the total capacity of 1.67 mtoe in 1994 from over 150 projects; with the falling price of electricity from these sources, the government announced expanded targets in 1998 for their future contribution. The interesting conclusion from those data is the dominance of biological sources (which exceed large-scale hydro power in the UK as a whole) and the small quantity produced by the most visible and controversial element, the wind farm. The latter, whose distribution in the mid-1990s is plotted in Figure 8.27, are mostly either coastal or upland and hence stir up argument. Some of the antagonism is on the grounds of noise but most is on the new element in the landscape, especially where moorland sites are chosen (Plate 8.7). In national parks and AONBs, the Countryside Commission has opined that 'We do not feel that it makes sense to tackle one problem by creating another . . .'[182] When sites are proposed in areas which are already settled and have a mixture of developments, then the noise level is perceived as low and the verticality of the pylons is seen as an extension of an existing roster of electricity and communications structures. Coasts and off-shore

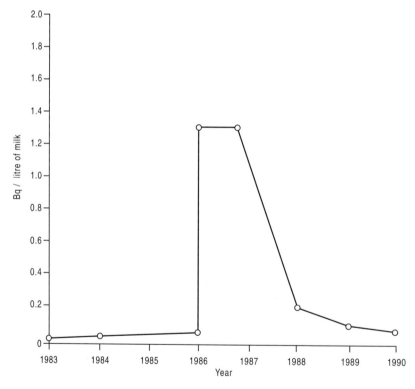

FIGURE 8.25 *An effect of the Chernobyl accident in 1986: the quantity of caesium-137 (half-life 30 years) in milk in England 1983–90, in Bq/litre of milk.*
Department of the Environment, Digest of Environmental Statistics 1996, *London: HMSO, 1996, p. 107, Table 5.2.*

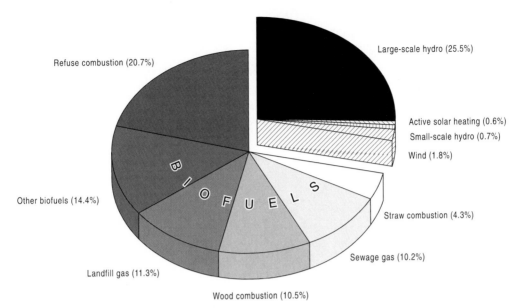

FIGURE 8.26 *The 1994 data for the proportions of energy in the UK supplied by renewable technologies. The amount of refuse combustion (vulnerable to changes in consumption habits and materials) is not perhaps a surprise, but the use of solar panels, even in Great Britain, seems very low, as if the technology and/or its associated costs have not yet taken off.*
G. Walker, 'Renewable energy in the UK: The Cinderella sector transformed', Geography 82, 1997, p. 72, Figure 5.

zones are also likely targets for would-be developers of this resource and conflict with scenic amenity seems inevitable.

The EU wants to see Europe getting 13 per cent of its electricity in 2010 from renewables (compared with about 5 per cent now): Britain could be one of the most favoured places to contribute to that aim but government help with the initial stages of introducing more solar energy use is painfully slow. Privatisation of energy supply has come about in the UK without any of the pressures to conserve energy that accompany utilities in, for example, the USA. This is probably one reason why such companies are eager to buy into the British industry. So for the foreseeable future, the pattern of use in which gas is pushed as the cheapest and least polluting fuel (and presumably the most profitable), allied to rather half-hearted government campaigns for conservation impelled by fears of global warming, seems likely to continue.[183] The curious situation in which the muscle tissues, if not the lifeblood, of an industrial society is in the hands of transnational companies and their shareholders rather than the people themselves, also seems unlikely to change.

POPULATION TRENDS

Though a few special interest groups exist with a Malthusian view of the population trends in the UK, there is more public concern about an ageing population than

FIGURE 8.27   *Location in 1997 of major wind power installations in the UK. The popularity of upland sites in the west is easily seen, as is the attraction of the upland areas like the Pennines. The potential for conflicts over landscape values is easily imagined.*
Courtesy of British Wind Energy Association.

about an expanding one.[184] The current levels and temporal movements of the population could, *ceteris paribus*, be applied to most other post-industrial nations. Scotland not only has rural emigration but also, more importantly, lower fertility levels and so its population is likely to fall to below 5 million during the next 25 years. One of the remarkable features is the persistence of the attraction of rural life, seen here in the movement to the countryside. In part, this is probably a simple consequence of

PLATE 8.7   *A wind farm in mid-Wales in 1997. The presence of grazing sheep suggests that leasing of land for this purpose may be a useful adjunct to falling farm incomes in the uplands. The towers are not environmentally significant in a place such as this, but are certainly an unfamiliar element in the landscape. Perhaps their main significance is to presage a different kind of human-environmental relation from those of the fossil-fuel era.*

the ubiquity and relative cheapness of the motor car, and the availability of reasonable roads on a year-round basis. Most inhabited parts of Britain are within daily reach of employment and shopping provided a car is available: there is scarcely an equivalent of *la France profonde*. At another level, nevertheless, it seems like yet another expression of the rural myth: that life will be better in all ways in the country. In practical terms, this meant that in three years of the early 1990s some 1.25 million people left six large metropolitan areas, taking part in a cascade of movement from cities to small towns and then to villages. The balance is only reversed by 19–29 year-olds. This is unsurprising, though even they try to go back once in the 30–44 age group; the rural areas comprised 10.4 per cent of the nation's people in 1995, up from 8.9 per cent in 1971, though scarcely more than the 1981 census figure. The mining and industrial areas housed 12.1 per cent of the population in 1995 compared with 13.0 per cent in 1971. It is calculated that in the 20 years after 1996, some 4.4 million new dwellings (another 11 Liverpools) would be needed to house an expected level of population in which single people would increase the number of households in a stable population.[185] The urbanised proportion of the nation would rise from 11 to 12.5 per cent. A key environmental question is whether the developments envisaged would be on green fields or on 'brown land', that is, in already urbanised areas, and be supplemented by other measures to reduce the actual proportion of greenfield housing to less than 25 per cent. A government-set target of 60 per cent on already developed (for urban uses) land was canvassed in 1996, though most commentators thought that 30 per cent was a more likely figure. In the public mind there is often confusion between greenfield sites and green belt sites in protected areas.

Administration

The story of the 1970s onwards in the environmental field has been one of increasing realisation that environmental management on an integrated basis is necessary but that achieving it is difficult (Table 8.11). Yet the regional splitting of Quangos such as the Nature Conservancy Council and the Countryside Commission was undertaken. Ostensibly this was to bring them closer to the people, but cynics pointed to the ability to fill them with placemen sympathetic to development rather than conservation. In 1996, an Environment Agency was created: while this enfolded the River Authorities and the Pollution Inspectorate into one, agriculture, fisheries and forestry still remained separate. The Labour government elected in May 1997 gave oversight of transport as well as those agencies to one minister, which was taken to signal a recognition of motor transport as a major environmental factor.

TABLE 8.11　UK: major conservation designations, 1996

Designation	Area covered	
	Million hectares	Percentage of UK total
National Parks	1.373	5.7
AONBs	3.390	14.1
National Nature Reserves	0.193	0.8
SSSIs	1.943	8.1
ESAs	3.160	13.1

AONB = Area of Outstanding Natural Beauty; SSSI = Site of Special Scientific Interest; ESA = Environmentally Sensitive Area
Source: Department of the Environment, *Digest of Environmental Statistics*, London: HMSO, 1996.

The time between the 1970s and the end of the century was generally a time of reluctance on the part of government to disturb market arrangements and vested interests, and so intervention was minimised. Several ministers tilted in favour of environmentally damaging developments in order not to offend powerful industries and businesses. In each sector, there were government players but they were only one set of actors and may not have played a leading role. In the case of acid rain, for example, an account published in 1995 asserted that,

> Policy has been determined as a result of negotiation between a small core group of players comprising the DoE, the DTI and the electricity industry. HMIP [the then Pollution Inspectorate] has played a peripheral role, while environmental NGOs have remained outside the policy network.[186]

One major influence, which can be readily documented, is the European Commission. In spite of the UK government's often reluctant compliance with directives, commentators have been able to write (in 1995) of 'a massive shift in power from national states to the EU which has occurred in environmental policy in recent years'.[187] The more Europe-friendly attitude of the post-1997 government may speed up the adoption of EU policies and standards but it seems unlikely that any UK government can in fact work up a national environmental policy when it controls so few of the key processes (the utilities, for example, being in private hands) and espouses an equivocal attitude to regulation.

THE INTERNATIONAL SCENE

To have lived in the UK in the 1990s was to experience a concern with 'Europe' that could only have been paralleled in the reign of Henry V or perhaps before that. Fuelled by a group of 'Eurosceptics' whose motivation was a xenophobia that found expression in the term 'loss of sovereignty', the debate rarely recalled the wider international webs into which the British were, and are, bound.

The withdrawal from empire was largely finished in this period and thus a major source of employment for environmental managers, who had fed many ideas into their home countries[188] was closed down as the newly independent nations sought such advice from a group wider than the colonial power, often located in an international agency or possibly in a NGO. Advice on environmental matters was more likely to be requested if it led to 'development': to growing crops rather than wildlife conservation, unless the latter were the basis of a tourist industry. The increasing per capita GDP at home led to more purchasing power and some of this has been funnelled into the rural south via export-oriented economies. One result of the power of large companies, coupled with high energy usage nearer home, has been the virtual elimination of seasonality in fresh produce. Any sense of the harvest seasons of our cool temperate climate has to be found in the creations of commerce (Christmas being the outstanding example) and not the ripening of crops: strawberries no longer belong with Wimbledon and Henley in the calendar of the year's progress. The perpetual presence of mangetout can symbolise our awareness of global linkages. These couplings are highly diverse in nature, ranging from the black letters of international treaties and conventions, through our visual perceptions of our place in the universe as photographed from space, to the electronic mesh of information that comprises the Internet. Nobody can say what the interaction of all these is, let alone will be; it is without doubt a non-linear system.

No excuse is needed, however, for raising the idea of visual representations of the earth and parts of it as being significant in the formation of our current environmental attitudes. Ever since the public availability of aerial photographs in the wake of World War II, the view from the air has exercised a fascination that has found its way into books of aerial photographs and satellite images which have had good sales.[189] In local and regional terms, such imagery has supplemented and certainly to some extent supplanted the Ordnance Survey map, whose symbols and contour patterns were once meaningful to every child who had survived school beyond the age of 14.[190] More conventionally, the BTA has issued a movie map of Great Britain, showing the locations for about 200 films and TV dramas, where people may search for the scenes of some of their favourite moments, though the famous buffet at Carnforth in Lancashire is no longer open for Brief Encounters.

Sensibilities and attitudes to environment

Even aspiring xenophobes have to admit that Great Britain is bound into a dynamic web of information, which comes at us in unremitting flows of ever greater magnitude. A great deal of it is ignored. Some of it is constructed into knowledge: a part of it by professionals in media, education and entertainment, other parts by poets and painters, for instance. Whether we reach the third stage, the distillation of wisdom, is usually only apparent in retrospect. We rarely look at wholes: an attempt to look at wastes in the whole of Wales, for example, constitutes an unusual piece of collation.[191] In environmental terms some of the linkages are global. The extension of the northern hemisphere ozone hole, for example, is beginning to affect Great Britain, bringing with it the higher UV levels that damage plant tissues and induce human cancers. One example of this type of linkage is our obligations under the Rio accords of 1992, where Great Britain agreed to stabilise and then reduce its CO_2 output. Much of the attention to such a shift comes in the form of reducing energy use overall, as in the shift to gas rather than oil in power stations and a series of promises to improve the quality and attractiveness of public transport. A lesser known element is rural land use: for 1988, it was calculated that agriculture and forestry account for no more than 3 per cent of all the UK emissions of 150 million tonnes. On balance, agriculture and forestry fix 9785 t/yr of CO_2 and emit 5032 t/yr, so their absorptive capacity is a positive, though quite small, contribution to the carbon balance.[192] The national response to being demonstrably part of a global system is of course tempered by the local expressions, and in many minds these are not easily distinguishable from home-grown pressures for change: a new MacDonald's in the High Street looks much the same as the conversion of Mrs Batty's café into a mock-American diner.

What clearly impinges upon sensibilities in this field is the pace of change. The rate of land conversion from rural to urban uses has been slower in recent years than in the inter-war period. Thus even in England, only 10 per cent of the land is classified as urban; little of it, however, is remote (Fig. 8.28). This belies some of the perceptual effects, however. To drive from Yorkshire to London along the Great North Road increasingly becomes a transit through a corridor of large service stations and diners run by major chains.[193] The native perception of Great Britain as being grounded in its rural scene is deeply implanted: a city-bred and city-living poet like Phillip Larkin took up this theme in 1972 in 'Going, going',

> before I snuff it, the whole
> Boiling will be bricked in
> Except for the tourist parts –

First slum of Europe: a role
It won't be so hard to win,

and there is a lot more of the same in this poem. Larkin was well known for his poet-ically sour exterior but there is no doubt that he touches a responsive chord here, for the level of concern about environmental matters is high in the public consciousness; around the time of the general election of 1997, broadsheet newspapers carried a cluster of articles about the countryside and its degradation; a legislation to restrict the grubbing out of hedgerows was early in the new government's programme. Yet, the rural scene was not high on the list of environmental concerns evoked by systematic sampling. In 1989, surveys showed that radioactive wastes, chemical emissions into rivers and the sea, and the contamination of beaches by sewage ranked the highest. Since most people got their information from television, these priorities presumably reflect what was being highlighted by that medium in that year. Most people, equally, will take action to improve the environment if something simple and clear can be done: three-quarters of the sample gave up using CFC-powered sprays, 40 per cent used bottle banks and 25 per cent recycled paper. More and more people joined environmentally related organisations: between 1971 and 1990 the membership of the National Trust rose from 278,000 to 2,032,000 and that of Friends of the Earth from 1,000 to 110,000. Even the more assertive Greenpeace acquired 372,000 members by the latter date. Interestingly enough, the new environmental feature of concern for one in five households has become noise. Mostly this is urban noise, from neigh-bours, children, dogs and the like. But rural areas are not that quiet, as refugees from the cities often discover: farm animals are careless of conventional hours of sleep, agricultural machinery is every bit as noisy as that on building sites, and commercial flight paths seem to have few lacunae over Britain. Walking in the hills, it is not uncommon to look down on a pair of military jets whose noise level is near to pain thresholds.[194]

But in general, the population was unenthusiastic about development of the economy being subjected to environmental priorities, a GDP of £440,000 million in 1983 rising to £570,000 in 1994 (at 1990 prices) was better news than the diminution of the number of calling bitterns to about 26. This suggests a kind of split-level attitude with a deeper affinity with the land (in spite of urbanism) overlain by the materialism which is encouraged by a capitalist economy. Although Larkin's gloom is widely shared, other creative artists resonate and celebrate British environmental qualities. This is no more so than in music, where the gentler inter-war tradition (see pp. 230–4) has been carried forward albeit into a more astringent harmonic. But, for example, much of Peter Maxwell Davies's output is reminiscent of his Orkney home even when the references are not explicit, and the poet George Mackay Brown (d. 1997) provides a parallel in words. In Harrison Birtwhistle's opera *Gawain* (1990) there is the explicit use of one of the most environmentally rich texts of English writing. More often work is suffused with an Englishness that is inclusive of its landscape: some of the settings of Wilfred Owen poems in Britten's *War Requiem* (1962) summon views of England just as Ralph Vaughan Williams's *Pastoral Symphony* had earlier recalled Flanders

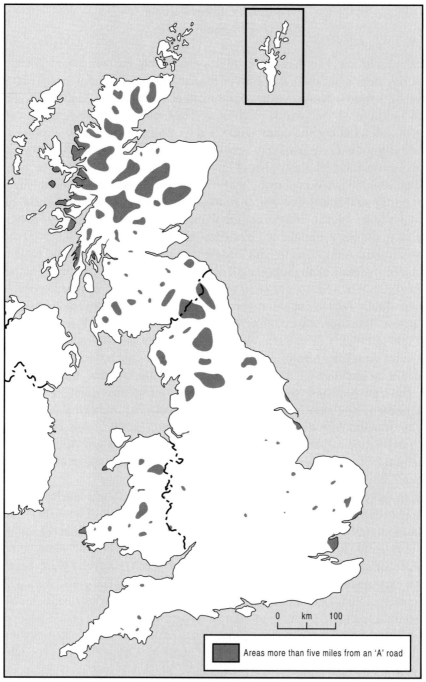

0   km   100

Areas more than five miles from an 'A' road

FIGURE 8.28   *Post-war Britain: areas more than five miles (8 km) from an 'A' class road. Apart from a few reclassifications (for example, the block in upper Weardale, County Durham will have been split) it is still much the same. In the original, Stamp was worried about access to land for production: the map is more likely now to be seen along with the 'tranquil areas' map (Figure 8.16). L. D. Stamp,* The Land of Britain, *3rd edn, London: Longman/Geographical Publications, 1962, p. 209, Figure 119.*

rather than Gloucestershire.[195] An identification with landscape which started in the first half of the century with the sculptures of Barbara Hepworth and Henry Moore has been carried on by Andy Goldsworthy in his re-arrangement of natural materials *in situ* which are photographed and then left to the elements: leaves, stone and timber pieces have all received this treatment. To some extent this is an extension of the practice started by the Forestry Commission at Grizedale in the Lake District, of having wooden sculptures installed at intervals along a walking trail.

The riches and pleasures of high culture have always appealed to a minority (though such is the social structure of Great Britain that it tends to be a decision-making class) whereas television is truly democratic. Advertising creativity is developed with the small screen in mind.[196] Therefore the treatment of the environment on television is very important, both in the programmes and the advertisements. That it reinforces the rural idyll is scarcely worth mentioning. Connections with an Edenic state abound, whether in programmes inhabited by rascally but characterful and basically good people with strong regional accents or in advertisements for 'wholesome' foods accompanied by the kind of music anthologised on CD as 'The Nostalgia Collection'. Backdrop is another use: a car can be shown to be a transport to heaven rather than Charon's ferry on wheels. In these scenes, the auto is nearly always shown by itself rather than, as is more normal, in the company of a great many other vehicles. Upland scenery of a rugged nature, such as the Scottish Highlands or the northern Pennines, is often used, though in somewhat uncharacteristic weather.

Of the programmes, there seems to be an inexhaustible flow; this presumably means a similar appetite. Wildlife, landscape, country pursuits, urban pollution, heritage conservation, quirky buildings and quirky characters are the daily fare, in peak time as well as in the hours any civilised nation would devote to a siesta. In the case of wildlife, most have in common the lack of acknowledgement of human involvement in producing the ecosystems containing the species studied, unless the programme is about peregrines nesting on tower blocks, for example. Curiously, there is the same lack of historical sense that allows the National Parks to be bracketed with Areas of Outstanding *Natural* Beauty. Most follow a standard format of a seasonal round followed by an indication of threats from human activity. Some of this, we can imagine, fuelled the rise in membership of the RSPB from 98,000 in 1971 to over a million at the end of the century.

CONTRADICTIONS OF THE TIME

How are these years different from those earlier in the century? Let us remember first that there are more of us. Growth may be very slow, but the population of Great Britain is not declining in absolute numbers. Nor are our environmental demands diminishing: our ecological impact both at home and abroad continues to expand, as expectations of access to energy and materials increase. The measure of Ecological Footprint, which gives a value of 1.7 ha/cap of biological production globally, stands at −3.5 ha/cap for the UK, showing our dependence on other ecosystems and our low area of carbon-absorbing forests. The Carbon Index (CI) views history as a means of capturing carbon since this also intensifies the footprint and reduces biodiversity. The world average CI in 1995 was 5.15 and the UK came out at 10.8 in a world where the USA was 24.72 and Ethiopia 1.07.[197]

Some forms of public concern for environment (even in the distinctly humanised form of National Trust properties) has never been greater.[198] Some kind of collective schizophrenia is clearly at work, though its contradictions are probably not recognised by most people. If they are, then they are shelved in face of the efforts of everyday living. At the heart, perhaps, is the binary division between humanity and nature which seems to be deeply embedded in the western world-view.

If, again, we want to single out something of singular importance then it must be our membership of the EU. On the one hand this has locked us into the intensifications of the CAP. On the other, some of its directives are supportive of better nature conservation and its fisheries policies might – just – save some stocks from extinction. Another binary division, another fragmentation.

CHAPTER NINE

Experience and meaning

INTRODUCTION: MANY PASTS

T. S. Eliot brings together the words 'experience' and 'meaning' in *The Dry Salvages*,[1] where he alerts us to the likelihood that 'We had the experience but missed the meaning'. We might profit from expanding this (without losing its poetic force) to think that neither experience nor meaning are homogenous and probably have never been so since the Neolithic, perhaps even before then. Diversity of experience according to place and occupation, for example, has been paralleled by the different ways of writing history which are alluded to in Chapter 1. In this book there has been some aim at being detached, and to try to extract regularity from the messiness of change. There has been no conscious attempt at a political history which aspires to the validation of the actions of a particular group such as government or tries to construct a sense of nationhood.

CRITICAL PERIODS

In attempting to give a detailed account of the years between the Palaeolithic and the present, it is easy to lose any sense of overall shape. At the risk of emphasising change rather than continuity, an effort is made in Table 9.1 to highlight a few periods and 'events' when the society–environment relationship seems in hindsight to have been particularly formative.

The main finding of this Table seems to be that rapid changes are less common than long periods of consolidation of a particular set of relationships. If we try to separate out turning points then the introduction of agriculture must be seen as one of them. The next single most crucial event is the move to an 'inorganic economy'[2] where the extra power (in many senses of the word) made available was used to manipulate the pre-existing systems of all kinds. The third and last, more difficult to isolate from an era of swift metamorphosis in so many areas of life, is the direction given by the lavish application of subsidies by governments. In the present context, those given directly to agriculture seem the most relevant, but there are many others: in forestry, the location of industry and the encouragement of certain types of fishing vessels, for instance. The underpinning of various environmental processes by public money thus becomes critical even though it is scarcely now a turning point (and the early examples of agricultural subsidies were meant to be temporary) in any

TABLE 9.1   Critical periods for environmental variables, with social context

Environmental Period	Social/economic context	Key 'events'
Growth of forests as climate ameliorated: 8000–6000 BC	Minor manipulations at upper edges and around openings by hunter-gatherers	Introduction of agriculture
Fragmentation and retreat of woodlands as agriculture expands: LBA–AS times, ± complete by Norman Conquest, end eleventh century	Ownership patterns are emplaced: monarch at the head of land-owning and managing hierarchies	
Maximum pre-industrial diversity: fields, common lands, mountains; water bodies altered but 'natural' states still often visible. Ends c.1750	Hierarchies still important but greater number of 'middle-class' owners with long-term interest in the land	Enclosure movements of all kinds
Industrialisation with effects well beyond cities 1750–1900. Intensive management of lowlands, control over water flows, land cover changes and pollution in urban-industrial centres	Profits from capitalist industry paramount in many areas, with spin-off into rural recreation patterns of both elites and masses	Introduction of the potato Substitution of coal for wood in iron-making 1750–1850. Stockton and Darlington Railway 1825; Alkali Act 1863. Imperial expansion
Seesaws between intensive use and weedy fields, moorland edge up and down. 1880s–present, cutting across next phase.	Subsidies as driving forces in rural production-based versus 'protectivist' balance	Scott Report 1942 Town and Country Planning Act 1947
Locking of some 'management boundaries' post-1950: NNRs, NPs, AONBs etc.	Spread of hitherto restricted values improving the standing of the non-human world	Accession to EU; ratification of global conventions, for example, Montreal, Rio Break-up of Nature Conservancy Public suspicion of agriculture after BSE, GM experiments, late 1990s

'short and sharp' interpretation of that phrase. However arguable these turns may be, there is no doubt that cumulatively they have shaped the present. We can view them from two main angles: the long-term cultural context and the practical consequences or pathways. These join together in a brief look at today's situation which is contrasted with the paths not taken; there is then a short coda.

CULTURAL CONTEXT

Most cultural historians find evidence of periodisations which relate to the outlook as well as the practices of the time in question. The all-pervasive religiosity of the Middle Ages is one such category of overview. That everything might be sacred did not prevent it being changed. To fell a woodland on a monastic estate could be presented as part of human participation in the divinely ordained work of completing the Creation. (Seven days had clearly turned out to be an underestimate, setting something of a precedent for the building industry.) The Renaissance saw 'man as the measure of all things' and so began a shift to ordering anything and everything to human estate, pragmatically using the rationalities disclosed by those sciences whose rise is epitomised by the founding of the Royal Society in 1662. The most recent of these cultural phases to be recognised without too much controversy is modernism. Made possible by the industrial revolution but following it at a decent interval, modernism is above all the embrace of technology as shown in the steel-framed, slab-sided building,[3] of free verse, abstract paintings and sculptures, and of the legitimacy of large-scale belief in, variously, capitalism, state socialism or the findings of the men (mostly *sic*) in white coats. There were concurrent reactions: the small-scale socialism of William Morris and the arts and crafts movement, the defiantly persistent representation of the rural scene as thatched cottages around village ponds, the pastoral music of Gerald Finzi (1901–56) and the verse of John Betjeman (1906–84). Much of this is memorably and sensitively traced in literary expression by Raymond Williams.[4] Whether post-modernism is simply the latest phase in modernism or represents an entirely new set of strands remains to be seen. In its abandonment of grand narratives and over-arching modes of explanation, however, there seems to be a good case for marking it off from its precursor. In particular, the validity of only local explanations (in time and space) appears to be isomeric with the kinds of fragmentations into social individualities and into environmental compartments that have been mentioned at various points in this book. Yet, in some creative contexts at least, fragments can be woven into a powerful story provided there is a warp, such as the land itself provides. W. G. Sebald's *The Rings of Saturn* provides an outstanding example.[5]

The construction of identity

It is probably always a mistake to derive culture and its values only from a material base. While connections undoubtedly exist, some developments float free of the material world. Thus the formation of British identity was greatly influenced by the way its Protestant inclinations to individuality and democracy contrasted with continental Catholic absolutism and hierarchy, but none of these can easily be related to forest management or fish catches. Nevertheless, the definition of British identity in terms of its mix of English, Scottish, Welsh and Irish elements and in terms of its imperial relationships clearly has its material constituents: maybe the peat content of Scottish water or the sugar crop of the West Indies are indeed related closely to the formation of identity. The creation of agricultural surpluses in feudal systems of rural economy or of money in the early phases of industrialisation was made possible by the resources of land and sea at home and abroad and in turn affected the ecology of land use systems within Great Britain. The upper limit of cultivation at the moorland edge has been significant since the Iron Age at least and the limitations imposed by climate still produce a type of livelihood that spills over into perceptions of human character as portrayed in popular entertainment. The same connecting strands can be seen in drama which emphasises the independence of small and isolated communities in Scotland: Local Heroes abound.

All in pieces . . .

John Donne lamented early in the seventeenth century that this world was 'spent', for,

> 'Tis all in pieces, all coherence gone;
> All just supply and all relation:

because the new cosmologies had forced both science and theology into new perspectives.[6] If we search for an analogous meta-trend in later centuries, then it seems as if technological determinism has gained in strength as time has gone by, rather as if we jumped onto its bus in the late eighteenth century and have not yet found the courage to ring the bell for a request stop.

It is possible at the same time to read the history of Great Britain as a process of individuation, from the probably cosmic identification of humans and nature, living and dead, in the Palaeolithic and Early Mesolithic, through the undoubted group territoriality of the later Mesolithic to medieval commonalities and their gradual disappearance into owned stretches of land. Each upward ratchet of high-energy technology has fragmented the terrestrial social and environmental patterns more and more, just as it has quite recently with the Saami reindeer herders and the recipients of the Green Revolution in rice growing in Asia. The nineteenth century in particular was a time of loosening of ties as fewer and fewer goods were re-used and recycled and there was a shift to emphasis on the new. But the idea of an 'organic society' which was lost in the nineteenth century and which had been some kind of kindly, rural, integral community which had a 'human naturalness or normality' or was 'right and inevitable' is the invention of literary critics who have never crouched out of the east wind behind the wall of a cow-byre eating fat bacon sliced off with a penknife onto greyish bread.[7]

No wonder, then, that the environment is itself fragmented. There is an apportionment of land and water to owners, who may exercise exclusivity, whether of production ('no public right of way') or protection ('permits to enter the reserve may be purchased from . . .') and it is assumed in both town and country that a car is available to proceed from one piece of fenced function to the other;[8] exclusion is a frequent result of not possessing that particular piece of technology because of age or income. It is easy to fall into a threnody for a non-existent golden age, but it does seem that the sense of human connection with the land (and water, via the seaside holiday) has been loosened by the ubiquity of the technologies of separation. Examples might be the Walkman, the private vehicle, television and video, and the easy transport to a foreign destination ('Foreign' because it is sunny, not because there is different food or an unfamiliar language). Safety equipment required for outdoor occupations isolates the worker. Even the disappearance of mixed farming and the decline of public broadcasting might be added to the list. Conventional views would look for a rebuilding of community in all these senses via social engineering, educational programmes, better public transport and the like.[9] A more radical perspective might point to the absolute inadequacy of prediction in such a complex system and to the fact that technological change has driven most of the metamorphoses that have been chronicled here. Given the rapid pace of innovation, it seems highly likely that it will be a piece of technology that will bring about the next set of social revolutions, not a politician's exhortation to be nice to each other and not to vandalise saplings.[10] More certain perhaps is the realisation that ecological integrity is dependent upon social, economic and political integrity.[11]

The wider cultural setting of twentieth-century environments can be described in more detail. There is, for example, the question of language: the ideas of individuals acting upon external objects was reinforced in medieval Europe when the pronoun 'I' (and its equivalents in other languages) becomes compulsory, especially in combination with 'shall' and 'will'. These verbs make life appear as a constant series of willed and directed activities. As Arran Gare puts it, 'this [outside] world is not spoken of as a world of momentary objective events of which people are part, but as a world of objects which is the substratum of will and duty, of planning and doing . . .'[12]

Even 'momentary objective events' implies a fragmentation rather than a flow. In pre-industrial times, an apportionment comes along with the division of the land and its allocation as private property. The distribution of the land in medieval Europe became a reflection of the capacity to till the soil using the deep plough, just as the power of the water mill became emblematic in all kinds of ways. The face of the land can be seen mirrored in the increasing privatisation of the individual: the lord and his family sequester themselves into the solar, the corridor introduces a separateness into the large house. This is exacerbated in industrial societies, which usually have a large urban population who tend to be arrayed in spatial homogeneities. After the celebration of the city by Claude Monet comes the retreat to his garden; after Impressionism, the dissections of Cubism. Isomeric creations include Webern's short sharp pieces, and the linguistic tesserae of Eliot's *Waste Land* and Joyce's *Ulysses*. Housing with walled enclosures is echoed by the bubble of the personal stereo; the cinema has been threatened by the VCR and TV, the concert hall (itself an eighteenth-century development) by the CD player. Public broadcasting has been subject to attrition in the face of commercial interests. Holidays are no longer a celebration within the community but an escape from it. There is still a seasonal round of festivals but these tend to celebrate private relationships and the individual's life cycle.[13]

Many features of the environment come to express the specialisation of human lifeways. The landscape becomes increasingly chopped up as land use systems which were interdependent lose those qualities and both ownership and purpose become more individualised. Thus there is land for intensive production, land for recreation and land for conservation but all are fenced and are managed with a single purpose in mind (Table 9.2). As population has grown and as consumption has increased in what can now be talked of as a 'full world', the parcelling of the environment has become more intensive: as if the less per head there was, the more obviously it had to be owned and devoted to an obvious purpose. This is encouraged by free-market enthusiasts who argue that resource management is the more effective if the resources are 'owned'; they say that it is common property resources which are over-used. The UN Law of the Sea conventions setting up Exclusive Economic Zones (EEZs) was a practical expression of this, as are permits to trade pollution.

TABLE 9.2   Processes of coalescence and of fragmentation: examples

	Coalescence and Integration	Fragmentation
Post-industrialist (after 1950)	ELECTRONIC COMMUNICATIONS Political and trade blocs Accession to global treaties and conventions	ELECTRONIC COMMUNICATIONS Private transport Income gap Privatisation of radio and TV
Industrialist (after 1800)	Public transport Theatre and cinema POPULATION MIGRATION Imperial relations The steam engine	Public radio Creative arts POPULATION MIGRATION Specialisation in science
Agriculturalist (after 3500 BC)	Inter-regional trade Cattle droves Integration of nation-state Overseas trade Adoption of Christianity	Enclosure Separation of production areas and 'waste'
Hunter-gatherer (after 10,000 BC)	Single mythology	Territoriality
All periods	INTRODUCTIONS of biota	Extinctions and INTRODUCTIONS

Processes in SMALL CAPITALS are simultaneously in both categories.

A parallel but somewhat different exposition of this in the context of environment comes in Niklas Luhmann's assertion[14] that, first of all, humans cannot communicate directly with nature (to which many twentieth-century rationalists will assent, though not all the Aquarians and 'New Age' believers) but secondly that, in talking to ourselves about it, we are incapable of formulating an accurately mimetic representation because our discourses are split into distinct channels. Given the history of Western culture, each channel's discourse tends towards one of two conditions: dualisms are writ large. Examples include legal/illegal, economic/uneconomic, employed/unemployed, religious/heathen, immanent/transcendent, reason/unreason and, eventually, good/evil.

Small wonder that the digital computer, which is either 1 or 0, has found such a welcome reception. One analogy might be the pipe organ: in each pipe there is a resonance and in the world of the Baroque and of Modernity there is a score, to which all the soundings are referred. In the post-modern world, the music is aleatory: the sense even of improvising on a theme is being rapidly lost and replaced by snatches of inter-pipe resonances which temporarily have meaning or make a nice sound or are sufficiently discordant to make people move away.

Fragmentation can be contrasted with coalescence. Today's outstanding cultural example is our national embeddedness in a world-system. The global simultaneity of electronic communications, for example, allows the transfer of money in nanoseconds, just as it can show the Internet freak an updated view from an Antarctic base every half minute, a scene which somehow lacks interest in the winter. Such networks were anticipated by the transnational companies (TNCs) who are structured so as to allow the polyfocal creation of profit, switching operations as advantage becomes apparent, in a kind of pinstripe update of shifting agriculture. Globalisation is a complex phenomenon but its effects in the economic sphere accelerate in the mid-1970s with, for example, the rapid spread of manufacturing industry into low labour-cost regions of the world.[15] Even before the TNCs, the industrial world was creating a global coalescence of enhanced levels of CO_2 in the upper atmosphere. All these are consequences of the emergence of a 'central civilisation' depending upon its connectedness for separation from its ancestors and co-evals, something which has only happened in the industrial and post-industrial phases of human history.

We are less in sole charge of our own relationships than we have ever been. Our membership in the EU brings in environmental legislation which is different from what we would have adopted (or declined to adopt) if left to ourselves.[16] Beyond the EU, we are becoming tied into a set of treaties and protocols which affect our impact upon the global environment. The seas and the atmosphere are global commons of concern here. These regulatory globalisations are, however, very slow in taking effect compared with the productive forces that are unleashed by capitalism. The ultra-rapid transmission of capital can be followed quite quickly by a new product from existing premises or the design–build–output sequence of a new factory. In these contexts, a UK planning mechanism little changed from that of 1947 tries to mediate between the largely conservative views of the informed and articulate part of the citizenry and the equally conservative desire for profit from the TNCs and the local entrepreneurs. The radicals in the environmentalist movement, who subscribe to neither of these outlooks, get practically nowhere in conventional terms and have a high profile only in major projects when they clash with the law by taking to the trees or to tunnels. The moderates can usually only modify location, rarely scale.

PATHWAYS

Interesting though our global relationships are, there are some down-home practicalities to be considered. If we have thought globally (not for very long, I suspect) in this context, then how have we and our representatives acted more locally?

State relations

The breakup of bodies like the Nature Conservancy Council in the 1980s was mostly a political desire to render them powerless so that they might not obstruct the onward progress of 'the market'. In a curious way, though, it presaged the debate of the 1990s over the degree of autonomy to be accorded to sub-state units within the UK, and the referenda of 1997. One end-point may well be the emergence of four nations, plus a degree of multiculturalism within them. If this then leads to some degree of devolution, what environmental implications might all these changes have, given the centrifugal effect of EU membership? That there is regional difference in landscape character produced by a combination of cultural and environmental factors is acknowledged but the charting of it has not so far produced anything that might govern, at any scale.[17] The map produced by the Countryside Commission in 1996 'charting the character of the entire English countryside' aims to define local character with a view to sensitive development rather than spreading 'bland uniformity'.[18] There are 181 regions on the map, each accompanied by a description of landscape, wildlife and natural features (Fig. 9.1). The (largely North American) post-modernist advocates of 'bioregionalism' might gain great encouragement from this map but even for them it would be an unlikely base for district council boundaries. English Nature in 1996 produced another regionalisation of England: of 'Natural Areas' with 'constituent (sic) Character Areas, which contain features of common interest for nature conservation and landscape . . .'[19] A 'Natural Area' is defined as a geographical unit derived from the characteristic wildlife, underlying geology, soils, land use and culture of the different parts of England. Hence, plans to reorganise the whole visitor experience at Stonehenge also envisage restoring arable land to calcareous flower-rich downland, including habitat for the stone curlew (*Burhinus oedicnemus*), now down to a few dozen pairs. What is missing from most of these attempts at drawing character on a conventional map is any sense of the dynamic: what are the intensities of land and water use, what are the pressures for the conversion of semi-natural habitats, for example?[20] The maps seem only reluctantly to acknowledge that almost every square centimetre of ground in these islands had been altered as a consequence of human activities.

At another spatial level, the UK is bound into a series of treaties concerned with the environment. Some are relatively local, such as the EU Directives on bird habitats;[21] others are much more wide-ranging and none more so than the Rio Accords of 1992, updated at various meetings thereafter, to stabilise and then reduce the emission of 'greenhouse gases'. Successful pursuit of that aim may well require some alterations in lifestyle rather than simply yet more technology, and the announcement in 1998 that the government was to promulgate an integrated public transport policy, so as to reduce motor car use, may have been a first step.[22] How to tackle in an integrated manner what the Chairman of Scottish Natural Heritage called in 1996 the current plagues of Scotland: red deer, greylag geese and windfarms, is a different kind of challenge.[23] Among other things, Nature Conservation's emphasis on areal protection is likely to undergo severe tests in a warming climate.

The overall impression at the end of the century seems to be that of an understanding of environment in London, Edinburgh and Cardiff that would have been at home in the nineteenth century. There is a collective sigh about big business and its

FIGURE 9.1 *'The character of England'. A map of regions of landscape, wildlife and natural features produced by the Countryside Commission and English Nature in 1996. Each region has a name, a list of which can be found in the original source. It is descriptive and to some extent explanatory but is essentially a snapshot in time rather than a picture of change.*
Source: The Countryside Agency, The State of the Countryside 1999, *Cheltenham: The Countryside Agency, 1999, p. 46.*

tumbling haste for profits, a faith that any technology must be embraced at once, and a distinct deafness to other choices. Only in the case of the 'greenhouse effect' is there any sign that our natural surroundings are important. This, to be repetitive, puts the future environment of Great Britain largely in the hands merely of technological change.

Does it matter?

I want to argue that it does, a great deal. There are two aspects to this: first, the way in which environment can be the scene for the practice of justice to these islands' inhabitants; secondly, our relations with it can be an example to the wider world community of how a *modus vivendi* between a post-industrial society and nature might be achieved. Just as we were early into the industrial revolution, we could try to lead into a different kind of revolution altogether.

Justice at home

The relationship of people to their land is bound to reflect the values of the society as well as the characteristics of the land itself. In Great Britain, the strong class division of society is apparent in the way that, for example, large areas of both town and country are fenced off into private status. That this need not be so can be seen from Sweden where most open land is accessible to all, subject to certain quite specific conditions of behaviour. It seems unrealistic to suppose that this could work in Great Britain except perhaps on moorlands and other unenclosed land. But it is not impossible for the people to have more say in all forms of land use, by bringing farming and forestry into the planning ambit, from which they have largely been excluded even after the 1947 Act. The way in which rural landowners assume that they have a perpetual right to determine their own land uses and other practices in a way that people in the towns do not is widely seen as unfair.

There is, though, another aspect of justice that must be considered. We often talk of justice as if it were a quality that can be applied only to humans. There is no denying that some jurists and philosophers would argue that only contracting parties have standing before the law and nature cannot enter such contracts. This rather overlooks the possibility that nature might be represented rather as a legal minor might be, so trees could have standing in courts of law. Underlying this discussion is the question of moral standing and moral concern. The treatment of animals, for example, altered considerably in the period 1600–1800, as Sir Keith Thomas has shown[24] and is still capable of change: the public protests over the export of veal calves in the early 1990s is one example and the proposals by the UK that Europe should phase out battery chicken production makes this clear. So the extension of moral concern beyond the human, started in the seventeenth century, carried on by Jeremy Bentham ('not, can they speak, but can they suffer?', he asked) and emplaced in legislation since the nineteenth century, is capable of still more extension. Hence, justice for the non-human components of our surroundings might allow them to flourish away from sequestered reserves and enjoy status as neighbours rather than as elements of the tourist gaze.

Setting an example

The position of the UK in the world is a difficult one. After losing our empire, we have had difficulty finding a political place in the world. Conservative governments have talked, in revealing metaphors, of 'punching above our weight' and expressed this in hostility to the EU and in a reliance (not always reciprocated) on the 'special relationship' with the USA. Labour governments have lacked the courage to stop spending money on nuclear weapons. Yet the UK continues to attract attention. It is still a strong partner in the Commonwealth of nations which were once colonies and dominions, an organisation which in the 1990s was still attracting new members. It is a major tourist destination, and while only a small proportion of foreign visitors stray beyond London, Edinburgh, Stratford and Oxford, they could carry home impressions of how things are done just as they remember the policemens' helmets and the awful state of the London Underground. Highly influential, too, is the status of English as *the* world language, even if the version that is most widespread is the dialect of one of our more powerful ex-colonies. The homeland of Shakespeare, though, can still be an object of respect.

I want to suggest that if Great Britain were in the vanguard of promoting a society–nature relationship that was shifted in favour of a greater esteem for the non-human, it would have wide repercussions. At present, the UK is not in the lead for adopting even the relatively mild 'green' measures which late capitalism has brought to, for example, the Netherlands and Germany. We stand nowhere in more radical measures. This is in spite of the commitments made to 'sustainability' in the Rio Accords of 1992 and their subsequent elaborations. Now, the concept of 'sustainability' is a enigmatic one to define and has become something of a shibboleth that has to be mentioned in contexts such as this book to show that the user is one of the good guys. No matter that the concept has all kinds of transhistorical and philosophical difficulties: if it proposes conserving the stock of natural assets, avoids damaging the regenerative capacity of semi-natural ecosystems, promotes social justice and avoids imposing added risks or costs on future generations, then it has a higher value than our current condition. It does not propose freezing practices or landscapes in their present state, nor does it desire change simply as a way for producers to make profits by manipulating the demands for fashion. The remarkable thing about such a different path is how un-radical it can look: the Town and Country Planning Association's report *Planning for a Sustainable Environment* of 1993 has no outlandish proposals and is concerned about the quality of people's lives and about social justice: there is no privileging of foxes here.[25] There is no reason why this should be 'the road not taken'.

Real consequences

In 10,000 years much has happened. Some of the changes discussed have come about irrespective of human activities, though none (we might guess) have gone unremarked by humans who have then made their own meanings of events like climatic change and the rises and falls of sea-level. This class of processes is still important: even if some climatic change at global scale is human-driven, other cycles are probably beyond our reach. There is no cosmic law that says that the glacial-interglacial fluctuations of the Pleistocene are over for ever: we might be living in an interglacial period with only a few thousand years to go before the polar bears close in on the northern suburbs of Glasgow and snowy owls perch on the runway light posts at Heathrow.[26]

For the moment, the view from all our windows is of a human-influenced scene, no matter whether we live in Upper Holloway or in Caithness. If we are fortunate there is something good from both nature and humanity; if we are less so, then probably the built environment dominates, though I would not like to live in one of the remoter islands of the Hebrides. The salmon leaping in the River Wear outside my study window, together with the occasional sight of a kingfisher, will do nicely, thank you, for the everyday, along with central heating and running water. The problems, then, are those of finding a route to long-term change which does not entail the adoption of a Utopia. Such visions of heaven nearly always involve trampling over somebody's life in order to impose a singular divination – and it is for their own good, of course. If you cannot see the summit of the hill, it is very easy to take a wrong route upwards but there is the further complication that the hilltop is in fact an illusion since it will have been itself subject to reconstruction (for example, by technological change and by the redefinition of our own goals) as we climb. So a radical openness to the future is needed. But then how do we tell which changes in our environmental relationships are creative and which are destructive? I can only repeat what is above: no Utopias of the de-industrialised, rural idyll, heritage-rich, local sufficiency-on-boiled-turnips kind: the time for those to be even possible has passed. Instead, there has to be a gradual shuffle through the muddle of conflicting values, poor information, good data not believed by decision-makers and the knowledge that in many cases there is only the least poor choice and the dignity, if any, resides in having a choice at all – something very poor people do not have. Equally, there may be the fortunate but paralysing choice between good things: if there is to be good public access to open land but at the expense of a fragile plant, or a fragile rural economy, who is to benefit and what are the bases for decision? In all this, the knowledge of other possibilities is critical, which is one of the reasons for having Universities, whose staff are or ought to be paid to think differently.

Above all, we need continually to be reminding ourselves that we are a rich nation and that we are members of a very affluent people's club. Thus we have the freedom to adopt a precautionary principle which can be encapsulated also as the 'least regrets' strategy: the avoidance or reduction of risks to the environment before specific environmental hazards are encountered.[27] This is, above all, a way of coping with uncertainty in which there are variables such as:

- changes in the natural systems;
- realisation by scientists that their models of these systems are non-linear and hence poor at prediction;
- variability in the ability of decision-makers to understand the information given to them;

- difficulties in distinguishing substitutable for non-substitutable environmental assets at a time of rapid cultural and technological change; and
- irregular distribution of the courage to adopt the seemingly rational course of action, as distinct from, for example, the politically acceptable.

We have sane and acceptable measures of the kind set out in the Town and Country Planning Association report[28] which could form the basis for this kind of policy. If we do not adopt them, it is because the people are the problem, not the environment. There might be two strands to this. The first is that the governments (at all levels from the parish council to the UN) are incapable of understanding the changes or of putting them into effect. The second is that the people are so devoted to their short-term material conditions that they resist changes in their habits. A spontaneous 'greening' of the British population seems unlikely, though a great deal can be done, as has been seen in, for example, Oregon, California, The Netherlands and Germany. We might suspect that people would shift their behaviour more if not subjected to the constant pressures to consume which emanate from large-scale industry. 'I shop, therefore I am' is the humorous version of a personal identity which is defined in terms of consumption of goods and services that leave a deep environmental footprint, at home and abroad.

If individuation and fragmentation are the distinct outcome of all these years of change, what might supersede it? It is difficult to imagine conventional politics doing so; 'environmentalism' is important but nowhere near crucial; religion has never appealed greatly to the British people as a guide to behaviour;[29] and to be again repetitive, that leaves technology as the critical mechanism. History gives us only partial clues about the reasons for the formation of economic identity. If it has lessons then they are of unforeseen change: hunter-gatherers in later Mesolithic Britain would have asserted that their way of life was 'sustainable'. In AD 1700 there would have been little sign that the agricultural dominance of the nation was to be overtaken so quickly and comprehensively by coal-based industrial life-ways. The actual Britain of the 1990s was little foreseen by the futurologists of the 1950s: 'electricity too cheap to meter' is not how many pensioners see it now. Not only do people's expectations change, but technological developments can spread very quickly as in the adoption rates of household appliances and TV, for example. We have only seen the tip of the iceberg of micro-electronics and biotechnology; nanotechnology has scarcely started to crystallise.

What might have been

It is always fascinating to think of the counterfactual history: what our land might have been like had certain events turned out differently. At a gross level, Great Britain would be very different indeed had the industrial revolution never taken off here. Even if we had been its cradle, but then exported it without adopting it then perhaps we would have remained pre-industrial.[30] Being conquered by France in the Napoleonic Wars and forced into a kind of Western European version of Algeria might (just) have brought that about. In contrast, suppose that we had had a Bolshevik Revolution in 1917 or 1918 and then adopted the 'total conquest of nature' attitude fostered by Lenin and Stalin. The legacy of industrial pollution, for example, would be much greater than we still now face. In response to the recreation needs of the *nomenklatura*, there might nevertheless be extensive wild areas where large mammals could be hunted. Much of Scotland might have been set aside as a forbidden zone for that purpose.

Those are broad-scale fantasies. Closer to actuality is the question of whether environmental history might have been different given the same basic economic legacy but with different standards of morality and legislation which were more environmentally tender all round. If, let us dream, the standards of the State of Oregon of today had been applied in Britain from 1500 onwards? Here are some 'could have beens', in no special order:

- The conservation of more woodland from the sixteenth century onward, so that the nineteenth century decline (to the lowest level in Europe) was not so devastating to the woodland stock. The surge in exotic plantations after World War I would not then have been so urgent.
- More protection for farmers from the economics of intensification in the twentieth century, especially after *c.*1960. This would perhaps have meant a nationalised loans scheme for equipment so that the debt burden would have been light enough to allow them to farm more extensively and employ more people.
- A sharper profile for statutory nature conservation (including the cash for compulsory purchase) after the 1949 Act.
- Control over the movement of open land into urban uses (especially housing) before the 1947 Act and its inadequate predecessors: say from 1918 onwards.
- Earlier recognition of the vulnerability of fish stocks and firm control of fishing effort and the by-catch: perhaps as early as the 1890s.
- The earlier loss of the social cachet of hunting and shooting (say immediately after 1918) so there was less manipulation of moorlands, woodlands and farmlands for sport and less cruelty to stags, foxes and hares together with less soil and nitrogen loss from upland vegetation.

There is perhaps some interest in the appearance of items from the years between 1880 and 1918, which is so often reckoned as a transition period in the UK and one in which many traditions were invented. Out of this and later periods, though, sprang most of the positive things which are chronicled through the relevant parts of the book and so we should not take away any kind of totally negative reading of our history any more than, quite clearly, we could impose a triumphalist view.

The consideration of the counterfactual does highlight the lack of the inevitability of any historical sequence and as such throws a sharply critical light on notions of 'sustainability'. At the most basic, an economy and a set of lifestyles which depend so much upon non-renewable supplies of fossil fuels cannot be maintained in the long run. The alternatives for maintaining our urban-industrial lifestyle are thus to adopt nuclear power to the level of, for example, France or Belgium, or to develop the 'alternative' energy sources as fast as possible and to accompany that with massive programmes of energy conservation. The latter, while seeming 'green' (and eventually contributing to lowering our carbon emissions, though most such technologies are energy and capital intensive at their beginnings) brings about environmental change: installations to harness the wind and the tides are large in size and foreign to most of our landscapes, for example.

Most discussions of sustainability have centred either on biodiversity or on some reckoning of the maintenance of the capital value of 'natural' assets. Biodiversity in Great Britain is rarely the basis of an economy the way it is in savanna Africa or could be in tropical lowland forests, so that it is unlikely to be costed and if it were would be of quite low value except perhaps as a regional recreation resource. Maintaining capital value tends to revolve around the further discoveries of oil and gas in the EEZ and so as much as anything is a measure of putting off the time of doing anything radical. In any case we cannot tell what is likely to come above the horizon of technological change in the medium term. So unless 'sustainability' has a time frame attached to it (and preferably not more than one of 25 years) then it is, in our case, more or less meaningless.

Any policy should be set in the frame of our global linkages (Plate 9.1). If, for example, there is a global need to reduce CO_2 emissions by 50 per cent by 2050 then the UK's per capita place is to reduce our current total emission of 550 million tonnes to 67 million tonnes, a reduction of 88 per cent. An interim 30 per cent cut by 2010 is deliverable by moving to 'alternative' energy sources such as wind, wave and solar power, making our homes at least 30 per cent more energy efficient and reducing the amount of energy expended in personal and commercial transport. The next 58 per cent seems to elude comment even by the more radical environmental reform groups. The reduction in energy use would have to be paralleled by cuts in the use of wood by 73 per cent, 15 per cent in water use and 27 per cent in production from agricultural land, to achieve the same share for other resources.[31]

Not fare well, but fare forward . . .

The main lesson of history, therefore, is change. Very little has remained constant for long: the main period of apparent stasis was that of the simpler societies of hunter-gatherer type. The more complex a society is, the more constant effort it has to put into maintaining stability and thus it is prone to unforeseeable transformations, people being human. There is no retreat from our present levels of complexity and so we have to work with them. It certainly seems as if since the nineteenth century we (like most of the West) have had most of our lifeways largely controlled by technological developments, albeit our metaphysics have been pre-programmed to accept rather than reject them. Yet the world as constituted by nature has limits as a human habitat. Few of them are obvious in Great Britain but they are there in the 'full world' of which we are an inescapable part, and from time to time they make themselves apparent: if not at home then certainly on the screen in the corner. Britain's long involvement with the

PLATE 9.1   *An oil platform in the North Sea. It is emblematic of many environmental linkages, not least those of global connections in the material world and of the Western acceptance of technology-based growth as the normal course of history.*
Photograph: Mark Tasker, courtesy of JNCC.

outside world makes it imperative that while we struggle to find a way of life within our own limits, we have also to fit into those confines which the rest of the world faces. John Donne's 'no man is an island' has acquired the status of cliché but few pieces of writing put it better.[32] And yet: nothing in history tells us that bells may only be used for mourning: they may ring in celebration, too. Donne's most recent heir is perhaps Dylan Thomas, and since Thomas was the first poet I ever seriously read, and as Eliot is too lacking in warmth at this point, I would prefer to set as our aspiration the shafts of sunlight that illuminate the resonances of his 'Prologue' to the *Collected Poems* of 1952:

> Hark! I trumpet the place
> From fish to jumping hill.[33]

The changing environment from the air

The aerial photograph has changed our perception of our environments, not least because it can bring into one frame so much of both past and present. The following pages present a small selection of easily interpreted pictures, each of which relates primarily to one of the chronological sections of the book. Behind that sharpest focus there is the history of the land and water up until that time and then there is the more recent change which has transformed some or sometimes most of it.

Phase	Picture number and location
Hunter-gatherer time	App. 1.1 The North York Moors
Solar-based agriculture (prehistoric)	App. 1.2 Holne Moor Reaves, Dartmoor
Solar-based agriculture (typical)	App. 1.3 Laxton open fields, Northamptonshire
Solar-based agriculture (consequences)	App. 1.4 Woburn, Bedfordshire
Solar-based agriculture (typical)	App. 1.5 Wicken Fen, Cambridgeshire
Industrialisation	App. 1.6 Ironbridge Gorge, Shropshire
Industrialisation	App. 1.7 Merthyr Tydfil, South Wales
Post-industrial uses	App. 1.8 The Isle of Rum, Hebrides, Scotland
Post-industrial economies	App. 1.9 The city of Edinburgh, Lothian, Scotland

APP. 1.1 THE NORTH YORK MOORS

The North York Moors, the most easterly upland in Britain, is based on limestone and sandstone rocks of Jurassic age. The summit plateau, about 1300ft OD, is covered by moorland which is now mostly managed for sheep and grouse. This is an entirely man-made landscape produced by a long history of human exploitation and management of the land. This is a vertical air photograph (June 1994) of part of the high central watershed of the North York Moors, with north towards the upper right-hand corner.

Between ten and five thousand years ago the North York Moors were covered firstly by coniferous forest and then by mixed deciduous woodland. The Mesolithic people who lived there were hunter-gatherers with a distinctive tool kit of microlithic flint

implements and weapons. The peat deposits on the moors preserve evidence, in the form of charcoal layers and flints of the Mesolithic period, which indicates that these people frequently fired the forests to drive game or to keep open areas to attract animals to the browsing and grazing. As on most uplands, the removal of the trees allowed peat growth. Since the North York Moors are relatively dry, the depth and spread of this blanket peat is restricted compared with, for example, the Pennines. Use of the unenclosed land in agricultural and industrial times has always involved combinations of burning and grazing so that the forest never had a chance to recolonise. If those influences were to stop then the upland would largely be covered with woodland after a few hundred years.

The stream between the road and the track traverses country shown by palaeoecological research to have been woodland subject to human manipulation during Mesolithic times. The present landscape shows many aspects of nineteenth-century environmental manipulation: the Rosedale railway (top left), and the patches of moor-burn which continue the methods of grouse moor management started in the mid-nineteenth century. To the north-east of the track (the line of the Lyke Wake Walk, a twentieth-century recreational phenomenon), more or less uniform regeneration of heather after the 1976 fire can be seen. White patches and streaks indicate soil erosion. So the basic lineament of the open environments is Mesolithic in origin but all subsequent phases of human use have added to the picture: agricultural, industrial and the most recent time of heavy recreational use and (since this is a national park) attempts to conserve the heather moorland.

Reference: I. G. Simmons, *The Environmental Impact of Later Mesolithic Cultures*, Edinburgh: Edinburgh University Press, 1996.
© Cambridge University Collection of Air Photographs.

App. 1.2 Holne Moor Reaves, Dartmoor

Reaves are long stone wall boundaries constructed in the mid-Bronze Age to divide pasture land on Dartmoor. Reave systems (patterns of parallel rows) vary in size, and the two largest on Dartmoor each exceed 3000 ha: the Dartmeet system of which Holne Moor is a part is 6.25 km long. Reaves are classed as coaxial field systems; that is, they have one dominant axial boundary which is clearly seen in all the axes and these are met at right angles by transverse boundaries. The number and density of transverse boundaries is variable. Reaves are 'terrain oblivious', that is, they run straight, regardless of physical features but where the line carries them along a steep valley side they, of necessity, follow the hillside contour. An outer, framing, boundary can also be seen, sweeping down almost to be touched by the intake end of the reservoir. All these are made visible by the presence of the remains of the walling structures that have been labelled 'reaves'. In a series of excavations from 1976 to 1986 of both reaves and their associated buildings Andrew Fleming established their date as mid-Bronze Age, 1700/1600 BC.

Although long known to the farmers of Dartmoor, reaves were 'discovered' by Revd J. H. Mason, Rector of Widecombe, and first described in 1825 by Thomas Northmore in Besley's *Exeter News*. Mason and Northmore identified the reaves as land divisions but subsequent writers confused them with trackways. The result was that their social and economic significance was obscured.

The extent of the reave systems on Dartmoor indicates a social system in which forward planning and large-scale construction programmes were both possible and sustainable over a long period of time. Fleming suggests that the reaves indicate land divisions among five distinct communities on the moor. Each system is associated with a round stone circle and would appear to belong to a large unit, possibly an extended family or larger group of people. There is pollen evidence for cultivation of cereals and beans and arable farming may have been both extensive and intensive, taking place alongside the pastoral economy. Dartmoor was not marginal farmland at this

time and Fleming suggests that the Bronze Age people there were part of a powerful elite grown rich on the trade in copper and tin which centred on the south-west.

The photograph shows a March 1985 vertical photograph of part of Holne Moor on Dartmoor. Venford Reservoir and the River Dart are conspicuous but the lattice of Bronze Age fields is also visible, especially to the left (that is, west) of the reservoir. They were built in an environment which inherited a great deal of open terrain from Mesolithic and Neolithic times. Holne Woods line the banks of the Dart: they are examples of woodland environments that have persisted on slopes too steep for clearance. That does not, however, mean that such woods were unmanaged. They were an integral part of the medieval economy, yielding underwood, fuel and timber as well as pasturage; now their biological diversity justifies their designation as part of a national nature reserve.

Reference: A. Fleming, *The Dartmoor Reaves: Investigating Prehistoric Land Divisions*, London: Batsford, 1988.
© Cambridge University Collection of Air Photographs.

App. 1.3 Laxton open fields, Nottinghamshire

This photograph is perhaps as much frozen in time as any of the sequence presented here. The pre-medieval agriculture is invisible and there have not been enormous changes since the picture was taken in 1949. The village of Laxton preserves a working remnant of the open-field communal farming system which became widespread in the early Middle Ages, in which the farming regime was agreed at the twice-yearly Manor Court. Laxton is on a heavy clay soil and it was often in areas of these more demanding soils that the open-field system lasted longest. At Laxton, although the meadows were enclosed in 1720 and four separate farms were laid out on the edges of the parish, the main three-field system survived. By 1906 Earl Manvers owned all of

the open-field system but although he altered the size and layout of the open fields he did not enclose them and the ancient communal farming system persisted. The vicar of that time, the Revd C. Collinson, publicised Laxton's open-field system in *Country Life* magazine. Interest in the archaic survival grew and by the 1920s its preservation was a matter of public debate. In 1938 C. S. and C. S. Orwin published *The Open Fields* and the importance of Laxton was further impressed upon the public mind.

The picture is an aerial oblique of part of the Laxton open fields in July 1949. The differences in crop between unenclosed portions of the fields give us hints of the farm environment of many areas of England that had open-field systems in the Middle Ages and, sometimes, beyond. Although the woodlands look very modern, they occupy the low watershed between Laxton and the next parish, which was a common place for areas of 'waste'. The stream now flows through enclosed and improved grassland and would in earlier times have been part of the common meadow which was a vital resource for stock feed: its hay crop was essential to feed the beasts through the winter. As such, use by stock was carefully regulated by the court and its boundaries carefully maintained. One lesson is that this was no paradise for wildlife, though fallow fields would have attracted lapwings, for example. On a bleak day in late December now, even a sparrow would excite a 'twitcher'.

The 6th Earl Manvers persuaded the Ministry of Agriculture to buy and maintain the fields in perpetuity. Despite this the government threatened to sell Laxton in 1979. The farmers were joined by a vigorous press campaign to protect the open fields and it was therefore sold to the Crown Estates Commission who promised to maintain the fields and the customs attached to them. There is a small visitor centre in the village and signboarded walks among the fields are open to the public.

Reference: J. V. Beckett, *A History of Laxton, England's Last Open-Field Village*, Oxford: Blackwell, 1989.
© Cambridge University Collection of Air Photographs.

APP. 1.4 WOBURN, BEDFORDSHIRE

Solar-powered agriculture can produce surpluses which allow the powerful to display their wealth in landscape and environmental alteration. Hence, Woburn Abbey, the seat of the Dukes of Bedford, represents one of the most developed expressions of English landscape gardening with a park of some 5000 acres in extent. Humphrey Repton, who landscaped the gardens at Woburn in 1804, described the delights of landscape gardening as depending upon 'Congruity, Utility, Order, Symmetry, Picturesque Effect, Intricacy, Simplicity, Variety, Novelty, Contrast, Continuity, Association, Grandeur, Approbation, Animation and The Seasons' E. Hyamns (1971), *Capability Brown and Humphrey Repton*, New York: Charles Scribner's Sons, pp. 115–207. This confused and self-contradictory list covers every aspect of his work at Woburn.

The designs for Woburn were apparently more elaborate and extensive than for any other garden he had ever done. Repton's Red Book (his original design book) for Woburn has not survived but it is believed that most if not all of his proposals for the

gardens were adopted. The house is surrounded by formal garden beds and a terrace. The austerity of Capability Brown's imitation of Nature, with the house standing isolated on an open lawn amid carefully planted parkland did not appeal to Repton. The grandeur, splendour and artistry of the house was to be extended into the garden and parkland; it was a gradual transition not an abrupt change from interior to exterior, artificial to natural. In Repton's own words, 'The gardens, or pleasure-grounds near a house, may be considered so many different apartments belonging to its state, its comfort and pleasure.' The 4th Duke had begun importing citrus fruit trees and exotic conifers in 1740. Repton popularised the term 'arboretum' to describe plantations of such exotic trees. He created a variety of diverse garden areas including an American garden, a Chinese garden and an animated garden (a menagerie). This latter is perhaps the aspect of Woburn which has been developed in order to allow the estate to survive financially. The photograph is from the summer of 1948, before Woburn's major development into a tourist attraction and indeed at a time when country houses were being pulled down daily. The 'apartments' near the house are apparent, with only the main block having clear vistas to the parkland with its ponded stream. A more domesticated pool separates the buildings from a formal hedge-maze; the foreground area shows the neglect (dead trees, for example) which the war years and death duties had forced on the owners of such properties. Nothing definitively American (nor indeed Chinese) can be easily identified in this picture, except perhaps a giant redwood.

Reference: T. Williams, *Polite Landscapes: Gardens and Society in 18th Century England*, Baltimore: Johns Hopkins University Press, 1995.

© Cambridge University Collection of Air Photographs.

APP. 1.5 WICKEN FEN, CAMBRIDGESHIRE

Wicken Fen, about 10 miles (16 km) north-east of Cambridge is a piece of relict fenland and a National Trust reserve. The Romans began land reclamation, but the vast network of drainage systems on which today's agriculture depends was mostly established in the seventeenth century. It has been improved and developed ever since. Fens naturally progress to woodland or to raised bog. But here the alkalinity of the water draining from the calcareous uplands generally has prevented the acid conditions necessary for raised bog. The succession to woodland was halted until the end of the nineteenth century by the harvesting of saw-sedge (*Cladium mariscus*) and reeds for thatching and kindling. While the land around Wicken was drained and reclaimed for agriculture this Fen owes its preservation to its use as a 'wash', an area into which excess water from the uplands was diverted to prevent flooding in the low-lying farm lands. At Wicken the development to scrub and to woodland has been arrested at various stages which are now maintained by the artificial management of water levels. This produces a variety of plant and animal habitats, from open water to fen-wood species. Despite the richness of the flora and fauna the area is completely human-made and depends on constant management to maintain the widest possible range of habitats.

Mid-afternoon in June of 1977 throws few shadows (north is to the right) but the texture of scrub and sedge vegetation is markedly different from the surrounding agricultural land near Wicken. The photograph should be studied in conjunction with Figure 4.2, where the use and some of the history of the compartments of the fen are identified. There is little difficulty in spotting the areas of scrub which derive from the cessation of sedge-cutting as an economic use that perpetuated many of the rare species (especially *Lepidoptera*) of the fen. The bushes also conceal much of the evidence for peat-cutting which was also a pre-industrial use of the area. Additional features of interest include the white streak to the west of the fen (that is, above it on the picture) which represents a former stream course, outlined by its silt load among the peats; and the clearly artificial lake with its island and marginal reed-beds. By such

means the Norfolk Broads were created, and linear baulks of peat there recall the arms of the island in this photograph. The lesson seems to be that pre-industrial use allowed habitat and species variety that became competed out under 'natural' succession in the nineteenth and twentieth centuries but which is being restored by active management undertaken in the light of the historical ecology.

Reference: H. Godwin, *Fenland: Its Ancient Past and Uncertain Future*, Cambridge: Cambridge University Press, 1978.
© Cambridge University Collection of Air Photographs.

App. 1.6 Ironbridge Gorge, Shropshire

The Ironbridge Gorge on the River Severn was a cradle of the Industrial Revolution. Its unique concentration of furnaces and foundries fascinated, appalled and attracted writers and artists throughout the eighteenth and nineteenth centuries. In 1708 Abraham Darby had taken over one of the charcoal blast furnaces which had operated

since the early seventeenth century among the many small coalmines around the River Severn. The following year he developed a method of smelting iron using coke, which was a much cheaper fuel. This single achievement contributed hugely to the advancement of industrialisation. The locality supplied all the materials and means of production: coal and iron from nearby mines, ample water power to drive machinery, limestone used as a flux in the smelting process and to make moulds for casting, and river transport to ship manufactures away to markets. The valley was also home to Maw's Tileworks, the largest in the world, and Coalport china.

The iron bridge itself was built under the supervision of Abraham Darby III. It was the first cast-iron civil engineering structure in the world. The bridge opened on New Year's Day 1781 and quickly became one of the wonders of the modern age. In the great Severn flood of February 1795 the Ironbridge was the only bridge on the river to remain unscathed. However, movement of its stone abutments was soon noticed and remedial work was undertaken on and off until 1931 when the bridge was closed to all vehicular traffic. In 1972 work began to insert a reinforced concrete slab on the bed of the river to hold the two abutments in place. The ironworks were gradually taken over by Allied Ironfounders who in 1959 first opened the Abraham Darby furnace site to the public. In 1968 The Ironbridge Gorge Museum Trust was set up: there is a visitor centre and several museum sites throughout about six square miles along the banks of the River Severn.

This 1947 view of the gorge looking westwards shows the gorge before some of the recent development and with some working barges on the river still. Coalbrookdale itself runs up to the right half-way up the photograph and Darby's original site is occupied by the works which have now become a museum. The picture exemplifies well many of the early industrial environments of Great Britain in the sense that it grew in a haphazard and untidy fashion with parcels of land bearing works in among woodland and farmland; houses, cottages and terraces mingled with gardens and smallholdings as is clear from the foreground of this picture. In such places, pollution of air and water was deleterious to most forms of life but the small fragments of scrub and waste would have harboured many birds and insects. There is little reason why most of the areas of woodland should not have been cleared for crops and pasture and so we suspect that ownership has had a considerable role in their continuation; they might perhaps be the descendants of woods managed for charcoal-producing coppice before Darby switched to coked coal.

Reference: N. Cossons and H. Sowden, *Ironbridge: Landscape of Industry*, London: Cassell, 1977. © Cambridge University Collection of Air Photographs.

APP. 1.7 MERTHYR TYDFIL, SOUTH WALES

In 1750 Merthyr Tydfil, on the River Taff, was a small rural village. By 1801 it had become the most populous town in Wales and remained so for the next 60 years. Its growth resulted from the development of the iron smelting industry using locally mined coal and iron ore. As well as being the iron and steel capital of the world Merthyr Tydfil was also the site of the first steam powered railway in 1804. The town exported iron and steel all over the world and the rails for the Trans-Siberian railway were produced here. The four largest ironworks, Cyfarthfa, Dowlais, Penydarren and Plymouth, dominated the town. Development was rapid and uncontrolled, and while a few ironfounders made their fortunes, wage levels for the mass of workers were very low with their housing being largely of poor quality, cramped terraces. Industrial output was in decline by the beginning of the twentieth century and although there was a slight boost to the local economy during World War I, Merthyr decayed so dramatically in the 1930s that it was designated a depressed area. Such was its economic and environmental poverty that complete depopulation was being considered when World War II gave the town a new lease of life.

Open-cast mine workings, colliery tips and derelict foundries still mark the land-scape. The valley sides show the intensity of environmental change brought about

by coal- and iron-based industrialisation. Many of the linear features on the far side are due to extraction, but the heaps near at hand are partially waste tips; the last coal mine closed in 1956. So what we see is the result of many years of gradual change but is dominated by 150 years of rapid, intensive metamorphosis in the service of an 'inorganic' economy. This picture dates from 1966, before the main thrust of reclamation legislation was under way but, equally, before there was much open-cast mining here. Since 1966 reclamation programmes have been carried out on some 514 ha which is about a quarter of the total derelict land at present in need of improvement. Merthyr Tydfil's local plan included 26 separate sites in need of reclamation and is expected to run until 2006. The area is still rich in coal and will be subject to open-cast extraction programmes for the foreseeable future. Many reclamation schemes have been carried through to 'green' the valley and from 1999 a project about 1.5 x 0.25 km will create open space around two lakes in the valley. Not quite a return to an organic economy, since the electricity will mostly come from hydrocarbons. Above the valley there is a typical moorland scene: 'island' farms of improved land, water impoundments and further traces of mining. The open nature of that environment is a prehistoric creation and has been agriculturally maintained, then industrialised; like many uplands it has been useful for water catchment. Beyond, to the south-east, are more of the ridges which separate the industrialised valleys of South Wales.

Reference: T. Bayliss-Smith and S. Owens (eds), *Britain's Changing Landscape from the Air*, Cambridge: Cambridge University Press, 1990.

© Cambridge University Collection of Air Photographs.

App. 1.8 The Isle of Rum, Hebrides, Scotland

The Isle of Rum, of 26,400 acres (10,692 ha), is wholly owned by Scottish Natural Heritage and since April 1957 has been a National Nature Reserve. The declared intention has been to improve the degraded land by restoring woody cover and increasing biological productivity. The white-tailed sea eagle was successfully reintroduced in the 1970s and bred on the island in 1985 for the first time since the nineteenth century. A study of the population dynamics of red deer has been going on since 1957 as has a programme to reintroduce species of trees and shrubs natural to the island. An enormous number of birds are found on Rum and it became a Special Protection Area under the EU directive on the conservation of wild birds. It is impossible to be sure if red deer are native to Rum or were introduced by humans.

Archaeological excavations from 1984 to 1986, at Kinloch, at the head of Loch Scresort, produced evidence of human settlement in Mesolithic times. At this time the island was covered by open heathland shrubs, juniper, bog myrtle and light woodland with copses of birch and hazel. How far the Mesolithic people modified the environment on Rum is impossible to say on present evidence. Certainly subsequent human occupation wrought great changes, including the pasturing of sheep, for which purpose a population of 400 people was 'cleared' in the early nineteenth century. The sheep were all removed in 1957, which makes Rum a unique upland environment.

Loch Scresort is the only large inlet on Rum and at its head is Kinloch Castle with amenity planting from the nineteenth century around it. A hint of the wildness of the coast which attracts the island's seabird populations is gained from the foot of the picture. The upland plateau is typical of many parts of Scotland which are not actually mountainous although the ploughing here was (August 1974) in the service of planting native trees rather than the conifers then favoured by the Forestry Commission. Since the castle and environs are virtually the only settlement on the island, it becomes an unusually flexible natural laboratory. To maintain a given level of plant and animal species it will probably be necessary always to manage the plant and animal communities on Rum. Since human beings have exploited its resources since Mesolithic times

there has probably never been a 'natural' self-regulating balance of plant and animal populations. Like many upland areas in Great Britain, there is visible or near-visible evidence of the environmental relationships of most periods since hunter-gatherer times.

Reference: F. Fraser Darling and J. Morton Boyd, *The Highlands and Islands*, New Naturalist, London: Collins, 1964.

© Cambridge University Collection of Air Photographs.

APP. 1.9 THE CITY OF EDINBURGH, LOTHIAN, SCOTLAND

Environmental change is not all about rural areas. The city (as shown in this vertical photograph of Edinburgh in 1999) represents a very thorough make-over of the pre-existing terrain. But few cities in Europe obliterate the underlying topography

completely. They may create their own distinctive urban climate and, within that frame, a series of microclimates and habitats for such biota as birds, insects and lichens. Here, the volcanic mass of Arthur's Seat (bottom right) has, like open space in cities such as Newcastle-upon-Tyne, stayed open for symbolic as well as physical reasons; its foot however has succumbed to the gentrification of the formal Palace garden. The rock mass of the Castle overshadows the reclaimed loch which has become gardens alongside the largely sunken track of the railway between Haymarket and Waverley stations; there is no indication though of what preceded the construction of the geometric New Town. Like its equivalent in Bath, this development appears to have been laid across the land as if unrolled like a carpet. But we can see that a river still runs along its margin, if not through it. The southern edge of the large open space which houses the Royal Botanic Garden and several sports grounds is visible in the top left corner: the presence of a cricket pitch suggests that no concessions to environmental determinism of the climatic kind are made in this city. Sheer topography however has preserved most of the open space in the city's central parts.

Reference: A. J. Youngson, *The Making of Classical Edinburgh*, Edinburgh: Edinburgh University Press, 1966.

© 2000 The XYZ Digital Map Company Ltd (www.xyzmaps.com) and Wildgoose.

Notes

1 Introduction

1. The act of writing takes place in time, of course. But it can try to look at all past times with an equal gaze, so to that extent it is 'timeless'. The author stands at the centre of an arc, viewing both prehistory and last month through the same set of instruments. Ideally, anyway.

2. There is an interesting contrast between two 'environmental histories' of Great Britain from the same year: H. J. Fleure, *A Natural History of Man in Britain*, London: Collins, 1951, uses historical periods as its framework, getting to AD 1066 only in Chapter 7. Jacquetta Hawkes, *A Land*, London: Cresset Press, 1951 (and Readers Union, 1953), is highly geology-determinist until Chapter 8.

3. D. Worster, *Dust Bowl: The Southern Plains in the 1930s*, New York: OUP, 1979, is the outstanding example. There are more detailed studies on, e.g., the Hudson River or Lake Tahoe, mostly for the late nineteenth and twentieth centuries.

4. G. Sumner, 'Wales', in D. Wheeler and J. Mayers (eds), *Regional Climates of the British Isles*, London and New York: Routledge, 1997, pp. 131–57.

5. D. Wheeler, 'North-east England and Yorkshire', in Wheeler and Mayers, *Regional Climates*, pp. 158–80.

6. K. Lambeck, 'Late Devensian and Holocene shorelines of the British Isles and North Sea from models of glacio-hydro-isostatic rebound', *Journal of the Geological Society* 152, 1995, pp. 437–48.

7. This world-view need not be the only one. In medieval Europe, the world was a book in which the word of God might be read. Every leaf might be sacred or might be a source of terror: the ambiguity of forests can be seen in that light. Hua-yen Buddhism sees all phenomena as temporary crystallisations in a cosmic net of energy. There is no simple causality of the A→B type and everything is connected to, and interpenetrative with, everything else.

8. H. Kearney, *The British Isles: A History of Four Nations*, Cambridge: Cambridge University Press, 1989; reprinted as Canto edition in 1995.

9. D. Coleman and J. Salt, *The British Population: Patterns, Trends and Processes*, Oxford: Oxford University Press, 1992.

10. See, e.g., N. Fenech, *Fatal Flight: The Maltese Obsession with Killing Birds*, London: Quiller Press, 1992.

11. A. Wilson, *Late Call*, London: Secker and Warburg, 1964; Penguin Books 1968. I was tempted to quote Betjeman on Slough but we Old Paludians feel (a little) touchy about that.

2 Hunter-gatherers and fisherfolk: 10,000 to 5500 BP

1. For example, M. Bell and M. J. C. Walker, *Late Quaternary Environmental Change*, London: Longman, 1992; N. Roberts, *The Holocene: An Environmental History*, Oxford: Blackwell, 1989. Journals include relevant numbers of *British Archaeological Reports* and

also *The Holocene, Journal of Archaeological Science* and *Vegetation History and Archaeo-botany*.

2. I. Shennan, 'Holocene crustal movements and sea-level changes in Great Britain', *Journal of Quaternary Science* 4, 1989, pp. 77–89.

3. I. Shennan, 'Sea-level and coastal evolution: Holocene analogues for future changes', *Coastal Zone Topics: Process, Ecology and Management* 1, 1995, pp. 1–9.

4. D. Hall and J. Coles, *Fenland Survey: An Essay in Landscape and Persistence*, English Heritage Archaeological Report 1, London: English Heritage, 1994.

5. The latest wild horse remains come from the years 9770–10,200 BP; reindeer from as late as 8300 BP on Assynt (Sutherland). Moose remains include an antler with a ^{14}C date of 3900 BP from Galloway, which contrasts with the English dates from nine millennia BP. One authority in the 1920s suggested that the moose survived beyond the ninth century AD in Scotland. See A. C. Kitchener and C. Bonsall, 'AMS radiocarbon dates for some extinct Scottish mammals', *Quaternary Newsletter* 83, 1997, pp. 1–11. There is even a polar bear skull from Assynt, dated to *c*.19,000 BP which was approximately the time of the glacial maximum of the Devensian.

6. B. Huntley, 'The post-glacial history of British woodlands', in M. A. Atherden and R. A. Butlin (eds), *Woodland in the Landscape: Past and Future Perspectives*, Leeds: Leeds University Press, 1997, pp. 9–25.

7. C. Smith, *Late Stone Age Hunters of the British Isles*, London and New York: Routledge, 1992.

8. P. Rowley-Conwy, 'Sedentary hunters: the Ertebølle example', in G. Bailey (ed.), *Hunter-Gatherer Economy in Prehistory: A European Perspective*, Cambridge: Cambridge University Press, 1983, pp. 111–26; Rowley-Conwy, 'Season and reason: The case for a regional interpretation of Mesolithic settlement patterns' in G. L. Peterkin, H. Bricker and P. A. Mellars (eds), *Hunting and Animal Exploitation in the Later Palaeolithic and Mesolithic of Eurasia*, Archaeological Papers of the American Anthropological Association 4, 1993, pp. 179–88; M. Zvelebil, 'Plant use in the Mesolithic and its role in the transition to farming', *Proceedings of the Prehistoric Society* 60, 1994, pp. 35–74.

9. This is *Alces alces* and is sometimes referred to as elk. North American common usage, however, uses 'elk' for the equivalent of the European red deer. 'Moose' is unequivocal.

10. The earliest excavations are detailed in J. G. D. Clark, *Excavations at Star Carr*, Cambridge: Cambridge University Press, 1971, 2nd edn; the most comprehensive re-interpretation of the bones is A. J. Legge and P. Rowley-Conwy, *Star Carr Revisited*, London: Birkbeck College, University of London, 1988; recent environmental work in P. Day, 'Preliminary results of high-resolution palaeoecological analyses at Star Carr, Yorkshire', *Cambridge Archaeological Journal* 3, 1993, pp. 129–33; and complications over the dating in P. A. Mellars, '"Absolute" dating of Mesolithic human activity at Star Carr, Yorkshire: New palaeoecological studies and identification of the 9600 BP radiocarbon "plateau"', *Proceedings of the Prehistoric Society* 69, 1994, pp. 417–22. The whole is gathered up in P. A. Mellars and P. Dark, *Star Carr in Context*, Cambridge: McDonald Institute Monographs, 1998 [Dark was née Day].

11. J. J. Wymer, 'Excavations at the Maglemosian sites at Thatcham, Berkshire, England', *Proceedings of the Prehistoric Society* 28, 1962, pp. 329–61.

12. J. M. Coles, 'Morton revisited', in A. O'Connor and D. V. Clarke (eds), *From the Stone Age to the 'Forty-Five'*, Edinburgh: John Donald, 1983, pp. 9–18.

13. P. A. Mellars, *Excavations on Oronsay*, Edinburgh: Edinburgh University Press, 1987.

14. C. Caseldine and D. Maguire, 'Late glacial and early Flandrian vegetation changes on northern Dartmoor, south-west England', *Journal of Biogeography* 13, 1986, pp. 255–64; R. L. Jones, 'The activities of Mesolithic man: Further palaeobotanical evidence from north-east Yorkshire', in D. A. Davidson and M. L. Shackley (eds), *Geoarchaeology: Earth Sciences and the Past*, London: Duckworth, 1976, pp. 355–67; M. B. Bush, 'Early

Mesolithic disturbance: A force on the landscape', *Journal of Archaeological Science* 15, 1988, pp. 453–62; A. G. Smith and E. W. Cloutman, 'Reconstruction of Holocene vegetation history in three dimensions at Waun-Fignen-Felen, an upland site in South Wales', *Philosophical Transactions of the Royal Society* series B, 322, 1988, pp. 159–219.

15. Day, 'Preliminary results'.

16. J. H. Tallis, 'Forest and moorland in the South Pennine uplands in the mid-Flandrian period. III. The spread of moorland – local, regional and national', *Journal of Ecology* 79, 1991, pp. 401–15.

17. J. H. Tallis and V. R. Switsur, 'Forest and moorland in the south Pennine uplands in the mid-Flandrian period. II. The hillslope forests', *Journal of Ecology* 78, 1990, pp. 857–83 (876).

18. Caseldine and Maguire, 'Late glacial and early Flandrian vegetation changes'.

19. P. D. Moore, 'The origin of blanket mires, revisited', in F. M. Chambers (ed.), *Climate Change and Human Impact on the Landscape*, London: Chapman and Hall, 1993, pp. 217–36.

20. A. U. Mallik, C. H. Gimingham and A. A. Rahman, 'Ecological effects of heather burning. I. Water infiltration, moisture retention and porosity of surface soil', *Journal of Ecology* 72, 1984, pp. 767–76.

21. This need not have been a very lengthy process. There is an eighteenth-century description of part of Scotland (Lochbrun) in which the author tells of the transformation of a 'Firr' wood into a bog: 'In the year 1651 . . . This little plain was at that time covered over with a firm standing wood; which was so very Old that the trees had no green leaves but the Bark was totally thrown off . . . the outside of these standing white Trees, and for the space of one Inch inward, was dead white Timber; but what was within that, was good solid timber, even to the very Pith, and as full of Rozin as it could stand in the wood. Some Fifteen years after . . . there was not so much as a Tree . . . but in place whereof . . . was all over a plain green ground, covered with a plain green Moss . . . the green Moss (there in the *British* language called Fog) had overgrown the whole timber . . . and they said none could pass over it, because the scurf of the Fog would not support them. I would needs try it; and accordingly I fell in to the Arm-Pits . . .' (George, Earl of Cromertie, FRS, 'An account of the mosses in Scotland', *Philosophical Transactions of the Royal Society of London*, XXVII, 1711, pp. 296–301. Most of the spelling has been modernised).

22. J. M. Coles and B. Orme, '*Homo sapiens* or *Castor fiber?*', *Antiquity* 57, 1983, pp. 95–102; J. M. Coles, 'Further thoughts on the impact of beaver on temperate landscapes', in S. Needham and M. Macklin (eds), *Alluvial Archaeology in Britain*, Oxbow Monographs 27, Oxford: Oxbow Books, 1992, pp. 93–99.

23. G. B. Corbet (ed.), *Animals of Western Europe*, London: G. T. Foulis and Co., 1966.

24. P. C. Buckland and K. J. Edwards, 'The longevity of pastoral episodes of clearance activity in pollen diagrams: The role of post-occupation grazing', *Journal of Biogeography* 11, 1984, pp. 243–49.

25. Smith and Cloutman, 'Reconstruction of Holocene vegetation history', p. 168.

26. D. L. Clark, 'Mesolithic Europe: The economic basis', in G. de G. Sieveking, I. H. Longworth and K. E. Wilson (eds), *Problems in Economic and Social Archaeology*, London: Duckworth, pp. 449–81.

27. K. J. Edwards and G. Whittington, 'Vegetation change', in K. J. Edwards and I. B. M. Ralston (eds), *Scotland: Environment and Archaeology 8000 BC–AD 1000*, Chichester: Wiley, 1997, pp. 63–82.

28. T. L. Affleck, K. J. Edwards and A. Clarke, 'Archaeological and palynological studies at the Mesolithic pitchstone and flint site of Auchareoch, Isle of Arran', *Proceedings of the Society of Antiquaries of Scotland* 118, 1988, pp. 37–59.

29. J. A. Fossitt, 'Late Quaternary vegetation history of the Western Isles of Scotland', *New Phytologist* 132, 1996, pp. 171–96.

30. R. L. Jones, 'The impact of early man in coastal plant communities in the British Isles', in

M. Jones (ed.), *Archaeology and the Flora of the British Isles: Human Influence upon the Evolution of Plant Communities*, Oxford University Committee for Archaeology Monograph no. 14; BSBI Conference Report no. 19, 1988, pp. 96–106.

31. R. Tipping, 'The determination of cause in the generation of major valley fills in the Cheviot Hills, Anglo-Scottish border', in Needham and Macklin, *Alluvial Archaeology in Britain*, pp. 111–21.

32. C. Tilley, *A Phenomenology of Landscape*, Oxford and Providence RI: Berg Press, 1994.

33. O. V. Ovsyannikov and N. M. Terebikhin, 'Sacred space in the culture of the Arctic regions', in D. L. Carmichael, J. Hubert, B. Reeves and A. Schanche (eds), *Sacred Sites, Sacred Places*, One World Archaeology 23, London and New York: Routledge, 1997, pp. 44–81.

34. R. Bradley, *The Significance of Monuments*, London and New York: Routledge, 1998.

3 Shafts of sunlight: agriculturalists

1. M. Aston and C. Gerrard, 'Unique, traditional and charming: The Shapwick Project, Somerset', *Antiquaries Journal* 79, 1999, pp. 1–58.

2. K. E. Barber, 'Peat-bog stratigraphy as a proxy climate record', in A. F. Harding (ed.), *Climatic Change in Later Prehistory*, Edinburgh: Edinburgh University Press, 1982, pp. 103–13.

3. J. J. Blackford, K. J. Edwards, A. J. Dugmore, G. T. Cook and P. C. Buckland, 'Icelandic volcanic ash and the mid-Holocene Scots pine (*Pinus sylvestris*) pollen decline in northern Scotland', *The Holocene* 2, 1992, pp. 260–65.

4. H. H. Lamb, 'Climate from 1000 BC to 1000 AD', in M. Jones and G. W. Dimbleby (eds), *The Environment of Man: The Iron Age to the Anglo-Saxon Period*, BAR British Series 87, Oxford: BAR, 1981, pp. 53–65.

5. D. D. Gilbertson, J.-L. Schwenninger, R. A. Kemp and E. J. Rhodes, 'Sand-drift and soil formation along an exposed North Atlantic coastline: 14,000 years of diverse geomorphological, climatic and human impacts', *Journal of Archaeological Science* 26, 1999, pp. 439–69.

6. E. Williams, 'Dating the introduction of food production into Britain and Ireland', *Antiquity* 63,1989, pp. 510–21.

7. A. Whittle, 'The first farmers', in B. Cunliffe (ed.), *The Oxford Illustrated Prehistory of Europe*, Oxford and New York: Oxford University Press, 1994, pp. 136–66.

8. R. Tipping, 'The form and fate of Scotland's woodlands', *Proceedings of the Society of Antiquaries of Scotland* 124, 1994, pp. 1–54; K. J. Edwards and I. B. M. Ralston (eds), *Scotland: Environment and Archaeology, 8000 BC–AD 1000*, Chichester: Wiley, 1997.

9. There is an excellent collection of plans and photographs in P. J. Fowler, *The Farming of Prehistoric Britain*, Cambridge: Cambridge University Press, 1983.

10. J. Richards, *The Stonehenge Environs Project*, London: Historic Buildings and Monuments Commission for England, 1990; C. Gingell, *The Marlborough Downs: A Late Bronze Age Landscape and Its Origins*, Devizes, Wilts.: Wiltshire Archaeological and Historical Society, 1990.

11. J. L. Davies, 'The early Celts in Wales', in M. J. Green (ed.), *The Celtic World*, London and New York: Routledge, 1995, pp. 671–700.

12. O. Rackham, 'Ancient woodland and hedges in England', in S. R. J. Woodell (ed.), *The English Landscape: Past, Present and Future*, Oxford: Oxford University Press, 1985, pp. 68–105.

13. B. W. Cunliffe, 'Man and Landscape in Britain 6000 BC–AD 400', in Woodell, *English Landscape*, 1985, pp. 48–67.

14. H. Cook, 'Groundwater development in England', *Environment and History* 5, 1999, pp. 75–96.

15. O. Rackham, *The History of the Countryside*, London: Dent, 1986.

16. C. Saunders, *The Forest of Medieval Romance: Avernus, Broceliande, Arden*, Cambridge: D. S. Brewer, 1993.

17. Fowler, *Farming of Prehistoric Britain*.

18. K. J. Edwards, 'Models of mid-Holocene forest farming for north-west Europe', in F. M. Chambers (ed.), *Climatic Change and Human Impact on the Landscape*, London: Chapman and Hall, 1993, pp. 133–45.

19. A. C. Kitchener and C. Bonsall, 'Further AMS radiocarbon dates for extinct Scottish mammals', *Quaternary Newsletter* 88, 1999, pp. 1–10. The aurochs survived until AD 1627 in continental Europe. There are claims that wild cattle were known in the Pembrokeshire Hills in the reign of Elizabeth I, just as there is a story of a wolf near Dolgellau in 1785. It seems best to stick to dates for which there is some firm evidence: the cattle might of course have been feral.

20. P. J. Reynolds, 'Rural life and farming' in Green, *Celtic World*, pp. 176–209.

21. P. J. Reynolds, *Iron Age Farm: The Butser Experiment*, London: Colonnade Books, 1979, p. 100.

22. L. Dumayne, 'Invader or native? – vegetation clearance in northern Britain during Romano-British time', *Vegetation History and Archaeobotany* 2, 1993, pp. 29–36; L. Dumayne and K. Barber, 'The impact of the Romans on the environment of northern England: Pollen data from three sites close to Hadrian's Wall', *The Holocene* 4, 1994, pp. 165–73; G. Whittington and K. J. Edwards, '*Ubi solitudinem faciunt pacem appellant*: The Romans in Scotland, a palaeoenvironmental contribution', *Britannia* 24, 1993, pp. 13–25.

23. M. Jones, *England before Domesday*, London: Batsford, 1986.

24. K. Dark and P. Dark, *The Landscape of Roman Britain*, Stroud, Gloucs.: Sutton Publishing, 1997.

25. M. G. Bell, 'The effects of land use and climate on valley sedimentation', in A. H. Harding (ed), *Climatic Change in Later Prehistory*, Edinburgh: Edinburgh University Press, 1982, pp. 127–42.

26. R. P. C. Morgan, *Soil Erosion and Conservation*, London: Longman, 1986.

27. M. G. Bell, 'The prehistory of soil erosion' in J. Boardman and M. G. Bell (eds), *Past and Present Soil Erosion*, Oxbow Monographs 22, Oxford: Oxbow Books, 1992, pp. 21–35.

28. F. Shotton, 'Archaeological inferences from the study of alluvium in the lower Severn-Avon valleys', in S. Limbrey and J. G. Evans (eds), *The Effect of Man on the Landscape: The Lowland Zone*, CBA Research Report 21, London: CBA, 1978, pp. 27–31.

29. M. Macklin and J. Lewin, 'Holocene river alluviation in Britain', *Zietschrift für Geomorphologie* Suppl Bd 88, 1993, pp. 109–22.

30. A. J. Howard and M. G. Macklin, 'A generic geomorphological approach to archaeological interpretation and prospection in river valleys: A guide for archaeologists investigating Holocene landscapes', *Antiquity* 73, 1999, pp. 527–41.

31. R. H. Squires, 'Flandrian history of the Teesdale rarities', *Nature* 229, 1971, pp. 43–4; Squires, 'Conservation in Upper Teesdale: Contributions from the palaeoecological record', *Transactions of the Institute of British Geographers* NS 3, 1978, pp. 129–50.

32. L. Dumayne and K. E. Barber, 'The impact of the Romans on the environment of northern England: Pollen data from three sites close to Hadrian's Wall', *The Holocene* 4, 1994, pp. 165–73.

33. J. Darrah, *Paganism in Arthurian Romance*, Woodbridge: Boydell, 1994.

34. M. G. Bell, 'People and nature in the Celtic world', in Green, *Celtic World*, pp. 145–58.

35. M. J. Green, 'The gods and the supernatural', in Green, *Celtic World*, pp. 465–88.

36. T. Darvill, 'The historic environment, historic landscapes, and space-time-action models in landscape archaeology', in P. J. Ucko and R. Layton (eds), *The Archaeology and Anthropology of Landscape*, One World Archaeology 30, London and New York: Routledge, pp. 104–18; the direct quotation is from R. Bradley, *The Significance of*

Monuments, London and New York: Routledge, 1998, p. 163.

37. C. Tilley, *A Phenomenology of Landscape: Places, Paths and Monuments*, Oxford and Providence, RI: Berg Publishers, 1994.
38. C. Tilley, *Phenomenology of Landscape*.
39. R. Hutton, *The Pagan Religions of the Ancient British Isles: Their Nature and Legacy*, Oxford: Blackwell, 1991.
40. R. Hutton, *The Stations of the Sun: A History of the Ritual Year in Britain*, Oxford: Oxford University Press, 1996. The spectacular Up-Helly Aa procession at Lerwick in Shetland was apparently started by the local Total Abstinence Society in 1870. A pity, really.
41. T. Hughes, *Gaudete*, London: Faber and Faber, 1977, p. 145.
42. Cunliffe, 'Man and Landscape in Britain'.

4 Closed and open systems, AD 550 to AD 1700

1. This is not the place for environmental theology. But many Christian traditions were hierarchical in which it was permissible for humans to try to control nature.
2. D. Coleman and J. Salt, *The British Population: Patterns, Trends, and Processes*, Oxford: Oxford University Press, 1992; R. M. Smith, 'Geographical aspects of population change in England 1500–1730', in R. A. Dodgshon and R. A. Butlin (eds), *An Historical Geography of England and Wales*, 2nd edn, London: Academic Press, 1990, pp. 151–80; S. G. Lythe and J. Butt, *An Economic History of Scotland 1100–1939*, Glasgow: Blackie, 1975.
3. Lamb, 'Climate from 1000 BC to 1000 AD'. Use of the occurrence of the pollen of *Empetrum nigrum* in upland peats as an indicator has suggested that 1100–1250 was especially dry. (J. H. Tallis, 'The pollen record of *Empetrum nigrum* in southern Pennine peats: Implications for erosion and climate change', *Journal of Ecology* 85, 1997, pp. 455–65.) Some interruptions occurred: *The Anglo-Saxon Chronicle* for 1032 records 'wildfire' (probably lightning without thunder) 'which did damage everywhere' and which might have been the consequence of a volcanic eruption elsewhere in the world. Caution over the identification of a medieval warm period is expressed by A. Ogilvie and G. Farmer, 'Documenting the medieval climate', in M. Hulme and E. Barrow (eds), *Climates of the British Isles: Present, Past and Future*, London and New York: Routledge, 1997, pp. 112–33.
4. J. H. Tallis, 'Climate and erosion signals in British blanket peats: The significance of *Rhacomitrium lanuginosum* remains', *Journal of Ecology* 83, 1995, pp. 1021–30.
5. H. H. Lamb, *Weather, Climate and Human Affairs*, London and New York: Routledge, 1988.
6. J. M. Grove, *The Little Ice Age*, London and New York: Methuen, 1988.
7. 9 Jan. 1685: 'I went across the Thames on the ice, now become so thick as to bear not only streets of booths, in which they roasted meat, and had divers shops of wares quite across as in a town, but coaches, carts and horses passed over . . . the fowls, fish and birds, and all our exotic plants and greens universally perishing. Many parks of deer were destroyed, and all sorts of fuel so dear that there were great contributions to preserve the poor alive.'
8. P. Brimblecombe, 'A meteorological service in fifteenth century Sandwich', *Environment and History* 1, 1995, pp. 241–49.
9. J. A. Steers, *The Coastline of Scotland*, Cambridge: Cambridge University Press, 1973.
10. B. M. S. Campbell, 'People and land in the Middle Ages, 1066–1500', in Dodgshon and Butlin, *Historical Geography of England and Wales*, pp. 69–122; H. E. Hallam (ed.), *The Agrarian History of England and Wales*, vol. 2, *1042–1350*, Cambridge: Cambridge University Press, 1988. J. Thirsk, 'The farming regions of England', in J. Thirsk (ed.), *The Agrarian History of England and Wales*, vol. 4, *1500–1640*, Cambridge: Cambridge University Press, 1967, pp. 1–112.
11. R. A. Dodgshon, 'Strategies of farming in the western highlands and islands of Scotland

prior to crofting and the clearances', *Economic History Review* 46, 1993, pp. 679–701; R. A. Dodgshon and E. G. Olsson, 'Productivity and nutrient use in eighteenth-century Scottish highland townships', *Geografiska Annaler* 70 B, 1988, pp. 39–51.

12. R. A. Dodgshon, 'The ecological basis of Highland farming 1500–1800 AD', in H. H. Birks, H. J. B. Birks, P. E. Kaland and D. Moe (eds), *The Cultural Landscape – Past, Present and Future*, Cambridge: Cambridge University Press, 1988, pp. 139–51.

13. I. Whyte and K. Whyte, *The Changing Scottish Landscape 1500–1800*, London and New York: Routledge, 1991.

14. M. Buchanan, 'Introduction', in M. Buchanan (ed.), *St Kilda: The Continuing Story of the Islands*, Edinburgh: HMSO/Glasgow Museums, 1995, pp. xi–xxiv.

15. C. Thomas, 'A cultural-ecological model of agrarian colonisation in upland Wales', *Landscape History* 14, 1992, pp. 37–50.

16. J. McDonnell, 'Medieval assarting hamlets in Bilsdale, north-east Yorkshire', *Northern History* 22, 1986, pp. 268–79.

17. R. Hodgson, 'Medieval colonization in northern Ryedale, Yorkshire', *Geographical Journal* 135, 1969, pp. 44–54.

18. R. Trow-Smith, *A History of British Livestock Husbandry to 1700*, London: Routledge and Kegan Paul, 1957.

19. Whyte and Whyte, *Changing Scottish Landscape*.

20. R. A. Donkin, 'Changes in the Early Middle Ages', in H. C. Darby (ed.), *A New Historical Geography of England before 1600*, Cambridge: Cambridge University Press, 1976, vol. 1, pp. 75–135.

21. J. V. Beckett, *A History of Laxton: England's Last Open-Field Village*, Oxford: Basil Blackwell, 1989.

22. E. Pollard, M. Hooper and N. W. Moore, *Hedges*, New Naturalist Series, London: Collins, 1974, Rackham, 'Ancient woodland and hedges'. The formula approach has been strongly criticised since it lumps togther all kinds of hedges and management regimes, not least the practice of buying shrubs (in the sixteenth and seventeenth centuries, for example) to create mixed hedges. See R. Muir, 'Hedgerow dating: A critique', *Naturalist* 121, 1996, pp. 59–64.

23. E. Salisbury, *Weeds and Aliens*, New Naturalist Series, London: Collins, 1961.

24. Shakespeare, *Coriolanus*, III, i, 69.

25. M. Jones, 'The arable field: A botanical battleground', in M. Jones (ed.), *Archaeology and the Flora of the British Isles*, Oxford: Oxford University Committee for Archaeology Monograph 14 and Botanical Society of the British Isles Conference Report 19, 1988, pp. 86–92.

26. R. Perry, *Wildlife in Britain and Ireland*, London: Croom Helm, 1978.

27. H. C. Darby, *The Changing Fenland*, Cambridge: Cambridge University Press, 1983.

28. H. Godwin, *Fenland: its Ancient Past and Uncertain Future*, Cambridge: Cambridge University Press, 1978.

29. Hall and Coles, *Fenland Survey*; H. E. Hallam, *The New Lands of Elloe: A Study of Early Reclamation in Lincolnshire*, Leicester: Leicester University Press, 1954; Hallam, *Settlement and Society: A Study of the Early Agrarian History of South Lincolnshire*, Cambridge: Cambridge University Press, 1965.

30. H. F. Cook and H. Moorby, 'English marshlands reclaimed for grazing: A review of the physical environment', *Journal of Environmental Management*, 38, 1993, pp. 55–72.

31. D. Woodward, 'Straw, bracken and the Wicklow whale: The exploitation of natural resources in England since 1500', *Past and Present* 159, 1988, pp. 43–76.

32. B. M. S. Campbell, 'Agricultural progress in medieval England: Some evidence from eastern Norfolk', *Agricultural History Review* 31, 1983, pp. 26–46.

33. J. N. Pretty, 'Sustainable agriculture in the Middle Ages: The English manor', *Agricultural History Review* 38, 1990, pp. 1–19. Further, the traditional manure heap was uncovered

and unfloored and would probably lose over half of its original nitrogen content before it was spread on the fields. See W. S. Cooter, 'Ecological dimensions of medieval agrarian systems', *Agricultural History* 52, 1978, pp. 458–77. The quantitative approach to nitrogen flows is taken up by R. S. Loomis, 'Ecological dimensions of medieval agrarian systems: An ecologist responds', *Agricultural History* 52, 1978, pp. 478–87.

34. H. S. A. Fox, 'Some ecological dimensions of medieval field systems', in K. Biddick (ed.), *Archaeological Approaches to Medieval Europe*, Kalamazoo, MI: Western Michigan University, Medieval Institute Publications, 1984, pp. 119–58.

35. Dodgshon, 'Strategies of Farming'.

36. O. Rackham, *Trees and Woodland in the British Landscape*, London: Dent, 1976.

37. D. Hooke, *Anglo-Saxon Landscapes of the West Midlands: The Charter Evidence*, Oxford: British Archaeological Reports British Series 95, 1981, p. 344.

38. C. Dyer, Hanbury: *Settlement and Society in a Woodland Landscape*, Occasional Papers 4 of the Department of English Local History, Leicester: Leicester University Press, 1991.

39. A. McDonald, 'Changes in the flora of Port Meadow and Picksey Mead, Oxford', in Jones, *Archaeology and the Flora*, pp. 76–85; J. Sheail, *Historical Ecology: The Documentary Evidence*, Huntingdon: Institute of Terrestrial Ecology, 1980.

40. H. F. Cook, 'Field-scale water management in Southern England to A.D. 1900', *Landscape History* 16, 1994, pp. 53–66; J. Sheail, 'The formation and maintenance of water-meadows in Hampshire, England', *Biological Conservation* 3, 1971, pp. 101–6.

41. B. Vesey-Fizgerald, *The Vanishing Wildlife of Britain*, London: MacGibbon and Kee, 1969.

42. R. A. Donkin, 'The Cistercian order and the settlement of northern England', *Geographical Review* 59, 1969, pp. 403–16.

43. In C. J. Glacken, *Traces on the Rhodian Shore*, Berkeley and Los Angeles: University of California Press, 1967, p. 309.

44. J. Sheail, *Rabbits and Their History*, Newton Abbott: David and Charles, 1971.

45. M. Bailey, *A Marginal Economy? East Anglian Breckland in the Later Middle Ages*, Cambridge: Cambridge University Press, 1989.

46. D. Stocker and M. Stocker, 'Sacred profanity – the theology of rabbit breeding and the symbolic landscape of the warren', *World Archaeology* 28, 1996, pp. 265–72; C. J. Bond, 'Settlement, land use and estate patterns on the Failand Ridge, North Somerset: A preliminary discussion', in D. Hooke and S. Burnell (eds), *Landscape and Settlement in Britain AD 400–1066*, Exeter: Exeter University Press, 1995, pp. 115–52.

47. M. L. Parry, *Climatic Change, Agriculture and Settlement*, Folkestone: Dawson, 1978.

48. I. Armit, *The Late Prehistory of the Western Isles of Scotland*, Oxford: British Archaeological Reports British Series 221, 1992; K. J. Edwards, 'Human impact on the prehistoric environment', in T. C. Smout (ed.), *Scotland since Prehistory*, Aberdeen: Scottish Cultural Press, 1993, pp. 17–27; T. C. Smout, 'Woodland history before 1800', in Smout, *Scotland since Prehistory*, pp. 40–9.

49. D. Hooke, 'Early Cotswold woodland', *Journal of Historical Geography* 4, 1978, pp. 333–41; Hooke, *The Anglo-Saxon Landscape: The Kingdom of the Hwicce*, Manchester: Manchester University Press, 1985.

50. D. Hooke, 'Pre-Conquest woodland: Its distribution and usage', *Agricultural History Review* 37, 1989, pp. 113–29.

51. W. Linnard, *Welsh Woods and Forests: History and Utilization*, Cardiff: National Museum of Wales, 1982.

52. R. Bromwich (ed. and trans), *Dafydd ap Gwilym: A Selection of Poems*, London: Penguin Books, 1985, 'New Edition'.

53. Rackham, *Trees and Woodland*, p. 65. The Great Hall of the Bishop of Hereford also has some fine timbers of medieval date, emulating a stone structure.

54. G. F. Peterken, *Natural Woodland: Ecology and Conservation in Northern Temperate Regions*, Cambridge: Cambridge University Press, 1996.

55. Rackham, *Trees and Woodland*, p. 73.

56. O. Rackham, 'The oak tree in historic times', in M. G. Morris and F. H. Perring (eds), *The British Oak: Its History and Natural History*, Faringdon, Berks: E. W. Classey for the Botanical Society of the British Isles, 1974, pp. 62–79.

57. O. Rackham, *Ancient Woodland: Its History, Vegetation and Uses in England*, London: Edward Arnold, 1980, p. 153.

58. R. Pott, 'The effects of wood pasture on vegetation', *Plants Today* September–October 1989, pp. 170–75.

59. J. Sheail, *Historical Ecology: The Documentary Evidence*, Huntingdon: ITE, 1980.

60. Quoted in Rackham, *Trees and Woodland*, 3rd edn, 1995, p. 62.

61. Anon, 'In search of veterans', *English Nature Magazine* 35, 1998, p. 11.

62. A. S. Mather, 'Pre-1745 land use and conservation in a Highland glen: An example from Glen Starthfarrar, North Inverness-shire', *Scottish Geographical Magazine* 86, 1970, pp. 159–70.

63. A. Dent, 'The last wolves in Yorkshire: And in England?', 43, 1982, pp. 17–26.

64. E. L. Jones, 'The bird pests of British agriculture in recent centuries', *Agricultural History Review* 20, 1972, pp. 107–25.

65. Glacken, *Traces on the Rhodian Shore*, p. 320.

66. *Sir Gawain and the Green Knight*, lines 740–46, trans. P. Stone, Harmondsworth: Penguin Classics, 1959. The locality of much of this poem can be traced directly from evidence in the verse, though Bertilak's castle is imaginary. The Green Chapel can be identified as a cleft in the Millstone Grit (Lud's Church Cave) near Gradbach in Staffordshire. See R. W. V. Elliott, *The Gawain Country*, Leeds Texts and Monographs NS 8, Leeds: University of Leeds School of English, 1984.

67. Geoffrey Chaucer, *The Book of the Duchess* (ed.) H. Phillips, Durham: Durham Medieval Texts no. 3, revised edition 1980. The very free translation is my own, with the help of the glossary from the Phillips edition.

68. D. Hooke, 'The Warwickshire Arden: The evolution and future of an historic landscape', *Landscape History* 10, 1988, pp. 51–9.

69. A great consolation to those of us who cannot see a large puddle without wanting to make a drainage channel with whatever footwear we currently have on.

70. N. D. G. James, *A History of English Forestry*, Oxford: Blackwell, 1981. The unpopularity of the Forests required this great number of enforcers, as did deer forests and grouse moors at their apogee before World War I.

71. D. J. Keene, 'Rubbish in medieval towns', in A. R. Hall and H. K. Kenward (eds), *Environmental Archaeology in the Urban Context*, London: CBA Research Report 43, 1982, pp. 26–30.

72. T. P. O'Connor, 'Pets and pests in Roman and medieval Britain', *Mammal Review* 22, 1992, pp. 107–13; P. L. Armitage, 'Small mammal faunas in later medieval towns', *Biologist* 32, 1985, pp. 65–71.

73. P. Hembry, *The English Spa 1560–1815: A Social History*, London: Athlone Press, 1990.

74. P. Brimblecombe, 'Early urban climate and atmosphere', in Hall and Kenward, *Environmental Archaeology in the Urban Context*, pp. 10–25.

75. Quoted in W. H. Te Brake, 'Air pollution and fuel crises in preindustrial London, 1250–1650', *Technology and Culture* 16, 1975, pp. 337–59. The diseases listed are from Evelyn's *Fumifugium*, also of 1661.

76. Te Brake, 'Air pollution'.

77. J. Eccleston and P. Eccleston, *A History and Geology of Staithes*, Staithes, Yorks: The Authors, 1998.

78. D. Woodward, '"Swords into ploughshares": Recycling in pre-industrial England', *Economic History Review* ser. 2, 38, 1985, pp. 175–91.

79. D. A. Jackson and R. F. Tylecote, 'Two new Romano-British iron-working sites in Northamptonshire', *Britannia* 19, 1988, pp. 275–98.

80. S. Barker, 'The history of the Coniston woodlands, Cumbria, UK', in K. J. Kirby and C. Watkins (eds), *The Ecological History of European Forests*, Wallingford and New York: CAB International, 1998, pp. 167–83.

81. G. T. Lapsley, 'The account roll of a fifteenth-century iron master', *English Historical Review* 14, 1899, pp. 509–29. The Bishop delegated the management of his woods in this part of Weardale to Robert Kirhous the 'Irynbrenner' for seven years. Robert might convert woods to charcoal, 'except ooke, esshe, holyn, wodapiltre and crabtre; and also except all the wood that wol be fynes or beemes the whuch alleway shall be fellyd thavys the consell of the forsaid Bysshop afore that the colyers make coke theer'.

82. Rackham, *Ancient Woodland*; he draws on the work of M. W. Flinn and of G. Hammersley, which can be found in summary form in J. R. Harris, *The British Iron Industry 1700–1850*, London: Macmillan Educational, 1988, ch. 2.

83. *Calendar of Close Rolls*, 1237, *Forest of Dean*.

84. E. J. T. Collins, 'Woodlands and woodland industries in Great Britain during and after the charcoal iron era', in J.-P. Metalie (ed.), *Protoindustries et Histoires des Forêts*, Toulouse: Les Cahiers de l'ISARD 3, 1992, pp. 109–20.

85. Whyte and Whyte, *Changing Scottish Landscape*; T. C. Smout and F. Watson, 'Exploiting semi-natural woods, 1600–1800', in T. C. Smout (ed.), *Scottish Woodland History*, Edinburgh: Scottish Cultural Press, 1997, pp. 86–150.

86. A. Fleming, 'Wood pasture and the woodland economy in medieval Swaledale', in M. A. Atherden and R. A. Butlin (eds), *Woodland in the Landscape: Past and Future Perspectives*, Leeds: Leeds University Press, 1998, pp. 26–42.

87. M. Cressey and G. Coles, 'The identification of non ferrous metal mining in N.E. Islay: An environmental, historical and archaeological perspective', *Quaternary Research Association Field Guide: The Quaternary of Islay and Jura*, 1997, pp. 46–55.

88. Woodward, '"Swords into Ploughshares"'.

89. R. B. Manning, *Hunters and Poachers: A Cultural and Social History of Unlawful Hunting in England 1485–1640*, Oxford: Clarendon Press, 1993.

90. J. S. Berrall, *The Garden: An Illustrated History*, Harmondsworth: Penguin Books, 1978.

91. J. Prest, *The Garden of Eden: The Botanic Garden and the Re-Creation of Paradise*, New Haven and London: Yale University Press, 1981.

92. S. Lasdun, *The English Park: Royal, Private and Public*, London: Andre Deutsch, 1991.

93. Hooke, 'Pre-Conquest woodland', p. 125, quotes charters and grants from the ninth century in which their recipients (dukes and monks) are exempted from entertainment of hunters and from the maintenance of hawks and hunting dogs for these *fæstingmen*. It seems quite likely that clearly recognised hunting areas in Anglo-Saxon times influenced the distribution of Royal Forests after the Norman Conquest: see D. Hooke, 'Medieval forests and parks in southern and central England', in C. Watkins (ed.), *European Woods and Forests: Studies in Cultural History*, Wallingford and New York: CAB International, 1998, pp. 19–52.

94. J. M. Gilbert, *Hunting and Hunting Reserves in Medieval Scotland*, Edinburgh: John Donald, 1979.

95. Celia Fiennes, *Through England on a side saddle in the time of William and Mary*. Christopher Morris (ed.), *The Journeys of Celia Fiennes*, London: Gresset, 1949.

96. O. Rackham, *The History of the Countryside*, London: Dent, 1986, p. 358.

97. B. Moss, 'Mediaeval man-made lakes: Progeny and casualties of English social history, patients of twentieth century ecology', *Trans Roy Soc S Afr* 45, 1984, pp. 115–28.

98. G. Whittington and K. J. Edwards, 'A Scottish broad: Historical, stratigraphic, and numerical studies associated with polleniferous deposits at Kilconquhar Loch', in R. A. Butlin and N. Roberts (eds), *Ecological Relations in Historical Times: Human Impact and Adaptation*, RGS/IBG Special Publications no 32, Oxford: Blackwell, 1995, pp. 68–87.

99. K. Walsh, D. O'Sullivan, R. Young, S. Crane and A. G. Brown, 'Medieval land use, agriculture

and environmental change on Lindisfarne (Holy Island), Northumbria', in Butlin and Roberts, *Ecological Relations in Historical Times*, pp. 101–21.

100. Coleman and Salt, *British Population*.

101. J. D. Post, *Food Shortage, Climatic Variability, and Epidemic Disease in Preindustrial Europe: The Mortality Peak in the Early 1740s*, Ithaca and London: Cornell University Press, 1985.

102. J. Thirsk, 'Farming in Kesteven, 1540–1640', in J. Thirsk (ed.), *The Rural Economy of England*, London: The Hambledon Press, 1984, pp. 129–47.

103. Glacken, *Traces on the Rhodian Shore*, p. 173. Emile Mâle, *The Gothic Image: Religious Art in France of the Thirteenth Century*, tr. Dora Nussey, Harper Torchbooks/Cathedral Library, New York: Harper Brothers, 1958.

104. K. Thomas, *Man and the Natural World: Changing Attitudes in England 1500–1800*, London: Allen Lane, 1983.

105. S. Schaffer, 'The earth's fertility as a social fact in early modern Britain', in M. Teich, R. Porter and B. Gustafsson (eds), *Nature and Society in Historical Context*, Cambridge: Cambridge University Press, 1997.

106. *Beowulf*, introd. and, trans. D. Wright, Harmondsworth: Penguin Books, 1957.

5 Building Jerusalem: the eighteenth century

1. J. Dibble, *C. Hubert H. Parry: His Life and Music*, Oxford: Clarendon Press, 1992, pp. 483–4. Parry wrote the setting for a patriotic organisation called 'Fight for Right', but in protest against jingoism and propaganda he withdrew support for that organisation and was delighted when it was proposed as the Women Voters' Hymn. (There is a parallel with Elgar and 'Land of Hope and Glory'). The orchestration of Parry's tune often heard was in fact written by Elgar in 1922. We used to sing 'Jerusalem' at school occasionally but I doubt it had much resonance in Slough, where the green fields were vanishing fast and the trading estate, though monotonous, could scarcely be described as dark and satanic. Then again, I was a regular promenader in the 1950s when it seemed quite reasonable to belt it out in the heady atmosphere of the Arena on the last night. Suez (1956) changed all that, though. The other contender for mass patriotism seems to be Holst's tune to the words 'I vow to thee my country' which was adapted in about 1921 from the central tune of 'Jupiter' (the Bringer of Jollity) in his orchestral suite *The Planets*, first played in its entirety in 1920. A. E. F. Dickinson calls the hymn 'that fantastic collocation' (*Holst's Music*, London: Thames Publishing, 1995, p. 123, and see fn. 4 on that page.)

2. D. S. Landes, *The Unbound Prometheus: Technological Change and Industrial Development in Western Europe from 1750 to the Present*, Cambridge: Cambridge University Press, 1969.

3. J. Grattan and M. Brayshay, 'An amazing and portentious summer: Environmental and social responses in Britain to the 1783 eruption of an Iceland volcano', *Geographical Journal* 161, 1995, pp. 125–34.

4. Rackham, *History of the Countryside*, pp. 90–2.

5. H. C. Prince, 'England circa 1800', in H. C. Darby (ed.), *A New Historical Geography of England after 1600*, Cambridge: Cambridge University Press, 1976, pp. 89–164; N. D. G. James, *A History of English Forestry*, Oxford: Blackwell, 1981.

6. C. R. Tubbs, *The New Forest: An Ecological History*, Newton Abbot: David and Charles, 1968.

7. W. Linnard, *Welsh Woods and Forests: History and Utilization*, Cardiff: National Museum of Wales, 1982.

8. J. M. Lindsay, 'The commercial use of woodland and coppice management', in M. L. Parry and T. R. Slater (eds), *The Making of the Scottish Countryside*, London: Croom Helm, 1980, pp. 271–89.

9.   S. Seymour, 'Landed estates, the "spirit of planting" and woodland management in later Georgian Britain: A case study from the Dukeries, Nottinghamshire', in Watkins, *European Woods and Forests*, pp. 115–34.

10.   M. Overton, 'The diffusion of agricultural innovations in early modern England: turnips and clover in Norfolk and Suffolk 1580–1740', *Transactions of the Institute of British Geographers* NS 10, 1985, pp. 205–21.

11.   Marshall wrote this in his *Rural Economy of Norfolk*, published in London in 1787. Quoted by H. C. Darby, 'The age of the improver', in H. C. Darby (ed.), *A New Historical Geography of England after 1600*, Cambridge: Cambridge University Press, 1976, p. 24.

12.   Rackham, 'Ancient woodland and hedges in England'.

13.   H. F. Cook and T. Williamson, *Water Management in the English Landscape – Field, Marsh and Meadow*, Edinburgh: Edinburgh University Press, 1999.

14.   H. F. Cook, 'Field-scale water management in Southern England', *Landscape History* 16, pp. 53–66.

15.   J. E. Denison, 'On the Duke of Portland's water meadows at Clipstone Park', *Journal of the Royal Agricultural Society* 1, pp. 359–70.

16.   Cook, 'Field-scale water management in Southern England'.

17.   Cook, 'Field-scale water management in Southern England'.

18.   E. Pollard, M. D. Hooper and N. W. Moore, *Hedges*, London: Collins, 1974.

19.   Jones, 'Bird pests'.

20.   Smout, *Scotland since Prehistory*; Smout, 'Highland land use before 1800: Misconceptions, evidence and realities', in Smout, *Scottish Woodland History*, pp. 5–23.

21.   F. Fraser Darling, *West Highland Survey*, Oxford: Oxford University Press, 1955, ch. 1.

22.   A. S. Mather, 'Land deterioration in upland Britain', *Progress in Physical Geography* 7, 1983, pp. 210–28.

23.   D. D. Gilbertson, M. Kent, J.-L. Schwenninger, P. A. Wathern, R. Weaver and B. Brayshay, 'The Machair vegetation of South Uist and Barra in the Outer Hebrides of Scotland: Its interacting ecological, geomorphic and historical dimensions', in Butlin and Roberts, *Ecological Relations in Historical Times*, pp. 124.

24.   A. Roberts, 'Midges in a changing Highland environment', in Smout, *Scotland since Prehistory*, pp. 113–24.

25.   M. Williams, 'The enclosure and reclamation of waste land in England and Wales in the eighteenth and nineteenth centuries', *Transactions of the Institute of British Geographers* 51, 1970, pp. 55–69.

26.   A. Bil, 'Transhumance economy, setting and settlement in Highland Perthshire', *Scottish Geographical Magazine* 105, 1989, pp. 158–67.

27.   M. L. Parry, 'Changes in the extent of improved farmland', in Parry and Slater, *Making of the Scottish Countryside*, pp. 177–202.

28.   J. Chapman, 'Parliamentary enclosure in the uplands: The case of the North York Moors', *Agricultural History Review* 24, 1976, pp. 1–17.

29.   J. R. Walton, 'Agriculture and rural society 1730–1914', in Dodgshon and Butlin, *Historical Geography of England and Wales*, pp. 323–50.

30.   An eighteenth-century landowner's view, quoted in M. Brayshay and A. Williams, 'A rough and cold country: Farming and landscape changes in northern Snowdonia since c. 1600', in Butlin and Roberts, *Ecological Relations in Historical Times*, pp. 122–45.

31.   J. A. Perkins, 'The prosperity of farming on the Lindsey uplands, 1813–37', *Agricultural History Review* 24, 1976, pp. 126–43.

32.   Prince, 'England *circa* 1800'. In Gainsborough's much-dissected painting of Mr and Mrs Andrews on their East Anglian estate, the hill in the background looks like a pretty rough scrub. In the original more than in reproductions Mrs A. looks pretty rough as well: 'nobody told me about *this* . . .'

33.   Whyte and Whyte, *Changing Scottish Landscape*.

34. M. Williams, *The Draining of the Somerset Levels*, Cambridge: Cambridge University Press, 1970.

35. B. K. Roberts, 'Rural settlements', in J. Langton and R. J. Morris (eds), *Atlas of Industrializing Britain 1780–1914*, London and New York: Methuen, 1986, pp. 54–9.

36. M. Palmer and P. Neaverson, *Industry in the Landscape, 1700–1900*, London and New York: Routledge, 1994.

37. M. Newsom, *Land, Water and Development: Sustainable Management of River Basin Systems*, 2nd edn, London and New York: Routledge, 1997.

38. J. Alfrey and C. Clark, *The Landscape of Industry: Patterns of Change in the Ironbridge Gorge*, London and New York: Routledge, 1993. Good illustrations of eighteenth- and nineteenth-century iron-making sites can be seen in B. Bracegirdle, *The Archaeology of the Industrial Revolution*, London: Heinemann, 1973, and in A. Briggs, *Iron Bridge to Crystal Palace: Impact and Images of the Industrial Revolution*, London: Thames and Hudson, 1979.

39. A. Burton, *The Canal Builders*, Newton Abbot: David & Charles, 1981.

40. C. Hadfield, *The Canals of Yorkshire and North-east England*, vol. 1, Newton Abbot: David and Charles, 1972.

41. C. Hadfield, *British Canals: An Illustrated History*, Newton Abbot: David and Charles, 1970.

42. N. Thrift, 'Transport and communication 1730–1914', in Dodgshon and Butlin, *Historical Geography of England and Wales*, pp. 453–86.

43. G. M. Binnie, *Early Dam Builders in Britain*, London: Thomas Telford, 1957.

44. J. U. Nef, *The Rise of the British Coal Industry*, 2 vols, London: Routledge, 1932. Re-issued in 1996 by Frank Cass, London.

45. G. Hammersley, 'The charcoal iron industry and its fuel, 1540–1750', *Economic History Review* 26, 1973, pp. 593–613. There is a mild irony in an editorial in *Farming and Conservation* 2 (2), 1995, p. 2, extolling the virtues of charcoal as a source of income for woodland owners. Oliver Rackham points out that simple calculations of the area needed to sustain an iron-works omits the problems of fragmentation of coppice lands and of the less than rational negotiations that might hinder the flow of wood products (Rackham, *Ancient Woodland*, p. 180).

46. J. D. Hunt, *The Figure in the Landscape: Poetry, Painting and Gardening during the Eighteenth Century*, Baltimore: The Johns Hopkins University Press, 1989.

47. S. Lasdun, *The English Park: Royal, Private and Public*, London: Andre Deutsch, 1991.

48. S. Daniels and S. Seymour, 'Landscape design and the idea of improvement 1730–1914', in Dodgshon and Butlin, *Historical Geography of England and Wales*, pp. 487–520.

49. R. Perry, *Wildlife in Britain and Ireland*, London: Croom Helm, 1978; T. Williamson, *Polite Landscapes: Gardens and Society in Eighteenth-Century England*, Stroud: Sutton Publishing, 1995.

50. P. Hembry, *The English Spa 1560–1815: A Social History*, London: Athlone Press, 1990.

51. G. Jackson, *The British Whaling Trade*, London: Adam and Charles Black, 1978.

52. D. Fairfax, 'Man and the basking shark in Scotland', in R. A. Lambert (ed.), *Species History in Scotland: Introductions and Extinctions since the Ice Age*, Edinburgh: Scottish Cultural Press, 1998, pp. 93–106.

53. See Dodgshon and Butlin, *Historical Geography of England and Wales*, p. 236.

54. C. W. J. Withers, 'How Scotland came to know itself: Geography, national identity and the making of a nation, 1680–1790', *Journal of Historical Geography* 21, 1995, pp. 371–97.

55. E. A. Wrigley, *Continuity, Chance and Change: The Character of the Industrial Revolution in England*, Cambridge: Cambridge University Press, 1988.

56. Until the 1880s, almost all petroleum products used in Britain came from the USA but by 1890 there was a serious challenge from Russia, with that nation being the chief supplier for a brief period around 1903; diversification then occurred with supplies from Romania, Dutch East Indies (Indonesia), Mexico and Persia arriving by 1914.

57. P. Bairoch, 'International industrialization levels from 1750 to 1980', *Journal of European Economic History* 11, 1982, pp. 269–333.

58. Glacken, *Traces on the Rhodian Shore*.

59. M. Andrews, *The Search for the Picturesque: Landscape Aesthetics and Tourism in Britain, 1760–1800*, Stanford, CA: Stanford University Press, 1989. The last quarter of the eighteenth century saw a remarkable number of 'tours' of, for example, the Lake District, the North of England, or Scotland which were illustrated with engravings derived from (usually) watercolours painted *in situ*. Turner was famous for these and the tradition continued into the nineteenth century with famous artists like John Sell Cotman (1782–1842). Pictures divide up 'the world' into separate chunks, usually rectangular.

60. J. Brewer, *The Pleasures of the Imagination: English Culture in the Eighteenth Century*, London: Harper Collins, 1987.

6 INDUSTRIAL GROWTH: MATERIAL EMPIRES, 1800–1914

1. P. Bairoch, 'International industrialization levels from 1750 to 1980', *Journal of European Economic History* 11, 1982, pp. 269–333.

2. See, for example, D. E. Allen, *The Naturalist in Britain: A Social History*, 2nd edn, Princeton, NJ: Princeton University Press, 1994. This account starts in the eighteenth century but devotes much of its space to the nineteenth century. The farm of the 1820s planned around a steam engine, the sea front at Blackpool and the magnate's country estate are all outgrowths of industrial development.

3. G. Manley, *Climate and the British Scene*, London: Collins, 1952.

4. H. H. Lamb, *Climate, History and the Modern World*, London: Methuen, 1982, ch. 13.

5. P. Brimblecombe, *The Big Smoke: A History of Air Pollution in London since Medieval Times*, London and New York: Routledge, 1987.

6. B. W. Clapp, *An Environmental History of Britain since the Industrial Revolution*, London and New York: Longman, 1994, ch. 2.

7. R. Hawes, 'The control of alkali pollution in St Helens, 1862–1890', *Environment and History* 1, 1995, pp. 159–72.

8. P. Haining, *The Great British Earthquake*, London: Robert Hale, 1976.

9. J. Winter, *Secure from Rash Assault*, Berkeley, Los Angeles, and London: University of California Press, 1999; K. L. Wallwork, 'Some problems of subsidence and land use in the mid-Cheshire industrial area', *Geographical Journal* 126, 1960, pp. 191–9.

10. J. T. Coppock, 'The changing face of England: 1850–*circa* 1900', in H. C. Darby (ed.), *A New Historical Geography of England after 1600*, Cambridge: Cambridge University Press, 1976, pp. 295–373.

11. P. R. Hobson, 'Patterns of the past: Ancient records and present evidence', in J. R. Packham and D. L. Harding (eds), *Ecology, Management and History of the Wyre Forest*, Wolverhampton: Wolverhampton Woodland Research Group, 1995, pp. 6–22.

12. B. M. S. Dunlop, 'The woods of Strathspey in the Nineteenth and Twentieth centuries', in Smout, *Scottish Woodland History*, pp. 176–89.

13. Rackham, *Ancient Woodland*, p. 324.

14. James, *History of English Forestry*, chs 8–9.

15. M. Jones, 'The rise, decline and extinction of spring wood management in south-west Yorkshire', in C. Watkins (ed.), *European Woods and Forests: Studies in Cultural History*, Wallingford and New York: CAB International, 1998, pp. 55–71.

16. J. Brown, *Agriculture in England: A Survey of Farming, 1879–1947*, Manchester and Wolfboro, NH: Manchester University Press, 1987.

17. J. D. Chambers and G. E. Mingay, *The Agricultural Revolution 1750–1880*, London: Batsford, 1966.

18. R. Grove, 'Coprolite mining in Cambridgeshire', *Agricultural History Review* 24, 1976, pp. 36–43.

19. D. W. Howell, *Land and People in Nineteenth Century Wales*, London, Henley and Boston, MA: Routledge and Kegan Paul, 1978.

20. T. Bayliss-Smith, *The Ecology of Agricultural Systems*, Cambridge: Cambridge University Press, 1978.

21. E. L. Jones, *Agriculture and the Industrial Revolution*, Oxford: Blackwell, 1974.

22. E. J. T. Collins, 'Agriculture and conservation in England: An historical overview, 1880–1939', *Journal of the Royal Agricultural Society of England* 146, 1985, pp. 38–46.

23. Howell, *Land and People*.

24. C. S. Orwin and R. J. Sellick, *The Reclamation of Exmoor Forest*, Newton Abbot: David and Charles, 1970 (first published by Oxford University Press in 1929).

25. M. Williams, 'The enclosure and reclamation of the Mendip Hills 1779–1870', *Agricultural History Review* 19, 1971, pp. 65–81; J. Winter, *Secure from Rash Assault: Sustaining the Victorian Environment*, Berkeley, Los Angeles and London: University of California Press, 1999, ch. 4.

26. T. D. Ford and J. H. Rieuwerts (eds), *Lead Mining in the Peak District*, Bakewell: Peak Park Planning Board, 1968.

27. R. T. Clough, *The Lead Smelting Mills of the Yorkshire Dales*, Leeds: The Author, 1962.

28. In Swaledale (N. Yorks.) grouse apparently replaced red deer as a noble quarry in the eighteenth century. They were hawked and netted until 1725 after which shooting was common, according to A. Fleming, *Swaledale: Valley of the Wild River*, Edinburgh: Edinburgh University Press, 1998, p. 76.

29. S. Tapper, *Game Heritage: An Ecological Review from Shooting and Gamekeeping Records*, Fordingbridge: Game Conservancy, 1992.

30. B. Vesey-Fitzgerald, *British Game*, London: Collins, 1946; P. J. Hudson, *The Red Grouse: The Biology and Management of a Wild Gamebird*, Fordingbridge: Game Conservancy, 1986.

31. D. N. McVean and J. D. Lockie, *Ecology and Land Use in Upland Scotland*, Edinburgh: Edinburgh University Press, 1969.

32. Clapp, *Environmental History of Britain*, p. 70.

33. J. Hassan, *A History of Water in Modern England and Wales*, Manchester and New York: Manchester University Press, 1998.

34. D. P. Butcher, J. Claydon, J. C. Labadz, V. A. Pattinson, A. W. R. Potter and P. White, 'Reservoir sedimentation and colour problems in Southern Pennine Reservoirs', *Journal of the Institute of Water and Environmental Management* 6, 1992, pp. 418–31. Peat erosion had, however, been going on since at least 1500–1700; there seems to have been a second major episode in the Pennines in the 1700s: A. C. Stevenson, V. J. Jones and R. W. Battarbee, 'The cause of peat erosion: A palaeolimnological approach', *New Phytologist* 114, 1990, pp. 727–35.

35. H. F. Cook, 'Groundwater development in England', *Environment and History* 5, 1999, pp. 75–96.

36. R. Perry, *Wildlife in Britain and Ireland*, London: Croom Helm, 1978.

37. T. A. Rowell and H. J. Harvey, 'The recent history of Wicken Fen, Cambridgeshire, England: A guide to ecological development', *Journal of Ecology* 76, 1988, pp. 73–90; H. Godwin, *Fenland: Its Ancient Past and Uncertain Future*, Cambridge: Cambridge University Press, 1978.

38. D. Francis, *A History of World Whaling*, New York: Viking Books, 1990.

39. G. Jackson, *The British Whaling Trade*, London: Adam and Charles Black, 1978.

40. M. Gray, *The Fishing Industries of Scotland, 1790–1914: A Study in Regional Adaptation*, Oxford: Oxford University Press, 1978.

41. *Fisheries Exhibition Literature*, 1887, vol. 4, p. 14.

42. A Royal Commission on Beam Trawling was set up which reported in 1885. Its conclusions on damage to other forms of netting and on stock declines mainly affected Scottish ports: see J. R. Coull, 'The trawling controversy in Scotland in the late nineteenth, and early twentieth centuries', *International Journal of Maritime History* 6, 1994, pp. 107–22.

43. R. Neild, *The English, the French and the Oyster*, London: Quiller Press, 1996.

44. The whole of continental European production amounted to about 3 million tons at that time.

45. R. Church, *The History of the British Coal Industry*, vol. 3: *1830–1913: Victorian Pre-eminence*, Oxford: Clarendon Press, 1986; M. W. Flinn, *The History of the British Coal Industry*, vol. 2, *1700–1830: The Industrial Revolution*, Oxford: Clarendon Press, 1984.

46. Clapp, *Environmental History of Britain*, pp. 14–15.

47. D. S. Landes, *The Unbound Prometheus: Technological Change and Industrial Development in Western Europe from 1750 to the Present*, Cambridge: Cambridge University Press, 1969.

48. B. Trinder, *The Making of the Industrial Landscape*, London: Dent, 1982: Phoenix, 1987.

49. D. Gregory, '"A new and differing face in many places": three geographies of industrialisation', in Dodghson and Butlin, *Historical Geography of England and Wales*, pp. 351–99.

50. A. Harris, 'Changes in the early railway age: 1800–1850', in Darby, *New Historical Geography of England*, pp. 165–226.

51. V. Robinson and D. McCarroll, *The Isle of Man: Celebrating a Sense of Place*, Liverpool: Liverpool University Press, 1990. J. P. Shimmin, 'The Port of Laxey', *The Isle of Man Natural History and Antiquarian Society* 8, 4, April 1978 to March 1980, pp. 378–84.

52. E. A. Wrigley, *Continuity, Chance and Change: The Character of the Industrial Revolution in England*, Cambridge: Cambridge University Press, 1988.

53. From T*he Condition of the Working Class in England 1844*, quoted in A. Clayre (ed.), *Nature and Industrialization*, Oxford: Oxford University Press/Open University Press, 1977, pp. 122–24. It was first published in German in 1845 and in English in 1877.

54. Quoted in Trinder, *Making of the Industrial Landscape*, 1987, p. 182. Trinder has the best selection of contemporary accounts of this type but says nothing about the original writers, whereas Clayre, *Nature and Industrialization*, has a biographical appendix. A. B. Reach was, it appears, a sociologist: an early member of that discipline. His language is rather heated: the pigs 'prey' upon the garbage when in reality they were probably doing a helpful job.

55. J. Sheail, 'Town wastes, agricultural sustainability and Victorian sewage', *Urban History* 23, 1996, pp. 189–210. I have the impression that bird recognition books were common users of the term 'sewage farms' as likely places to find waders, for example.

56. B. Trinder, *The Making of the Industrial Landscape*, London: Phoenix Books, 1997, p. 207.

57. E. Hobusch, *Fair Game: A History of Hunting, Shooting and Animal Conservation*, New York: Arco Publishing, 1980.

58. Anthony Trollope, *Phineas Redux*, 1876, ch. 16.

59. Traditional fox hunts killed about 20,000 foxes in the 1990s, which represents about 2.5 per cent of the UK fox population. The hunts also kill between 8000 and 10,000 hounds per year.

60. A. Vandervell and C. Coles, *Game and the English Landscape*, New York: Viking Press, 1980.

61. John Betjeman, *Collected Poems*, London: John Murray, 1973, p. 45.

62. The Alan Bennett film *A Day Out* (BBC2, 1972) captures this perfectly in the account of a cycling club's day ride from Halifax in 1911 and its post-1918 aftermath.

63. W. Kime, *Skegness in Old Photographs*, Stroud: Sutton Publishing, 1992.

64. A. J. Ludlam, *The East Lincolnshire Railway*, Oxford: The Oakwood Press, 1991.

65. J. Sheail, *Nature in Trust: The History of Nature Conservation in Britain*, Glasgow and London: Blackie, 1976. The last known British Great Auk was killed on St Kilda in 1840 by men who believed it to be a witch: see R. A. Lambert, 'From exploitation to extinction, to environmental icon: Our images of the Great Auk', in R. A. Lambert (ed.), *Species*

History in Scotland: Introductions and Extinctions since the Ice Age, Edinburgh: Scottish Cultural Press, 1998, pp. 20–37.

66. J. Sheail, 'Nature protection, ecologists and the farming context: A UK historical context', *Journal of Rural Studies* 11, 1995, pp. 79–88.

67. J. Sheail, 'Wild plants and the perception of land-use change in Britain: An historical perspective', *Biological Conservation* 24, 1982, pp. 129–46.

68. 'The plume trade', in P. R. Ehrlich, D. S. Dobkin, D. Wheye and S. L. Pimm, *The Birdwatcher's Handbook*, Oxford: Oxford University Press, 1984, pp. 23–7.

69. A. D. Middleton, *The Grey Squirrel*, London: Sidgwick and Jackson, 1931.

70. C. Lever, *The Naturalised Animals of the British Isles*, London: Paladin Books, 1979.

71. E. Salisbury, *Weeds and Aliens*, London: Collins, 1961.

72. Clapp, *Environmental History of Britain*, p. 45. I assume that 'Kyrle' is after John Kyrle (1637–1724) who lived in Ross on Wye and devoted his surplus income to works of charity and the improvement of the town and countryside. He was, says the *DNB*, a bachelor, and was well-built, red-faced and hearty.

73. J. Sheail, 'The sustainable management of industrial watercourses: An English historical perspective', *Environmental History* 2, 1997, pp. 197–215.

74. R. Grove, *Green Imperialism: Colonial expansion, tropical island Edens and the origins of environmentalism 1600–1860*, Cambridge: Cambridge University Press, 1995.

75. There are no records so far as I know of travellers demanding to be blindfolded, like Adam of Usk crossing the Alps. I suppose that today's equivalent is the consumption of brandy and Valium at airports.

76. William Blake, *The Marriage of Heaven and Hell*, 1790, plate 14.

77. M. Drabble, 'The Romantics', *A Writer's Britain: Landscape in Literature*, London: Thames and Hudson, 1979, pp. 147–92.

78. William Wordsworth, *Lines Composed a few miles above Tintern Abbey on revisiting the banks of the Wye during a tour*, ll. 93–102.

79. J. Bate, *Romantic Ecology: Wordsworth and the Environmental Tradition*, London and New York: Routledge, 1991, ch. 1.

80. Bate, *Romantic Ecology*.

81. J. R. Gold and M. M. Gold, *Imagining Scotland: Tradition, Representation and Promotion in Scottish Tourism since 1750*, Aldershot: Scolar Press, 1995.

82. A. Bermingham, *Landscape and Ideology: The English Rustic Tradition 1740–1860*, Berkeley and Los Angeles: University of California Press, 1986.

83. S. Daniels, *Fields of Vision: Landscape Imagery and National Identity in England and the United States*, Oxford: Polity Press, 1993, ch. 7; P. Bishop, *An Archetypal Constable: National Identity and the Geography of Nostalgia*, Madison, WI, and Teaneck, NJ: Farleigh Dickinson University Press, 1995.

84. Both this and *Rain, Steam and Speed* are in the National Gallery in London. On one visit I have to say that I thought the 'ghost ship' effect looked rather vulgar (The *Temeraire* fought at Trafalgar and so there is some irony in the picture of her last voyage being in a gallery which fronts onto Trafalgar Square.)

85. Daniels, *Fields of Vision*, ch. 4. I used Slough station daily on the way to and from school for six years but the atmosphere of pioneering and royalty had by then worn off rather. The full penetration of industrialisation and the steam engine into the national consciousness was accomplished by the 1850s, according to S. Hill, *The Tragedy of Technology*, London: Pluto Press, 1988.

86. On the other hand, the railway may simply not have been fenced at that time. The picture may be seen in the Courtauld Gallery in London, a small but chronologically inclusive collection that richly fills a couple of hours on a quiet morning in the winter. I am convinced that Pissaro's pictures of autumn in France do smell: *Autumn morning, Eragny* from the Pushkin Museum in Moscow perfumed the corner of a crowded room at the

Gemeentemuseum in Den Haag in the summer of 1996 with fruit, mushrooms and fallen vegetation.

87. The roots and development of which are explored for Great Britain and for North America in M. Bunce, *The Countryside Ideal: Anglo-American Images of Landscape*, London and New York: Routledge, 1994.

88. Thomas Hardy, *Return of the Native*, Book 1, 'The Three Women', 1.

89. A. Enstice, *Thomas Hardy: Landscapes of the Mind*, New York: St. Martin's Press, 1979; P. Porter (ed. and intro.), *Landscape Poets: Thomas Hardy*, London: Weidenfeld and Nicholson, 1981; M. Drabble, *Writer's Britain*, pp. 91–8. A. Alvarez, 'Introduction', in Thomas Hardy, *Tess of the D'Urbervilles*, Penguin: Penguin Classics edition, London, 1985, p. 11. The quotation about the engineman is from ch. xlvii, p. 405, in the Penguin edition. I can remember exactly the same kind of rig (steam engine and threshing 'drum' but with a lifter to take the straw up to form another stack) in Lincolnshire in the 1940s.

90. By 1911, Britain had seven major conurbations while no other European country had more than two. In order of size, the list read Greater London (7.3 million), S.E. Lancs (2.1 million), West Midlands (1.6 million), W. Yorkshire and Clydeside (1.5 million each), Merseyside (1.2 million) and Tyneside (0.8 million) (see n. 87).

91. Hobsbawm's list includes people like Stravinsky and Picasso, Matisse, Schönberg, Proust, Yeats, Joyce, Mann, Kafka and Pound (E. Hobsbawm, *Age of Extremes: The Short Twentieth Century 1914–1991*, London: Michael Joseph, 1994, p. 179 fn.).

92. For an imaginative recreation of part of Elgar's later life, including its melancholies, see the novel by James Hamilton-Patterson, *Gerontius*, London: Macmillan, 1989. In his middle years, Elgar and his wife moved house frequently and he is reputed to have said that this was because he had 'used up' the landscape for his compositions. The theme of the Introduction and Allegro for strings (Op. 47) was apparently 'plucked out of the air' on a walk on the Welsh coast. The literary equivalent of Elgar is perhaps A. E. Housman, though his scale is usually smaller. His evocation of the Welsh Marches in, for example, 'A Shropshire Lad' (1896) is the more remarkable for being written in Highgate and Hampstead by a Worcestershire man.

93. Trinder, *Making of the Industrial Landscape*.

7 'A fit country for heroes', 1914–50

1. 'What is our task? To make Britain a fit country for heroes to live in.' Speech by David Lloyd George (then Prime Minister) at Wolverhampton, 25 November 1918. However, the continuities between pre-1914 and post-1918 were strong at the macro-level: a liberal capitalist economy was accepted by almost everybody except a few socialists, for example. See J. Stevenson, *British Society 1914–1945*, London: Pelican Books, 1984, reprinted in Penguin Books, 1990.

2. P. Deane and W. A. Cole, *British Economic Growth 1688–1959*, 2nd edn, Cambridge: Cambridge University Press, 1967.

3. W. S. Humphrey and J. Stanislaw, 'Economic growth and energy consumption in the UK, 1700–1975', *Energy Policy* 7, 1979, pp. 29–42. 'tce' = tonnes of coal equivalent. There are data problems since these are based on British Isles figures, with the Irish Republic subtracted after 1920, but this overall trend seems firmly based. The later date includes other energy sources, such as oil and hydropower.

4. T. J. Chandler and S. Gregory (eds), *The Climate of the British Isles*, London and New York: Longman, 1976.

5. Coventry held a referendum in 1948 on the idea of smokeless zones; the vote was 5 to 2 in favour.

6. Brimblecombe, *The Big Smoke*.

7. M. A. Arber, 'Dust storms in the fenland round Ely', *Geography* 31, 1946, pp. 23–6. Ms Arber thinks the earliest dust storms in the area were in 1929 and that the bad storms of 1943 were not reported in the newspapers because of the war.

8. K. J. Hilton (ed), *The Lower Swansea Valley Project*, London: Longman Green, 1967.

9. P. N. Grimshaw, *Sunshine Miners: Opencast Coal Mining in Britain 1942–1992*, Mansfield, Notts.: British Coal Opencast, 1992.

10. Council for Environmental Conservation, *Scar on the Landscape? A Report on Opencast Coal Mining and the Environment*, London: CoEnCo, 1979.

11. R. Evans, 'Sensitivity of the British landscape to erosion', in R. S. G. Thomas and R. J. Allison (eds), *Landscape Sensitivity*, Chichester: Wiley, 1993, pp. 189–210.

12. M. Macklin, 'Metal pollution of soils and sediments: A geographical perspective', in M. Newsom (ed.), *Managing the Human Impact on the Natural Environment: Patterns and Processes*, London and New York: Belhaven Press, 1992, pp. 172–95; M. Macklin and J. Lewin, 'Sediment transfer and transformation of an alluvial valley floor, the river South Tyne, Northumbria, U.K.', *Earth Surface Processes and Landforms* 14, 1989, pp. 233–46.

13. J. N. Hutchinson, 'The record of peat wastage in the East Anglian Fenlands at Holme Post, 1848–1978 A.D.', *Journal of Ecology* 68, 1980, pp. 229–49.

14. S. J. Richardson and J. Smith, 'Peat wastage in the East Anglian fens', *Journal of Soil Science* 28, 1977, pp. 485–9.

15. A 1996 biography of Gladstone by Roy Jenkins shows that felling trees was a favourite leisure activity of the grand old man in the last third of the nineteenth century. Gladstone's diary records such an event on many of the days he spent on his own, and others', estates. He helped bring about the 90 per cent dependence on imports that pertained just before World War I with Russia as the biggest supplier (5 million loads), and thereafter Sweden (1.7 million), France (0.9 million, from a country that was in fact a net importer), Canada and Newfoundland (0.9 million) and the USA (0.5 million). Dependence on sea transport and on unpredictable supplies is obvious: 'The Russian forests are something of a mystery' says the source of these data: Advisory Committee on Forestry, *Report. July to October, 1912*, London: HMSO, 1913, Cmd 6713.

16. James, *History of English Forestry*. This forms the source for most of the administrative material in this section, with a few details from H. L. Edlin, *Trees, Woods and Man*, New Naturalist, London: Collins, 1956.

17. Known as the Acland Report, after its Chairman, Lord Acland, it was the *Report of the Forestry Sub-Committee of the Reconstruction Committee*, London: HMSO, 1918, Cmd 8881.

18. The 'nuclear winter' scenarios of the 1980s made much of the fact that forests would burn and thus add to the total breakdown of photosynthesis in the northern hemisphere.

19. M. L. Anderson (ed. C. J. Taylor), *A History of Scottish Forestry*, vol. 2, *From the Industrial Revolution to Modern Times*, London: Nelson, 1967.

20. E. J. T. Collins, *The Economy of Upland Britain 1750–1950: An Illustrated Review*, Reading: University of Reading Centre for Agricultural Strategy, 1978.

21. D. Woolfenden, 'Forestry in the Borders', *Forestry* 62, 1989, pp. 297–314.

22. G. Ryle, *Forest Service: The First Forty-Five Years of the Forestry Commission of Great Britain*, Newton Abbot: David and Charles, 1969.

23. CPRE stands for the Council for the Protection of Rural England now; 'Preservation' in the 1930s. Some discussion of this and similar organisations is on pp. 284–91.

24. It can be argued that the way the Commission went about afforestation was heavily influenced by the German model of 'Hochwald' based on even-aged, monocultural, high forest plantations managed by clear-felling: see G. F. Peterken, *Woodland Conservation and Management*, 2nd edn, London: Chapman and Hall, 1993.

25. Sheail, *Nature in Trust*.

26. More likely, possibly, for Venice, which can be closed off at the Causeway. 'Name three ways in which Giorgione broke with tradition in *La Tempesta*'; 'What are the cognate

nouns for *Doge* in English and French?' 'Was San Marco a cathedral during the period of the Republic?' 'Does the Arsenal still have a football team?' And so on. A number of people in Britain would have us leave the EU, and presumably Umberto Bossi's 'Padania' in what is now Northern Italy would want to form a league with the more racist parts of France.

27. F. Fraser Darling (ed.), *West Highland Survey: An Essay in Human Ecology*, Oxford: Oxford University Press, 1955, pp. 303–05.
28. Extracted from the tables in L. D. Stamp, *The Land of Britain: Its Use and Misuse*, 3rd edn, London: Longman Green/Geographical Publications Ltd, 1962.
29. D. Grigg, 'Types of farming in England and Wales', *Geography Review* 1(4), 1988, pp. 20–4.
30. D. Grigg, *English Agriculture: An Historical Perspective*, Oxford: Basil Blackwell, 1989. Most significant data for English agriculture from 1800 to 1980 can be found in tabular or graphical form in this book.
31. Grigg, *English Agriculture*. One sheep is 1 livestock unit; 1 bovine is 5 units.
32. As an employee of the Ford Motor Company, my father spent long hours in World War II assembling tractors at Dagenham, and many of the rest as an ARP Warden.
33. J. Brown, *Agriculture in England: A Survey of Farming 1870–1947*, Manchester: Manchester University Press, 1987. He is citing P. E. Dewey, 'Food production and policy in the United Kingdom, 1914–18', *Transactions of the Royal Historical Society* ser 5, 30, 1980, pp. 71–89.
34. Stamp, *Land of Britain*, p. 424.
35. A. Burrell, B. Hill and J. Medland, *Agrifacts: A Handbook of UK and EEC Agricultural and Food Statistics*, London: Harvester Wheatsheaf, 1990, p. 26; A. Edwards and A. Rogers (eds), *Agricultural Resources*, London: Faber and Faber, 1974.
36. To some degree, of course, they were right: a less intensive (that is, less dependent upon fossil fuels in the shape of machines and chemicals) agriculture is more 'organic' in the usual definition of that word in this context. But any agriculture, anywhere, involves considerable manipulation of the environment.
37. R. J. O'Connor and M. Shrubb, *Farming and Birds*, Cambridge: Cambridge University Press, 1986. The wryneck is a summer visitor and confined to the Chalk and Greensand areas of the south-east of England.
38. (Lord) Porchester, *A Study of Exmoor: A Report to the Secretary of State for the Environment and to the Minister of Agriculture, Fisheries and Food*, London: HMSO, 1977.
39. An account of the LUS's (Land Use Survey) procedure (as well as a discussion of the results) is in the various editions of Stamp, *Land of Britain*, 1948, 1950 and 3rd edn 1962.
40. A. S. Mather, 'The alleged deterioration in hill grazings in the Scottish Highlands', *Biological Conservation* 14, 1978, pp. 181–95.
41. J. Allaby, *The Changing Uplands*, CCP 153, Cheltenham: Countryside Commission, 1983. This is a 'popular' version of the *Report of the Upland Landscapes Study*, by G. Sinclair, S. Bell et al. Plas Machynlleth: Environment Information Services, 1983.
42. P. H. Armstrong 'The heathlands of the East Suffolk sandlings', *Suffolk Natural History* 15, 1972, pp. 417–30; Armstrong, 'Changes in the land use of the Suffolk sandlings: A study of the disintegration of an ecosystem', *Geography* 58, 1973, pp. 1–8.
43. D. N. McVean and J. D. Lockie, *Ecology and Land Use in Upland Scotland*, Edinburgh: Edinburgh University Press, 1969.
44. Fraser Darling, *West Highland Survey*.
45. The YHA originated in the 1929–1931 period; in the latter year it had 73 hostels and 6439 members; the Ramblers' Association came from a federation of Rambling Clubs which changed its name in 1933.
46. T. Stephenson, *Forbidden Land: The Struggle for Access to Mountain and Moorland*, Manchester: Manchester University Press, 1989.
47. One element has changed: in 1911 there were about 23,000 gamekeepers in Great Britain and in 1951 the census counted 4391, some (perhaps most) of whom were part-time. In the 1990s there are probably about 3000.

48. As early as 1911, there was a government-sponsored enquiry into grouse population fluctuations, chaired by Lord Lovat.

49. S. Tapper, *Game Heritage: An Ecological Review from Shooting and Gamekeeping Records*, Fordingbridge, Hants.: Game Conservancy 1992.

50. M. M. Bower, 'The distribution of erosion in blanket peat bogs in the Pennines', *Trans Inst Brit Geogr* 29, 1961, pp. 17–30; F. A. Barnes, 'Peat erosion in the Southern Pennines: Problems of interpretation', *East Midland Geographer* 3, 1963, pp. 216–22; A. C. Stevenson, V. J. Jones and R. W. Battarbee, 'The cause of peat erosion: A palaeolimnological approach', *New Phytologist* 114, 1990, pp. 727–35; J. H. Tallis, 'Climate and erosion signals in British blanket peats – the significance of *Rhacomitrium lanuginosum* remains', *Journal of Ecology* 83, 1995, pp. 1021–30.

51. L. B. Wood, *The Restoration of the Tidal Thames*, Bristol: Adam Hilger, 1982; see also F. T. K. Pentelow, *River Purification: A Legal and Scientific Review of the Last 100 Years*, London: Edward Arnold, 1953, for a review of the legal measures to that date.

52. Much of the labour for Slough came from South Wales, even into the 1950s: an immigrant then meant somebody from Wales. Later waves included people from post-1956 Hungary as well as families from the Caribbean and South Asia. Slough's first black woman mayor was called Simmons.

53. PEP (Political and Economic Planning), *Report on the Location of Industry*, London: PEP, 1939, p. 107.

54. K. Warren, *Mineral Resources*, Harmondsworth: Penguin Books, 1973.

55. We might wonder what Warren would have made of the Eton boating lake, carved out of the Thames flood-plain near the college.

56. O. Lancaster, *Pillar to Post: English Architecture without Tears*, 2nd edn, London: John Murray, 1956, originally published in 1938. 'Notice the skill . . . that insures that the largest possible area of countryside is ruined with the minimum of expense . . . and with what studied disregard of the sun's aspect the principal rooms are planned' (p. 68). Actually, I lived in one from 1950 to 1962 and although the siting in the middle of open country was scarcely sensitive at the time (it was built in the 1930s), the actual structure was not at all bad.

57. G. M. Boumphrey, 'Shall the Towns Kill or Save the Country?', in Clough Williams-Ellis (ed.), *Britain and the Beast*, London: Readers' Union/Dent, 1938. Famous contributors included Maynard Keynes, E. M. Forster and G. M. Trevelyan. In 1930, Williams-Ellis had published a book entitled *England and the Octopus*.

58. J. B. Cullingworth, *Environmental Planning 1939–1961*, vol. 3, *New Towns Policy*, London: HMSO, 1979.

59. It can still be seen on biscuit tins, calendars and in tourist board promotional material, and on TV advertisements. Villages with such a reputation are heavily visited for recreational purposes. Even in 1996, problem pages of broadsheet newspapers deal with the question of whether children should be brought up in the country rather than the city. The answer seems to be that if freedom from crime, vandalism and drugs is sought then the differences are not very great: it is the parents' image that is being catered to. The youngsters hate having to be taken and fetched everywhere for a start.

60. P. Scupham, 'Transformation Scenes, I, Evening in the park', in *Winter Quarters*, Oxford and New York: Oxford University Press, 1983, p. 29. It echoes the well-known descriptions of a country house after wartime depredations in Waugh's *Brideshead Revisited*.

61. These numbers mislead in the sense that the Air Ministry's acquisitions were mostly new, as was aviation; the army was adding to an already established estate on Salisbury Plain and in the Aldershot region, for example.

62. Though the British dimension is emphasised here, there was very great loss of life and damage in The Netherlands.

63. The journal *Geography* 38, 1953, pp. 132–89, contains a number of papers on the 1953 floods, including A. H. W. Robinson, 'The storm surge of 31st January–1st February, 1953

and the associated meteorological and tidal conditions', pp. 134–41; F. A. Barnes and C. A. M. King, 'The Lincolnshire coastline and the 1953 storm flood', pp. 141–60; A. T. Grove, 'The sea flood on the coasts of Norfolk and Suffolk, pp. 164–70; A. H. W. Robinson, 'The sea floods around the Thames estuary', pp. 170–6. The historic context is given by H. A. P. Jensen, 'Tidal inundations past and present', *Weather* 8, 1953, pp. 85–9 and 108–12. Cf. Pepys's *Diary* for 7 December 1663, 'there was last night the greatest tide that ever was remembered in England to have been in this river, all Whitehall having been drowned'. Another flood on the Thames in January 1928 drowned 14 people in basements and rendered 4000 people temporarily homeless.

64.	R. Robinson, *Trawling: The Rise and Fall of British Trawl Fishing*, Exeter: Exeter University Press, 1996.

65.	G. Jackson, *The British Whaling Trade*, London: Adam and Charles Black, 1978.

66.	Humphrey and Stanislaw, 'Economic growth'.

67.	G. Humphrys, 'Power and the industrial structure', in J. W. House (ed.), *The UK Space: Resources, Environment and the Future*, London: Weidenfeld and Nicholson, 1973, pp. 219–82.

68.	Though I remember the 1940s on a remote farm in the Lincolnshire fens: the radio was powered by a very large and heavy but non-recyclable battery and by an accumulator which had to be taken by bicycle to the shop every week for renewal. My grandparents' house in St Albans had gas lighting well into the 1950s.

69.	P. L. Payne, *The Hydro*, Aberdeen: Aberdeen University Press, 1988.

70.	Any second-hand bookshop will show what an outpouring of books about the British (and especially the English) scene there was; if the shop deals in remainders then the current fascination will also be demonstrated. At the highly literary end, one of the best is C. E. Montague, *The Right Place*, London: Chatto and Windus, 1924. To be fair, two of its 15 essays are about towns. One, inevitably, is about the Alps and comparisons with Italy abound. See P. Gruffudd, 'Selling the countryside: Representations of rural Britain in J. R. Gold and S. V. Ward (eds), *Place Promotion*, Chichester: Wiley, 1994, pp. 247–63; M. Bunce, *The Countryside Ideal: Anglo-American Images of Landscape*, London and New York: Routledge, 1994, especially ch. 1 and 2.

71.	For much of this section I have relied upon B. Green, *Countryside Conservation*, The Resource Management Series, 3, London: Allen and Unwin, 1981; G. E. Cherry, *Environmental Planning*, vol. 2, *National Parks and Recreation in the Countryside*, London: HMSO, 1975; Sheail, *Nature in Trust*; Sheail, *Rural Conservation in Inter-War Britain*, Oxford: Clarendon Press, 1981.

72.	The overall position and development of the Trust, the RSPB and wildife trusts is discussed by J. Dwyer and I. Hodge, *Countryside in Trust: Land Management by Conservation, Recreation and Amenity Organisations*, Chichester: Wiley, 1996. But the actual management of land and water is not part of the book.

73.	Sheail, *Nature in Trust*, p. 171. Sheail's books are by far the most readable and authoritative accounts of the history of conservation in the UK.

74.	J. T. R. Sharrock, *The Atlas of Breeding Birds in Britain and Ireland*, Berkhampstead: Poynser, 1976, p. 448. This species is one of the only two birds lost from the breeding fauna this century without subsequently recolonising; the other is the white-tailed eagle (*Haliaëtus albicilla*).

75.	T. Stephenson, *Forbidden Land: The Struggle for Access to Mountain and Moorland*, Manchester: Manchester University Press, 1989.

76.	J. B. Cullingworth, *Environmental Planning 1939–1969*, vol. 1, *Reconstruction and Land Use in Planning*, London: HMSO, 1975, has the fullest minute-by-minute account of these committees. The reports are: C. Addison, *Report of the National Park Committee*, Cmnd 3851, London: HMSO, 1931; Lord Justice Scott, *Report of the Committee on Land Utilisation in Rural Areas*, Cmnd 6378, London: HMSO, 1942; J. Dower, *National Parks in England and*

Wales, Cmnd 6628, London: HMSO 1945; A. Hobhouse, *National Parks Committee*, Cmnd 7121, London: HMSO, 1947.

77. This worked both ways, of course. I was lucky.

78. J. Pope-Hennessy, *Life of Queen Mary*, London: George Allen & Unwin, 1959, ch. 8. It cannot be pretended, however, that any move to increase public access to the countryside was not met with opposition from landowners and farmers; it did then and still does.

79. R. Price, *Scotland's Golf Courses*, Aberdeen: Aberdeen University Press, 1989.

80. Attributed without date to 'Lord Mancroft' in F. Metcalf (ed.), *The Penguin Dictionary of Humorous Quotations*, London: Penguin Books, 1987, p. 63, though in this case 'British' is used: surely not.

81. D. Coleman and J. Salt, *The British Population: Patterns, Trends and Processes*, Oxford: Oxford University Press, 1992. As they pitch it on p. 93: 'generally the regional batting order remains much as it was in the days of W. G. Grace'. Dr W. G. Grace lived from 1848 to 1915.

82. Sir Gwilym Gibbon, a civil servant who retired from the Ministry of Health (then the body charged with planning) in 1935. Quoted by Sheail, *Rural Conservation in Inter-War Britain*, p. 132.

83. P. Abercrombie, *Greater London Plan 1944*, London: HMSO, 1945. There are 221 pages 12 x 9½ inches and colour maps, at a cost of £1 5s 0d.

84. In 1952, R. Hill (*The Concerto*, Harmondsworth: Pelican Books, p. 260) says of it, 'One can only hope that in future we may have more frequent opportunities of hearing this great work.' Indeed we have.

85. The Thomas poem was written in May 1915; he died on active service in 1917.

86. A somewhat harsh judgement on work which does indeed have its more fluid moments. But Vaughan Williams served in World War I with an ambulance corps at the front and his *Pastoral Symphony* is about the fields of Flanders, not those of the West Midlands. The music of Gerald Finzi (1901–56) is more conventionally pastoral in its direction if not its harmonies. J. Burke, *Musical Landscapes*, Exeter: Webb & Bower, 1983, is mostly about rural England and mostly about music before 1950, with the exception of Britten.

87. Auden and mining in northern England are brought together in A. Myers and R. Forsythe, *W. H. Auden: Pennine Poet*, Nenthead, Cumbria: North Pennines Heritage Trust publication no. 7, 1999. The quotations are to be found in 'New Year Letter', 'Woods' and 'Ode to Gaea', found in *Selected Poems*, London: Faber and Faber, 1968, and *Collected Shorter Poems 1927–1957*, London: Faber and Faber, 1966.

88. The first stretch of the M6 (near Preston) was actually opened in 1958, and the initial phase of the M1 in 1959.

8 A POST-INDUSTRIAL WORLD, 1950 TO THE PRESENT

1. Though one of the most influential books about Britain of the 1990s, Will Hutton's *The State We're in* (London: Cape, 1995, with subsequent revised editions) has virtually nothing on environment at all.

2. The year 1998 saw a number of government proposals to reduce the level of car usage in the UK. It is too soon to see if they are having any effect.

3. A. Wilson, *Hemlock and After*, London: Penguin Books,1956; first published in 1952. The previous sentence mentions the Cold War, which was an undoubted ingredient of the uncertainties of the time. This novel, *Anglo-Saxon Attitudes* (1956) and *Late Call* (1964) seem to me to be the best examples of his work in the present context. The biography by Margaret Drabble (*Angus Wilson: A Biography*, London: Secker and Warburg, 1995) suggests that most of Wilson's creative writing about Britain was in fact done while he was abroad.

4. There is a nice commentary upon this theme in the novel by John Mortimer, *Paradise Postponed*, London and New York: Viking, 1958; reissued by Penguin Books in 1986 to coincide with a television adaptation. Elgar's Cello Concerto was used as theme music.

5. Though I have read somewhere the serious proposition that the climate has been the reason for British pre-eminence in painting with water colours.

6. P. Jones and M. Hulme, 'The changing temperature of "central England"', in M. Hulme and E. Barrow (eds), *Climates of the British Isles: Present, Past and Future*, London and New York: Routledge, 1997, pp. 173–96. The 'Central England' instrumental record is continuous from 1659, which allows some very accurate comparisons to be made.

7. C. Kidson, 'The Exmoor storm of 15th August, 1952', *Geography* 38, 1953, pp. 1–9.

8. G. F. Peterken, *Natural Woodland: Ecology and Conservation in Northern Temperate Regions*, Cambridge: Cambridge University Press, 1996, pp. 328–31. On the tenth anniversary of the great storm, Oliver Rackham pointed out that in public places a horizontal tree is 'an object of curiosity and delight – a relief from thousands of boring upright, unclimbable trees.' (*The Guardian*, 15 October 1997, pp. 4–5). The use of heavy machinery in clearing-up operations often compacted soils so that regeneration of trees was impeded.

9. M. Dukes and P. Eden, '"Phew! What a scorcher": Weather records and extremes', in Hulme and Barrow, *Climates of the British Isles*, pp. 262–98.

10. Brimblecombe, *The Big Smoke*, ch. 8. I remember the smog chiefly as the source of disrupted train services in the Thames Valley which made it difficult to get to school and so it had a sort of silver lining. The provisions of the Clean Air Act which created smokeless zones in towns came to the city of Durham in 1996.

11. I. Lowles and H. ApSimon, 'The contribution of sulphur dioxide emissions from ships to coastal acidification', *International Journal of Environmental Studies* 51, 1996, pp. 21–34.

12. A. Davison and J. Barnes, 'Patterns of air pollution: Critical loads and abatement strategies', in M. Newsom (ed.), *Managing the Human Impact on the Natural Environment: Patterns and Processes*, London and New York: Belhaven Press, 1992, pp. 109–29; P. Brimblecombe and G. Bentham, 'The air that we breathe: Smogs, smoke and health', in Hulme and Barrow, *Climates of the British Isles*, pp. 243–61.

13. F. Pearce, 'Is there life after death for British lichens?', *New Scientist*, 1 March 1997, p. 7.

14. DoE, *Digest of Environmental Statistics* 18, London: HMSO, 1996, pp. 41–2.

15. First Report of the Quality of Urban Air Review Group, *Urban Air Quality in the United Kingdom*, London: Department of the Environment, 1993.

16. K. Muirhead and A. P. Cracknell, 'Straw burning over Great Britain detected by AVHRR', *International Journal of Remote Sensing* 6, 1985, pp. 827–33.

17. A. P. Cracknell and M. R. Saradjian, 'Monitoring of straw burning in the U.K. using AVHRR data – Summer 1995', *International Journal of Remote Sensing* 17, 1996, pp. 2463–6.

18. M. Parry and R. Duncan (eds), *The Economic Implications of Climate Change in Britain*, London: Earthscan, 1995. The loss of the Gulf Stream due to cold fresh water flooding down from the Arctic is behind the prognoses of colder conditions altogether.

19. R. P. C. Morgan, 'Soil erosion and conservation in Britain', *Progress in Physical Geography* 4, 1990, pp. 24–47.

20. I. Foster, S. Harrison and D. Clark, 'Soil erosion in the West Midlands: An Act of God or agricultural mismanagement?', *Geography* 82, 197, pp. 231–9.

21. J. Boardman, 'Damage to property by runoff from agricultural land, South Downs, southern England, 1976–93', *Geographical Journal* 161, 1995, pp. 177–91.

22. A. Warren, 'Conservation and the land', in A. Warren and F. B. Goldsmith (eds), *Conservation in Perspective*, Chichester: Wiley, 1983, pp. 19–39.

23. Department of the Environment, *The Investigation and Management of Erosion, Deposition and Flooding in Great Britain*, London: HMSO, 1995.

24. L. J. A. Munro, E. C. Penning-Rowsell, H. R. Barnes, M. H. Fordham and D. Jarrett, 'Infant mortality and soil type: A case study in south-central England (with discussion)',

European Journal of Soil Science 48, 1997, pp. 1–17.

25. Department of the Environment, *The UK Environment*, London: HMSO, Government Statistical Service, 1996. The actual run of data for river flow and groundwater levels is only about 30 years long; rainfall has been measured since 1823 (DoE, *Digest of Environmental Statistics* 18, London: HMSO, 1996, p. 45).

26. Thames Water, *Environmental Review*, London: HMSO, 1997, p. 5.

27. About the only diversions not seriously tried have been cloud seeding and desalination, though the possibility of a desalting plant on the coast of Norfolk or Lincolnshire was floated in 1997.

28. S. Gregory, *The Price of Amenity*, London and Basingstoke: Macmillan, 1971, ch. 4; T. J. Bines, J. P. Doody, I. H. Findlay and M. J. Hudson, 'A retrospective view of the environmental impact on Upper Teesdale of the Cow Green Reservoir', in R. D. Roberts and T. M. Roberts (eds), *Planning and Ecology*, London and New York: Chapman and Hall, 1984, pp. 395–421.

29. K. Smith, *Water in Britain*, London: Macmillan, 1972.

30. H. Cook, 'Groundwater development in England', *Environment and History* 5, 1999, pp. 75–96.

31. For example, the process of burning sewage sludge as a method of disposal. Using it as fertiliser sounds environmentally attractive but is complicated by the metal content of the sludge. In 1997, some possible 'hot spots' (at field size) of cadmium were reported.

32. M. Newsom, 'Patterns of freshwater pollution', in M. Newsom (ed.), *Managing the Impact on the Natural Environment: Patterns and Processes*, London and New York: Belhaven Press, 1992, p. 49. Newsom notes that it is possible that groundwater aquifers can become contaminated with organic chemicals. He also estimates (p. 131) that storm water drainage in the UK has to cope with 17 g/m^2/yr of dog faeces. Given the low density of dogs in the Scottish Highlands, the quantity thus predicted for towns seems to accord with experience.

33. A-J. Beer, 'Something in the water', *Biologist* 44, 1997, p. 296. Friends of the Earth claimed (*Earth Matters* 43, 1999, p. 9) that ICI on Teesside released more than 8 tonnes of alkylphenols in 1998.

34. J. G. Jones, 'Windermere – lake under pressure', *Biologist* 44, 1997, pp. 369–73.

35. N. Arnell, 'Implications for water supply and water management', in M. Parry and R. Duncan (eds), *The Economic Implications of Climate Change in Britain*, London: Earthscan, 1995, pp. 28–45.

36. In the period 1950–1965, 506,000 ha were lost to agriculture; woodland gained 266,000 ha and urban development 253,000 ha. Some of the woodland came from reclamation schemes rather than agricultural land.

37. Peterken, *Natural Woodland*.

38. J. Tsouvalis-Gerber, 'Making the invisible visible: Ancient woodlands, British forest policy and the social construction of reality', in C. Watkins (ed.), *European Woodlands and Forests: Studies in Cultural History*, Wallingford and New York: CAB International, 1998, pp. 215–29.

39. 'Coppicing dividends', *English Nature Magazine* 42, 1999, p. 8. Small-leaved lime (*Tilia cordata*) is the main coppiced tree.

40. This is being undertaken by a private group, the Borders Forest Trust. Details in 1998 were to be found at a website: www.scotweb.co.uk/environment/wildwood/. If sheep-farming in the uplands gets even less profitable, succession to scrub and perhaps woodland may become frequent.

41. G. F. Peterken, 'Applying natural forestry concepts in an intensively managed landscape', *Global Ecology and Biogeography* 8, 1999, pp. 321–28.

42. Forest Enterprise, *Annual Report 1997–1998*, London: HMSO, 1998.

43. E. C. Mackey, M. C. Shewry and G. J. Tudor, *Land Cover Change: Scotland from the 1940s to the 1980s*, Edinburgh: The Stationery Office, 1998.

44. A. S. Mather and N. C. Murray, 'The dynamics of rural land use changes: The case of

private-sector afforestation in Scotland', *Land Use Policy* 5, 1987, pp. 103–21; A. S. Mather, 'The effects of afforestation on agriculture in Scotland', *Journal of Rural Studies* 11, 1995, pp. 187–202; A. S. Mather, 'The inter-relationship of afforestation and agriculture in Scotland', *Scottish Geographical Magazine* 112, 1996, pp. 83–91.

45.   Peterken, *Natural Woodland*; C. Lavers and R. Haines-Young, 'The impact of afforestation on upland birds in Britain', in C. Watkins (ed.), *Ecological Effects of Afforestation: Studies in the History and Ecology of Afforestation in Western Europe*, Wallingford: CAB International, 1993, pp. 127–52; K. J. Kirby, 'The effects of plantation management on wildlife in Great Britain: Lessons from ancient woodland for the development of afforestation sites', in Watkins, *Ecological Effects*, pp. 15–29. Dormice, it seems, are reluctant to cross large clearings.

46.   Nature Conservancy Council, *Nature Conservation and Afforestation in Britain*, London: HMSO, 1986; S. J. Petty, *Ecology and Conservation of Raptors in Forests*, Forestry Commission Bulletin 118, London: HMSO, 1998.

47.   P. Lusby, 'On the extinct plants of Scotland', in R. A. Lambert (ed.), *Species History in Scotland: Introductions and Extinctions since the Ice Age*, Edinburgh: Scottish Cultural Press, 1998, pp. 45–62.

48.   In 1963 the landscape architect Dame Sylvia Crowe was appointed to give advice. One narrative suggests that it was a cosmetic exercise and was not to interfere with timber production. See J. Gerber, 'The social construction of nature: the case of forestry in Great Britain since the turn of the 20th century', University of Oxford D.Phil. thesis, 1997.

49.   J. Rodwell and G. Patterson, *Creating New Native Woodlands*, Forestry Commission Bulletin 112, London: HMSO, 1994.

50.   D. A. Stroud, T. M. Reed, M. W. Pienkowski and P. A. Lindsay, *Birds, Bogs and Forestry: The Peatlands of Caithness and Sutherland*, Peterborough: Nature Conservancy Council. Between 1985 and 1996, 91 per cent of all afforestation in Britain was in Scotland and much of that was a way of avoiding tax by the creative use of Schedules B and D, a practice finally ended in 1988. But it meant that the seriously rich became very high-profile owners of forests in circumstances of considerable controversy. See S. Tompkins, *Forestry in Crisis*, London: Croom Helm, 1989. The Treasury had concluded by 1972 that government investment in forestry was only justifiable in terms of creating rural employment but the annual reports of the Commission at that time boast of the increasing efficiency due to mechanisation.

51.   D. Taylor, 'Land availability for future afforestation', in G. R. Hatfield (ed.), *Farming and Forestry*, Farnham: Forestry Commission, 1988, pp. 40–3; the price of land is the major determinant of what is bought for planting up. C. J. Cadbury, G. Williams and R. Green, 'Towards an upland habitat action plan', *RSPB Conservation Review* 7, 1993, pp. 5–11; R. Green and G. Williams, 'The ecology of the corncrake *Crex crex* and action for its conservation in Britain and Ireland', in E. M. Bignal, D. I. McCracken and D. J. Curtis (eds), *Nature Conservation and Pastoralism in Europe*, Peterborough: JNCC, 1994, pp. 69–74. E. M. Bignal and D. I. McCracken, 'Low-intensity farming systems in the conservation of the countryside', *Journal of Applied Ecology* 33, 1996, pp. 413–24.

52.   G. F. Peterken, 'Woodland conservation in Britain', in Warren and Goldsmith, *Conservation in Perspective*, pp. 83–100; Peterken, *Woodland Conservation and Management*, 2nd edn, London: Chapman and Hall, 1993; C. Watkins, *Woodland Management and Conservation*, Newton Abbot and London: David and Charles, 1990; D. C. Malcolm, J. Evans and P. N. Edwards, *Broadleaves in Britain: Future Management and Research*, Farnham: Forestry Commission/Institute of Chartered Foresters, 1984.

53.   Forestry Commission, *Broadleaves in Britain*, London: HMSO, 1984; G. W. Dimbleby, 'Natural regeneration of pine and birch on the heather moors of North-east Yorkshire', *Forestry* 26, 1953, pp. 41–52; Dimbleby, *Experiments with Hardwoods on Heathlands*, Oxford: Imperial Forestry Institute, 1958.

54. K. J. Kirby and R. C. Thomas, 'Fragmentation patterns of ancient woodland in England', in J. Dover (ed.), *Fragmentation in Agricultural Landscapes*, Aberdeen: International Association for Landscape Ecology, 1994, pp. 71–8.

55. R. C. Thomas, K. J. Kirby and C. M. Reid, 'The conservation of a fragmented ecosystem within a cultural landscape – the case of ancient woodland in England', *Biological Conservation* 82, 1997, pp. 243–52.

56. S. Kay, 'Factors affecting severity of deer browsing damage within coppiced woodlands in the south of England', *Biological Conservation* 63, 1993, pp. 217–22.

57. B. Boag and S. Tapper, 'The history of some British gamebirds and mammals in relation to agricultural change', *Agricultural Zoology Reviews* 5, 1992, pp. 273–311.

58. G. Cox, C. Watkins and M. Winter, *Game Management in England: Implications for Public Access, the Rural Economy and the Environment*, Rural Research Monograph series no. 3, Gloucester: Countryside and Community Press, G. Cox, 'Shooting a line? Field sports and access struggles in Britain', *Journal of Rural Studies* 9, 1993, pp. 267–76; R. Sidaway, *Birds and Walkers: A Review of Existing Research on Access to the Countryside and Disturbance to Birds*, London: Ramblers' Association, 1990.

59. M. R. G. Connell and R. C. Dewar, 'The carbon sink provided by plantation forests and their products in Britain', *Forestry* 68, 1995, pp. 35–48.

60. M. Whitby, 'The United Kingdom', in M. Whitby (ed.), *The European Environment and CAP Reform*, Wallingford: CAB International, 1996, pp. 186–205. See also A. W. Gilg, *Countryside Planning: The First Half Century*, 2nd edn, London and New York: Routledge, 1996.

61. P. Hetherington and M. Walker, 'Farm subsidies', *The Guardian*, 9 December 1997, p. 15.

62. Department of the Environment, *Countryside Survey 1990 Main Report*, London: DOE, 1993; R. Westmacott and T. Worthington, *Agricultural Landscapes: A Third Look*, CCP 521, Gloucester: Countryside Commission, 1997.

63. P. Ambrose, 'The rural/urban fringe as battleground', in B. Short (ed.), *The English Rural Community: Image and Analysis*, Cambridge: Cambridge University Press, 1992, pp. 175–94.

64. R. S. Thomas, 'Cyndyllan on a Tractor', *Song at the Year's Turning*, London: Rupert Hart-Davis, 1955.

65. In 1999, the Countryside Agency said that only 13 per cent of dry stone walls were in good condition.

66. M. D. Hooper, 'Hedgerows and small woodlands', in J. Davidson and R. Lloyd (eds), *Conservation and Agriculture*, Chichester: Wiley, 1977, pp. 45–57.

67. Ministry of Agriculture, Fisheries and Food, *Modern Farming and the Soil*, London: HMSO, 1970.

68. D. J. Briggs and F. M. Courtney, *Agriculture and the Environment: The Physical Geography of Temperate Agricultural Systems*, London and New York: Longman, 1985.

69. B. W. Ilbery and I. R. Bowler, 'Horticultural change and the Horticultural Buildings and Orchard Replanting Scheme in England', *Environment and Planning C: Government and Policy* 13, 1995, pp. 67–78.

70. K. Mellanby, *Waste and Pollution: The Problem for Britain*, London: HarperCollins, 1992, p. 116. Mellanby gives no date for the data in his Table 10 but the text implies that it is a year at the time of writing, so the late 1980s can be inferred.

71. Events will no doubt have overtaken this paragraph between manuscript and press. BBC News websites are good places to begin since they have links to both the government statements and those of the NGOs.

72. A. E. B. Taylor, P. W. O'Callaghan and S. D. Probert, 'Energy audit of an English farm', *Applied Energy* 44, 1993, pp. 315–35.

73. If the 'greenhouse' value of a molecule of CO_2 is 1.0, then that for nitrous oxide is 185 and methane 27. Even so, the CO_2 emissions are by far the largest.

74. Going back to the part of east Lincolnshire where I spent much of my childhood, I notice that the landscape has 'filled up' with a lot more buildings: housing for machinery,

glasshouses and bigger dwellings, for example. By contrast, a number of the labourers' cottages have disappeared: this is not a very popular region for second homes.

75. P. Lowe, G. Cox, D. Goodman, R. Munton and M. Winter, 'Technological change, farm management and pollution regulation: The example of Britain, in P. Lowe, T. Marsden and S. Whatmore (eds), *Technological Change and the Rural Environment*, London: David Fulton, 1990, pp. 53–80.

76. P. J. Johnes and T. P. Burt, 'Nitrate in surface waters', in T. P. Burt, A. L. Heathwaite and S. T. Trudgill (eds), *Nitrate: Processes, Patterns and Management*, Chichester: Wiley, 1993, pp. 269–317.

77. T. P. Burt and P. J. Johnes, 'Managing water quality in agricultural catchments', *Transactions of the Institute of British Geographers* NS 22, 1997, pp. 61–8.

78. Pollard, Hooper and Moore, *Hedges*.

79. English Nature/RSPB, *The Indirect Effects of Pesticides on Birds*, Totnes: Natural History Books, 1997.

80. R. K. Murton and N. J. Westwood, 'Birds as pests', *Applied Biology* 1, 1976, pp. 89–181; R. J. O'Connor and M. Shrubb, *Farming and Birds*, Cambridge: Cambridge University Press, 1986.

81. M. D. Hooper, 'What are the main recent impacts of agriculture on wildlife? Could they have been predicted, and what can be predicted for the future?', in D. Jenkins (ed.), *Agriculture and the Environment*, Cambridge: Institute of Terrestrial Ecology, 1984, pp. 33–7.

82. O. Tickell, 'Paradise postponed', *New Scientist* 157, 15 January, 1998, pp. 18–19.

83. R. S. Smith, 'Farming and the conservation of traditional meadowland in the Pennine Dales Environmentally Sensitive Area', in M. B. Usher and D. B. A. Thompson (eds), *Ecological Change in the Uplands*, British Ecological Society Special Publication no. 7, Oxford: Blackwell Scientific Publications, 1988, pp. 183–99.

84. This is set out in an ADAS report to MAFF: *Historical Monitoring in the Pennine Dales ESA 1987–1995*, found in this case as http://www.maff.gov.uk/environ/envsch/ESA/STAGE1/PENHIST.DOC downloaded on 21 May 1999.

85. Quite a lot of farmers are not keen, either. See the quotes in C. Morris and C. Potter, 'Recruiting the new conservationists: Farmers' adoption of agri-environmental schemes in the UK', *Journal of Rural Studies* 11, 1995, pp. 51–63.

86. W. M. Adams, 'Places for nature: Protected areas in British nature conservation', in F. B. Goldsmith and A. Warren (eds), *Conservation in Progress*, Chichester: Wiley, 1993, pp. 185–208; Adams, *Future Nature: A Vision for Conservation*, London: Earthscan, 1996, pp. 132–3.

87. T. P. Burt and N. E. Haycock, 'Catchment planning and the nitrate issue: A UK perspective', *Progress in Physical Geography* 16, 1992, pp. 379–404.

88. S. Laing, 'Images of the rural in popular culture 1759–1950', in B. Short (ed.), *The English Rural Community: Image and Analysis*, Cambridge: Cambridge University Press, 1992, pp. 133–51.

89. Bignal and McCracken, 'Low-intensity farming systems'.

90. J. Thirsk, *Alternative Agriculture: A History from the Black Death to the Present Day*, Oxford: Oxford University Press, 1997.

91. The Church Commissioners own about 67,000 ha, almost as much as the National Trust in England and Wales. But it includes the freehold of places like the Metrocentre at Gateshead, not normally associated with the term 'sacred'.

92. C. Chippendale, *Stonehenge Complete*, rev. edn, London: Thames and Hudson, 1994.

93. J. Lowerson, 'The mystical geography of the English', in B. Short (ed.), *The English Rural Community: Image and Analysis*, Cambridge: Cambridge University Press, 1992, pp. 152–74.

94. E. J. P. Marshall, P. M. Wade and P. Clare, 'Land drainage channels in England and Wales', *Geographical Journal* 144, 1978, pp. 254–63.

95. J. A. Taylor, 'The peatlands of Great Britain and Ireland', in A. J. P. Gore (ed.), *Mires:*

Swamp, Bog, Fen and Moor. B. Regional Studies, Amsterdam: Elsevier, 1983, pp. 1–46.

96. B. Moss, 'The Norfolk Broadland: Experiments in the restoration of a complex wetland', *Biological Reviews* 58, 1983, pp. 521–61; M. George, *The Land Use, Ecology and Conservation of Broadland*, Chichester: Packard Publishing, 1992; E. Maltby, *Waterlogged Wealth*, London: Earthscan, 1986.

97. L. Parkyn, R. E. Stoneman and H. A. P. Ingram (eds), *Conserving Peatlands*, Wallingford and New York: CAB International, 1997, is a collection of essays many of which are relevant to these themes. There is something of an emphasis on Scotland, Ireland and the Borders.

98. G. Gilbert, M. Painton and K. W. Smith, 'An inventory of British reedbeds in 1993', *RSPB Conservation Review* 10, 1996, pp. 39–44.

99. Royal Commission on Common Land 1955–8, *Report*, Cmnd 462, London: HMSO, July 1958.

100. L. D. Stamp and W. G. Hoskins, *The Common Lands of England and Wales*, New Naturalist series no. 45, London: Collins, 1963. An appendix lists all the commons and village greens known to Stamp at the time, with their area and brief comments on their use.

101. D. R. Denman, R. A. Roberts and H. J. F. Smith, *Commons and Village Greens*, London: Leonard Hill, 1967.

102. O. Wilson and J. A. Wilson, 'Common cause of common concern? The role of common lands in the post-productivist countryside', *Area* 29, 1997, pp. 45–58.

103. In the mid-1970s, there were 1043 hill sheep holdings in Scotland, 619 in northern England and 1720 in Wales. In Scotland and Wales, upland and hill farming accounted for about 30 per cent of the gross output by value of the agricultural sector (Jenkins, *Agriculture and the Environment*).

104. Jenkins, *Agriculture and the Environment*; Jenkins, *Upland Land Use in England and Wales*, Cheltenham: Countryside Commission, 1978; D. F. Ball, J. Dale, J. Sheail and O. W. Heal, *Vegetation Change in Upland Landscapes*, Bangor: ITE, 1982; A. Hopkins, J. Wainwright, P. J. Murray, P. J. Bowling and M. Webb, '1986 survey of upland grassland in England and Wales: Changes in age structure and botanical composition since 1970–72 in relation to grassland management and physical features', *Grass and Forage Science* 43, 1988, pp. 185–98.

105. D. N. McVean and J. D. Lockie, *Ecology and Land Use in Upland Scotland*, Edinburgh: Edinburgh University Press, 1969.

106. G. R. Miller and A. Watson, 'Heather moorland in Northern Britain', in Warren and Goldsmith, *Conservation in Perspective*, pp. 101–17.

107. R. T. Smith and J. A.Taylor (eds), *Bracken: Ecology, Land Use and Control Technology*, Carnforth, Lancs.: Parthenon Publishing, 1986.

108. Mackey, Shewry and Tudor, *Land Cover Change*.

109. J. A. Taylor, 'The bracken problem: A local hazard and global issue', in Smith and Taylor, *Bracken*, pp. 21–42.

110. R. Brown, 'Bracken on the North York Moors: Its ecological and amenity implications in national parks', in Smith and Taylor, *Bracken*, pp. 77–86; J. A. Taylor, 'Bracken: Global weed and environmental issue', *Geography Review* 11(5), 1998, pp. 12–15.

111. D. S. Allen, 'Habitat selection by whinchats: A case for bracken in the uplands?', in D. B. A. Thompson, A. J. Hester and M. B. Usher (eds), *Heaths and Moors: Cultural Landscapes*, Edinburgh: HMSO, 1995, pp. 200–5.

112. R. H. Marrs and P. J. Pakeman, 'Bracken invasion – lessons from the past and prospects for the future', in Thompson, Hester and Usher, *Heaths and Moors*, pp. 180–93.

113. L. Dudley Stamp (ed.), *The Land of Britain: The Report of the Land Utilisation Survey of Britain*, Part 44, London: Geographical Publications Ltd, 1941; R. Pollard, 'The Role of the Department of Agriculture, Fisheries and Forestry in the Management of the Manx Hill Lands', in *Proceedings of the Manx Hill-Land Seminar 11th–13th April 1995*, Isle of Man: Manx Conservation Trust, 1997, pp. 9–16; J. Harris, 'Hill sheep in the Isle of Man (in Comparison with Northern England)' in *Proceedings of the Manx Hill-Land Seminar 11th–13th April 1995*, pp. 29–30.

114. Not this century, at least. Some of it had probably been cereal land during the Napoleonic Wars, for instance.

115. (Lord) Porchester, *A Study of Exmoor*, London: HMSO, DoE/MAFF, 1977. 'Warring' is scarcely an exaggeration for the conflicts within the park planning committee and among the committee, its officers and other bodies and individuals. One complication was the advocacy of keeping land open for stag hunting, a recreational activity itself bound into a web of controversy.

116. D. Baldock and G. Beaufoy, *Nature Conservation and New Directions in the EC Agricultural Policy*, London and Arnhem: Institute for European Environmental Policy, 1993.

117. A. J. A. Stewart and A. N. Lance, 'Moor-draining: A review of impact on land use', *Journal of Environmental Management* 17, 1983, pp. 81–99.

118. P. J. Hudson, *Grouse in Space and Time*, Fordingbridge, Hants.: The Game Conservancy, 1992; Hudson, 'Ecological trends and grouse management in upland Britain', in Thompson, Hester and Usher, *Heaths and Moors*, pp. 282–93. According to *The Times* on 11 August 1999, a pair of top-quality 12-bore shotguns would auction for £50–60,000.

119. D. B. A. Thompson, A. J. MacDonald, J. H. Marsden and C. A. Galbraith, 'Upland heather moorland in Great Britain: A review of international importance, vegetation change and some objectives for nature conservation', *Biological Conservation* 71, 1995, pp. 163–78.

120. B. White, 'Natural resource management: The case of heather moorland', in P. Allanson and M. Whitby (eds), *The Rural Economy and the British Countryside*, London: Earthscan, 1996, pp. 62–82. The critical density above which heather declines seems to be 1.5 ewes/ha. Another relevant statistic is that 40 per cent of upland farms have a net income of less than £10,000 p.a. (and which has fallen by 60 per cent in the last few years) and that half the families questioned by the NFU said that their children would not be carrying on farming (*English Nature Magazine* 40, 1998, p. 14). An upland farm devoted to sheep will be getting about 60 per cent of its income from subsidies; a lowland arable unit in Cambridgeshire about 6 per cent.

121. A. C. Stevenson and D. B. A. Thompson, 'Long-term changes in the extent of heather moorland in upland Britain and Ireland: Palaeoecological evidence for the importance of grazing', *The Holocene* 3, 1993, pp. 70–6.

122. D. N. McVean and J. D. Lockie, *Ecology and Land Use in Upland Scotland*, Edinburgh: Edinburgh University Press, 1969.

123. B. W. Staines, R. Balharry and D. Welch, 'The impact of red deer and their management on the natural heritage in the uplands', in Thompson, Hester and Usher, *Heaths and Moors*, pp. 294–308.

124. C. Sydes and G. R. Miller, 'Range management and nature conservation in the British uplands', in M. B. Usher and D. B. A. Thompson (eds), *Ecological Change in the Uplands*, BES Special Publication no. 7, Oxford: Blackwell Scientific Publications, 1988, pp. 323–37.

125. M. Toogood, 'Representing ecology and Highland tradition', *Area* 27, 1995, pp. 102–9. He quotes the capital value of stalking rights as £10,000–20,000 per stag in the early 1990s. At the same time the Red Deer Commission proposed a initial reduction of 100,000 in the total red deer population of about 300,000. The idea of 'sustainable' tourism in Scotland as yet another myth is discussed in G. Hughes, 'Tourism and the environment: A sustainable partnership', *Scottish Geographical Magazine* 112, 1996, pp. 107–13. The altogether differing outlook of crofters is reported in F. MacDonald, 'Viewing Highland Scotland: Ideology, representation and the "natural heritage"', *Area* 30, 1998, pp. 237–44.

126. R. S. Thomas, *Pietà*, London: Rupert Hart-Davis, 1966.

127. T. Hughes, *Remains of Elmet: A Pennine Sequence*, London and Boston: Faber and Faber, 1979. It is wonderfully illustrated with black and white photographs by Fay Godwin.

128. A. S. Mather, 'Land use, physical sustainability and conservation in Highland Scotland', Land Use Policy 9, pp. 99–110. The paper ranges rather more widely than its title suggests.

129. (Lord) Nugent, *Report of the Defence Land Holdings Committee 1971–73*, London: HMSO, 1973.

130. T. C. E. Wells, J. Sheail, D. F. Ball and L. K. Ward, 'Ecological studies on the Porton Ranges: Relationships between vegetation, soils and land use history', *Journal of Ecology* 64, 1976, pp. 589–626.

131. F. M. Szasz, 'The impact of Word War II on the land: Gruinard Island, Scotland, and Trinity Site, New Mexico as case studies', *Environmental History Review* 19, 1995, pp. 15–30.

132. P. Marren, 'The siege of the NCC: Nature conservation in the eighties', in Goldsmith and Warren, *Conservation in Progress*, pp. 283–99. We now have 'English Nature' and its Welsh equivalent, and 'Scottish Natural Heritage', which combined with the Countryside Commission for Scotland.

133. P. Collen, 'The reintroduction of beaver (*Castor fiber* L.) to Scotland: An opportunity to promote the development of suitable habitat', *Scottish Forestry* 49, 1995, pp. 206–16; J. W. H. Conroy, A. C. Kitchener and J. A. Gibson, 'The history of the beaver in Scotland and its future reintroduction', in Lambert, *Species History in Scotland*, pp. 107–28.

134. D. Evans, *A History of Nature Conservation in Britain*, 2nd edn, London and New York: Routledge, 1997, ch. 8.

135. Anon, '90% grounds for optimism', *English Nature Magazine* 34, 1997, p. 8; there were in that year 4000 SSSIs covering 955,000 ha (2.36 million acres).

136. *Sustainable Development – the UK Strategy*, Cmnd 2426, London: HMSO, 1994.

137. Anon, 'New life for old heath', *English Nature Magazine* 98, 1998, p. 10; C. Evans, R. Marrs and G. Welch, 'The restoration of heathland on arable farming at Minsmere RSPB nature reserve', *RSPB Conservation Review* 7, 1993, pp. 80–4.

138. N. Cooper, 'A sanctuary for wildlife', *Biologist* 44, 1997, pp. 417–19; F. Greenoak, *Wildlife in Churchyards*, London: Little Brown, 1993; N. S. Cooper, *Wildlife in Church and Church-yard: Plants, Animals and Their Management*, London: Church House Publishing, 1995.

139. W. M. Adams, *Future Nature: A Vision for Conservation*, London: Earthscan, 1996. Even in the present system, government actions are sometimes difficult to understand. In response to an EU request for SAC (Special Area of Conservation) status the government submitted 331 areas (2.8 per cent of the national territory) whereas WWF's 'shadow list' comes to 1000 areas or 8 per cent, which is more like other European countries. Designation has to be complete by 2004.

140. D. B. A. Thompson, A. J. MacDonald, J. H. Marsden and C. A. Galbraith, 'Upland heather moorland in Great Britain: A review of international importance, vegetation change and some objectives for nature conservation', *Biological Conservation* 71, 1995, pp. 163–78.

141. There had been intervening enquiries and reports but none had resulted in significant change.

142. In perhaps 1963, I heard one of the National Parks Commissioners say that the parks were certainly not national and they weren't parks either. In 1999, one public debate enquired as to whether they were national parks or car parks.

143. See for instance C. Warren, 'Scottish National Parks: Overdue or unnecessary?', *Geography Review* 12(3), 1999, pp. 17–19. As it happened, the first business of the new Scottish Parliament in June 1999 included a bill for a Loch Lomond National Park.

144. The controversies detailed in the summer of 1997 in issue 21 of *Viewpoint* (the newsletter of the Friends of National Parks) were mostly, however, of the more precisely located kind: military expansion in Northumberland, quarry extension in the Peak District and North York Moors, an oil spill off Pembrokeshire and the summit of Snowdon. Road issues were confined to a report that the Okehampton bypass on the edge of Dartmoor had failed to take lorries out of the town because its gradients across the moorland edge were too steep.

145. P. D. Lowe, 'Values and institutions in the history of British nature conservation', in Warren and Goldsmith, *Conservation in Perspective*, pp. 329–52.

146. A. S. Mather, 'Protected areas in the periphery: Conservation and controversy in northern Scotland', *Journal of Rural Studies* 9, 1993, pp. 371–84.

147. Adams, *Future Nature*, ch. 10.

148. T. Champion, 'Studying counterurbanisation and the rural population turnaround', in P. Boyle and K. Halfacree (eds), *Migration into Rural Areas: Theories and Issues*, Chichester: Wiley, 1998, pp. 21–40; the net migration in 1990–1 for Inner London, for instance, was –1.24 per cent but for remote rural districts it was +0.61 per cent and for the most remote rural districts +0.77 per cent. In absolute numbers these equated to 31,009 people, 10,022 and 36,450 respectively. The 'urban rural mixed' and 'industrial' were the largest net sources of immigrants (24.9 per cent and 13.2 per cent respectively) suggesting that some 9000 people had put their toes into the rural water, as it were, before relocating to remoter areas.

149. R. Strong, *Country Life 1897–1997: The English Arcadia*, London: Country Life Books and Boxtree Ltd, 1999.

150. J. Pretty, *The Living Land: Agriculture, Food and Community Regeneration in Rural Europe*, London: Earthscan, 1998. For a comprehensive view of most rural matters in a political and institutional framework, see M. Winter, *Rural Politics: Policies for Agriculture, Forestry and the Environment*, London and New York: Routledge, 1996.

151. P. Allanson and M. Whitby, *The Rural Economy and the British Countryside*, London: Earthscan, 1996, pp. 146–9.

152. T. K. Marsden, J. Murdoch, P. Lowe, R. J. C. Munton and A. Flynn, *Constructing the Countryside*, London: UCL Press, 1993.

153. P. Lowe, G. Cox, D. Goodman, R. J. C. Munton and M. Winter, 'Technological change, farm management and pollution regulation: The example of Britain', in P. Lowe, T. K. Marsden and S. Whatmore (eds), *Technological Change and the Rural Environment*, London: David Fulton, 1990, pp. 53–80. The Environment Protection Agency now deals with pollution incidents in rural areas.

154. The fact that there are fewer farmers and farm workers in rural areas may well be a factor which is implicated in the high suicide rates: 993 farmers and farm workers killed themselves in 1979–90 and suicide is the second commonest form of death in farmers under 45. This represents a rate about twice that of the 'average' male. A more meaningful figure perhaps is that 27 farmers killed themselves in Shropshire in 1992.

155. P. Cloke and M. Goodwin, 'Conceptualizing countryside change: From post-Fordism to rural structured coherence', *Transactions of the Institute of British Geographers* NS 17, 1992, pp. 321–36. In 1996, the *Guardian* cartoonist Steve Bell made the suggestion that the Falklands War was an eruption of the rural idyll. There is a good history of attitudes to town and country in both Great Britain and North America in M. Bunce, *The Countryside Ideal*, London and New York: Routledge, 1994.

156. Another set of data puts the Swiss in the lead at 42 per cent and the Dutch lower down at 28 per cent. Either way, Britain is below 10 per cent. See the *Guardian*, 10 June 1998, and Friends of the Earth's *Earth Matters* 43, 1999, p. 9.

157. A study in Lancashire showed that 3 per cent of gulls carried *Escherichia coli* 0157, a virulent food-poisoning bacterium. It is possible that they transfer it from sewage slurry and landfill sites to farmland and thus infect cattle. 'Seabirds' dirty diet spreads disease', *New Scientist*, 22 March 1997, p. 11.

158. R. Mabey, *Flora Britannica*, London: Sinclair Stevenson, 1996.

159. P. Kivell and D. Lockhart, 'Derelict and vacant land in Scotland', *Scottish Geographical Magazine* 112, 1996, pp. 177–80.

160. J. A. Harris, P. Birch and J. Palmer, *Land Restoration and Reclamation: Principles and Practice*, London: Longman, 1996.

161. In the l990s, the national percentage of household waste recycled was very variable: Switzerland at 42 per cent headed the list, and countries like Finland, Canada, the USA and

Denmark were in the 20–30 per cent category; the UK's figure was 6 per cent. The general feeling has been that only a direct tax on wastes that need collection will alter attitudes.

162. R. Walford, *Land Use – UK*, Sheffield: The Geographical Association, 1997. It includes Northern Ireland, which the previous surveys did not.

163. P. J. Smart, B. D. Wheeler and A. J. Wheeler, 'Plants and peat cuttings: Historical ecology of a much exploited peatland – Thorne Waste, Yorkshire, UK', *New Phytologist* 104, 1986, pp. 731–48.

164. For a scholarly exploration of some of the complexities see M. Crang, 'On the heritage trail: Maps of and journeys to olde Englande', *Environment and Planning D: Society and Space* 12, 1994, pp. 341–55.

165. The whole question is explored throughly in terms of both history and sociology by Marion Shoard in her books. The most recent is *A Right to Roam: Should We Open Up Britain's Countryside*, Oxford: Oxford University Press, 1999. Although her depth of knowledge is impressive, the purpose is persuasive rather than detached-objective: 'shamelessly pugnacious', said the *New Statesman*, 10 May 1999.

166. A Mori poll in 1997 found that 84 per cent were in favour of the abolition of deer hunting. The equivalent figures for fox and hare were 76 per cent and 74 per cent.

167. These populations are of course different from deer which are kept in parks, of which there are perhaps 125 in Great Britain. Many of them carry the imported fallow deer rather than a native species like red deer.

168. D. M. Evans and S. Warrington, 'The effects of recreational disturbance on wintering water-birds on a mature gravel pit lake near London', *International Journal of Environmental Studies* A 53, 1997, pp. 167–82; C. H. Tuite, P. R. Hanson and M. Owen, 'Some ecological factors affecting winter wildfowl distribution on inland waters in England and Wales, and the influence of water-based recreation', *Journal of Applied Ecology* 21, 1984, pp. 41–62.

169. F. B. Goldsmith, 'Ecological effects of visitors and the restoration of damaged areas', in Warren and Goldsmith, *Conservation in Perspective*, pp. 201–14. The references have some useful systemic material by recreation type, though not all are for Great Britain.

170. A 1996 survey found an average 1482 items of debris per km of beach (sample = 196 stretches of beach). The most frequently found items were pieces of plastic, followed by cotton bud sticks, rope and cord, crisp and sweet packets, and lids. These items reflected both land-based and marine sources.

171. It emerged in 1997 that about 2 tonnes of low-level radioactive material was dumped there as well in the 1950s and 1960s.

172. R. B. Clark, *Marine Pollution*, 2nd edn, Oxford: Clarendon Press, 1989.

173. M. J. Tooley, 'The effects of sea-level rise', in M. Parry and R. Duncan (eds), *The Economic Implications of Climate Change in Britain*, London: Earthscan, 1995, pp. 8–27.

174. G. A. Macay, 'The UK fishing industry and EEC policy', *Three Banks Review* 132, 1981, pp. 48–62.

175. D. Symes, 'UK demersal fisheries and the North Sea: Problems in renewable resource management', *Geography* 76, 1991, pp. 131–42.

176. 'New fish farm pesticides to flood Scottish lochs', *New Scientist*, 22 March 1997, p. 10. The chemicals are azamethiphos (an organophosphate) and cypermethrin (a synthetic pyrethroid).

177. Evidence to the House of Lords Select Committee on Science and Technology, *Fish Stock Conservation and Management*, London: HMSO, 1995, HL Paper 25-I.

178. House of Lords Select Committee on Science and Technology, *Fish Stock Conservation and Management: Government Response*, London: HMSO, 1996, HL Paper 75.

179. [UK] Department of Trade and Industry, *Digest of United Kingdom Energy Statistics 1996*, London: HMSO, 1996.

180. W. A. Kerr and S. Mooney, 'A system disrupted – the grazing economy of North Wales in the wake of Chernobyl', *Agricultural Systems* 28, 1988, pp. 13–27.

181.   G. Walker, 'Renewable energy in the UK: The Cinderella sector transformed', *Geography* 82, 1997, pp. 59–74. In 1998, the price of electricity from all sources to the supplier companies was 2.5p per kilowatt, and, from renewables, 3.2p. In early 1997, the renewable price was 3.68p.

182.   Anon, 'Fears over beauty and the wind beasts', *Countryside* 85, winter 1997–8, p. 1.

183.   G. Manners, 'Energy conservation policy', in T. S. Gray (ed.), *UK Environmental Policy in the 1990s*, Basingstoke: Macmillan Press, 1995, pp. 144–58.

184.   The 1996 OPCS projection for England and Wales was 54.4 million in 2024, compared with an actual level of 51.6 million in 1994.

185.   A DoE consultation paper of 1996 forecast that the number of households will rise from 19.2 million in 1991 to 23.1 million in 2016, with projected rises the highest in a NE–SW belt from Cambridgeshire to Somerset, with outliers in Cornwall and Shropshire (*Guardian*, 26 November 1996). The total absolute population loss 1981–91 was highest from London (–628,000) and the West Midlands (–220,510) and lower in, for example, Tyne and Wear (–46,364), as reported of an ESRC-funded study in the *Guardian*, 6 December 1996.

186.   J. Skea, 'Acid rain: A business-as-usual scenario', in Gray, *UK Environmental Policy*, pp. 189–209.

187.   N. Haigh and C. Lanigan, 'Impact of the EU on UK policy making', in Gray, *UK Environmental Policy*, 1995, pp. 18–37.

188.   R. H. Grove, *Ecology, Climate and Empire: Colonialism and Global Environmental History 1400–1860*, Cambridge: Cambridge University Press, 1997.

189.   My impression (no more than that) is that there are fewer now in the bookshops than 10 years ago and that this has peaked in terms of novelty and become more of a routine item than a wonder: CD-ROM has taken them up but not with any air of breathlessness.

190.   Though not any longer, to judge from the current generation of undergraduates in Geography (*sic*).

191.   J. N. M. Firth, S. J. Ormerod and H. J. Prosser, 'The past, present and future of waste management in Wales – a case-study of the environmental problems in a small European region', *Journal of Environmental Management* 44, 1995, pp. 163–79.

192.   M. Whitby and A. Adger, 'UK land use and the global commons', in S. Harper (ed.), *The Greening of Rural Policy: International Perspectives*, London and New York: Belhaven, 1993, pp. 67–81.

193.   In 1998, it was reported that tankers full of toxic liquids from Holland and Germany were being ferried to Dover and then driven round the M25 until their cargo had all dribbled out; they were then driven home.

194.   It is an interesting, if possibly marginal, point that many people who work in rural areas are greatly isolated from the ecology of their surroundings by the safety equipment needed in an era of mechanisation. The forestry worker is a good example but think also of the recommended gear for the worker engaged in dipping sheep. Cyndyllan should now wear ear muffs.

195.   The list of composers influenced by Vaughan Williams can be very long and includes Birtwhistle. Also: Finzi, Hadley, Ireland, Bax, Howells, Rubbra, Lloyd, Brian, Tippett, Britten, Davies and the American Lou Harrison (W. Mellers, 'Postlude: A valediction forbidding mourning', *Vaughan Williams and the Vision of Albion*, London: Barrie and Jenkins, 1989, pp. 244–59). Britten is more complicated, not least by the suggestion that after his visit to the Belsen concentration camp soon after its liberation, every note he wrote was coloured by that memory.

196.   The small screen is clearly a validating medium: see how many TV advertisements portray a computer screen to increase our confidence in the product. Likewise, people in documentaries usually have a PC in the background to tell us that they are up-to-date. They also appear in libraries, thus carrying on the tradition of Renaissance painting where

some saints always carry books. (We know when it is a library rather than a private collection – we can see the catalogue labels on the spines.)

197. See, for example, M. Wackernagel and W. E. Rees, *Our Ecological Footprint: Reducing Human Impact on the Earth*, Philadelphia: New Society Publishers, 1996; J. C. J. M. Van den Bergh and H. Verbruggen, 'Spatial sustainability, trade and indicators: An evaluation of the "ecological footprint"', *Ecological Economics* 29, 1999, pp. 61–72.

198. It might have been a bigger body: in the years 1945–55, one 'stately home' was demolished about every 5 days. By the 1960s, some 600 great houses (not all National Trust properties) were open to the public.

9 EXPERIENCE AND MEANING

1. In the second 'movement' of *Dry Salvages*.

2. This is Sir Tony Wrigley's term and citations will be found in material on the nineteenth century.

3. Note also that the twentieth century has meant an emphasis (or at any rate a strong reintroduction) on verticality in the landscape: pylons, hi-rise buildings, windfarm towers and communication masts are examples. The recent addition of masts (and the conversions of other structures like water towers) for mobile phone networks has continued the theme.

4. R. Williams, *The Country and the City*, London: Chatto and Windus, 1973.

5. W. G. Sebald, *The Rings of Saturn*, London: The Harvill Press, 1998. First published in German in 1995 as *Die Ringe des Saturn*. The blurb (accurate for once) says, 'Its narratives are unfolded . . . like its {Saturn's} rings, [and] created from the fragments of shattered worlds'.

6. 'An anatomy of the world: the first anniversary', lines 213–14, first published in 1611. In his notes, A. J. Smith (*John Donne: The Complete English Poems*, Harmondsworth: Penguin Books, 1971, p. 600) says of the pages of the long poem which contain these and some similar lines that 'Donne points to a double lapse into chaos, in the heavens themselves and in our understanding of them.'

7. The quotations are from one of the most famous of all: F. R. Leavis and D. Thompson, *Culture and Environment*, Westport, CT: Greenwood Press, 1977, reprint of the original Chatto and Windus volume of 1933. See the chapters on 'The Organic Community' and 'The Loss of the Organic Community'. The eighteenth century was probably a time of enormous individuation, when many erstwhile public matters became private. Death, carving at table, punishment and sleeping all moved from openness to seclusion. See P. Spierenburg, *The Broken Spell: A Cultural and Anthropological History of Preindustrial Europe*, New Brunswick, NJ: Rutgers University Press, 1991. And I have crouched with a farm labourer in the circumstances described: his name was Harold Smith.

8. Around 50 per cent of Scotland is owned by 608 landlords, mostly as tax write-offs or sporting estates: see A. Wightman, *Who Owns Scotland?*, Edinburgh: Canongate Press, 1996. A plan to use the Territorial Army to guard little tern nests in north-east England was under discussion in 1999.

9. A sense of the complexity experienced by people living in the countryside in the face of social change brought about by immigration from the towns can be read in P. Cloke, P. Milbourne and C. Thomas, 'Living lives in different ways? Deprivation, marginalization and changing lifestyles in rural England', *Transactions of the Institute of British Geographers* NS 22, 1997, pp. 210–30. The notes give an interesting insight into the ways in which differences between the conventional categories of, for example, 'deprivation' and the lived experience of it are problematic for, for example, government agencies with well-defined sectoral responsibilities.

10. Not very successful unless nature becomes less segregated and more a part of our lived and felt surroundings. See Adams, *Future Nature*, and 'Rationalization and conservation:

Ecology and the management of nature in the United Kingdom', *Transactions of the Institute of British Geographers* NS 22, 1997, pp. 277–91.

11.   There were some apocalyptic statements around 1970. P. R. Ehrlich famously averred in his *Population Bomb* (New York: Ballantine Books, 1969) that England would not exist in the year 2000, and there is a collection of essays edited by E. Goldsmith, *Can Britain Survive?* (London: Tom Stacey, 1971/Sphere Books, 1972), both driven along by the neo-Malthusianism of the time. Ehrlich's statement has been thrown in his face ever since by growth-friendly opponents.

12.   A. Gare, *Nihilism Incorporated: European Civilization and Environmental Destruction*, Bungendore, NSW: Eco-logical Press, 1977.

13.   The escape may of course be to a Mediterranean resort with plentiful English (*sic*) beer and breakfast. The difference in holidays is made in R. Hutton, *The Stations of the Sun: A History of the Ritual Year in Britain*, Oxford: Oxford University Press, 1996.

14.   N. Luhmann, *Ecological Communication*, Cambridge: Polity Press, 1989.

15.   Eric Hobsbawm suggests that the ubiquity-indicators are plastic sheeting, artificial fibres, Coca-cola and digital watches. He also points out that only in the 1960s were Nobel Prizes awarded outside Europe and North America and that the atrium hotel is currently for bourgeois society what the opera house was in the late nineteenth century (E. Hobsbawm, *Age of Extremes: The Short Twentieth Century, 1914–1991*, London: Michael Joseph, 1994).

16.   The best example of this in the near future may well be the effect of the phased withdrawal of agricultural subsidies on hill farming. The ecological and scenic consequences could be very far-reaching.

17.   There are various UK and European data holdings of interest and a collection is indexed by the Centre for Ecology and Hydrology at Monks Wood, Cambridgeshire, whose web pages are a good gateway (www.ceh.ac.uk).

18.   *Countryside* 81, Winter 1996/97, p. 1. This was the freesheet 'Newspaper of the Countryside Commission'. In 1998, the Commission was amalgamated with the Rural Development Commission to form the Countryside Agency. One of the new body's first publications does not mention 'sustainability' except in the context of agriculture (The Countryside Agency, *Tomorrow's Countryside – 2020 vision*, no date nor place of publication but probably Cheltenham 1999).

19.   A pamphlet from 1998, *Working Today for Nature Tomorrow*, Peterborough: English Nature.

20.   See the interesting if diffuse discussion in S. Owens and R. Cowell, 'Lost land and limits to growth: Conceptual problems for sustainable land use change', *Land Use Policy* 11, 1994, pp. 168–80. A different perspective on the cognition of valued places is M. Crang, 'On the heritage trail: Maps of and journeys to olde Englande', *Environment and Planning D: Society and Space* 12, 1994, pp. 341–55.

21.   There is an interesting footnote to this in the finding that between 1982 and 1996, the average date for the first kestrel eggs in the Chiltern hills of Buckinghamshire fell inexorably (though not smoothly) from 14 May to 17 April (P. Burton, 'Birds of prey on farmland', *The Raptor* 24, 1997, pp. 18–20.

22.   The Secretary of State was at pains to point out that he wished to reduce levels of use, not ownership. More energy is used in making a car than in running it for the rest of its life but energy is not the only consideration.

23.   In the course of a Conference lecture in Glasgow in 1996.

24.   K. Thomas, *Man and the Natural World*, London: Allen Lane, 1983.

25.   A. Blowers (ed.), *Planning for a Sustainable Environment*, London: Earthscan, 1993.

26.   There is a nice fantasy piece about the reversion of London to 'natural' conditions under a temperate climate (assuming desertion after a Chernobyl-type accident or in the face of an epidemic) in L. Spinney, 'Return to paradise', *New Scientist*, 20 July 1996, pp. 26–31.

27.   M. Newsom, 'Planning, control or management?', in M. Newsom (ed.), *Managing the*

Human Impact on the Natural Environment: Patterns and Processes, London and New York: Belhaven Press, 1992, pp. 258–79.

28. Blowers, *Planning for a Sustainable Environment*.
29. After hearing an evangelical sermon, Lord Melbourne (1779–1848) is reputed to have said, 'Things have come to a pretty pass when religion is allowed to invade the sphere of private life' (*Oxford Dictionary of Quotations*, 3rd edn, Oxford: Oxford University Press, 1979).
30. Kingsley Amis's novel *The Alteration* (London: Cape, 1978) is based on the supposition that Britain was pre-industrial and Roman Catholic.
31. *UK Environmental Accounts 1998*, London: HMSO, 1998; D. McLaren and S. Bullock, 'Tomorrow's World', *Earth Matters*, 36, Winter 1997, pp. 8–9; T. Jackson, N. Marks, J. Ralls and S. Stymne, *Sustainable Economic Welfare in the UK 1950–1996*, London: New Economics Foundation, 1997; J. Pearce, *Measuring Social Wealth*, London: New Economics Foundation, 1996. See also the book-length treatment of D. McLaren, S. Bullock and N. Yousuf, *Tomorrow's World: Britain's Share in a Sustainable Future*, London: Earthscan, 1998.
32. It comes from *Meditation* XVII [with the spelling modernised], 'No man is an island, entire of itself; every man is a piece of the continent, a part of the main; if a clod be washed away by the sea, Europe is the less . . . any man's death diminishes me, because I am in involved in Mankind; And therefore never send to know for whom the bell tolls; It tolls for thee.'
33. Dylan Thomas, 'Author's Prologue', *Collected Poems 1934–1952*, London: Dent, 1952, p. viii. The *Adagio* of Carl Neilsen's 5th Symphony (op. 50) seems like a suitable accompaniment.

Biographies

BACON, FRANCIS (1561–1626)

Viscount St Albans, Baron of Verulam (1618) and Lord Chancellor of England from 1618 to 1621. Bacon was a lawyer, politician, philosopher and man of letters. He became Lord Keeper of the Seal in 1617 and the following year was made Lord Chancellor. In 1621 Sir Edward Coke brought a charge of corruption against Bacon who retired in disgrace. He spent much of his time thereafter writing. He had a clear analytical mind and rejected theoretical and hypothetical (Aristotelian) structures of knowledge in favour of empirical observation of nature. He published *Novum Organum*, in 1620, in which he described new scientific methods of organising knowledge, and a Utopian study called *New Atlantis* was published posthumously in 1627.

BENTHAM, JEREMY (1748–1832)

Philosopher, moralist, writer on jurisprudence and reformer. He was called to the Bar in 1817 but did not practise his profession. Instead he devoted his life to composing civil and criminal codes of law and all aspects of reform of the legal system. He coined many words now in standard use; codify, international and minimise, for example, and developed a philosophy of 'utility' which sought to value events and situations for the pleasure they gave without giving intellectual or fashionable weight to one recreation over another. It was said of him that he found the philosophy of law a chaos and left it a science. He campaigned for the reform of Parliament, for democracy, and for poor law and prison reform. He is seen as a founding father of University College, London.

BLAKE, WILLIAM (1757–1827)

Printmaker, painter, poet and prophet and visionary, Blake developed a personal mythology which he published in the form of combined printed art and text he called 'illuminated printing'. For him art of every kind offered an opportunity for redemption through insight into the spiritual world. He produced a series of mythological epics such as *The Book of Thel* (1789), *The Song of Los* (1795) and *The Book of Urizen* (1794) which explored the conflict between restrictive morality and anarchic liberty. Blake believed revolution had both a creative and destructive role and was the inevitable result of the separation of humanity from the spiritual dimension. His last two books, *Milton: A Poem* (1803–8) explains his life-work as a response to the poetic and religious legacy of Milton, while *Jerusalem* (1804–20) deals with the fall of man and ends with a vision of redemption.

CONSTABLE, JOHN (1776–1837)

Landscape painter. He studied at the Royal Academy and began his career as a portrait painter but in 1802 he sent a landscape to the Royal Academy Exhibition. This exhibition was a catalyst in the formation of his philosophy of art, putting his work into a context which enabled him to see what was lacking. He resolved thereafter to portray nature in a pure and unaffected manner. Constable created a completely new style and method of painting and chose as his subjects the humble, ordinary beauties of the English landscape. He obeyed the creative demands of his own talent and artistic expression, a new departure in the role of artists who were largely the journeymen servants of their patrons. *The Hay Wain* of 1824 is now an icon of Englishness.

Darwin, Charles Robert (1809–82)

Naturalist who first made his name as a geologist, later famous for describing the role of natural selection in the evolution of species. His appointment as naturalist for the five-year-long expedition of HMS *Beagle* in 1831 allowed him to observe the differences in isolated animal and bird populations and led him to develop his theory. In 1859 he published *On the Origin of Species by Means of Natural Selection* and in 1871 *The Descent of Man and Selection in Relation to Sex*. His interests as a naturalist were vast. His last book, *The Formation of Vegetable Mould through the Action of Worms* (1881) was the result of an enduring interest in earthworms and did much to explain their importance to the health of soil.

Elgar, Sir Edward (1857–1934)

A composer, principally of symphonic orchestral music of great emotional force. He also composed songs, incidental music for the theatre, music for royal occasions, oratorios and concertos. From 1924 Elgar was Master of the King's Musick. His music was imbued with a great national feeling, but also had a huge worldwide appeal. His sympathy with and understanding of the problems of orchestral musicians made him an excellent and effective conductor. The result of his patient and detailed interpretation of his music has meant that it is still played today as he intended. His most famous pieces are the *Pomp and Circumstance* marches, *The Enigma Variations* and the Concerto for Cello and Orchestra.

Engels, Friedrich (1820–95)

German socialist philosopher and co-author with Karl Marx of the *Communist Manifesto* published in 1848. Engels edited volumes 2 and 3 of Marx's *Das Kapital* after the latter's death in 1883 and was his main source of financial support during their years of collaboration. Engels was convinced that the communist revolution would take place in an industrially advanced country, such as England, where traditional society was breaking down and class divisions were being strengthened. Engels left his native Prussia for the Manchester branch of his family's business in 1842 and remained there for nearly all of the rest of his life. In 1845 he published *The Condition of the Working Class in England in 1844*.

Evelyn, John (1620–1706)

Polymath, writer, diarist, authority on architecture and landscape gardening; co-founder, Fellow, and sometime secretary of the Royal Society. Evelyn joined the Royalist Army in 1642 but was a prudent and cautious royalist who lived quietly at his Sayes Court estate during the Interregnum (1649–60). Here he cultivated his garden and a circle of friends of an intellectual and scientific mind. He avoided political intrigue, but devoted himself to the public good, accepting appointments to commissions on street improvement, the management of the Royal Mint and care of the sick, wounded and POWs of the second Dutch War among other duties. His diaries (from 1641) are an invaluable historical source.

Fiennes, Celia (1662–1741)

Traveller who, initially for the sake of her health, but largely to satisfy her curiosity, travelled in every county in England, visiting relations and keeping notes of her observations of the urban and industrial growth of the country. In 1697 she rode more than 600 miles in 6 weeks through northern England. In 1702 she wrote up her notes which provide the first eyewitness account of English life since Elizabethan times.

Fraser Darling, Sir Frank (1903–79)

Ecologist and conservationist. He became an early campaigner for conservation when, during his residence on Rona he became convinced of the need to live in harmony with nature. He was Director of the West Highland Survey (1944–50), Senior Lecturer in Ecology and Conservation at Edinburgh (1953–8) and vice-president of the Conservation Foundation in Washington DC, 1959–72. He advised government bodies on the creation of national parks and conservation areas and carried out official surveys in Alaska and East Africa. He also served on the Royal Commission on Environmental Pollution.

Hardy, Thomas (1840–1928)

Poet and novelist whose writing displays a stoic pessimism. His plots frequently make use of cruel coincidence and his main characters are invariably tragic, though not without some nobility or at least dignity. The novels are set in 'Wessex' a fictional county based upon his native Dorset. His most productive years were from 1878 to 1895 when he wrote, among other books, the *Return of the Native, Tess of the D'Urbervilles* and *Jude the Obscure*. Most of his output remains popular today and many of his novels have been made into films.

King, Gregory (1648–1712)

Genealogist, engraver, statistician and writer; King was the son of a surveyor in Lichfield. He left school at 14 to assist his father in the business of map making during the land-market boom of the post-Restoration period. He later worked for William Dugdale, Norroy King at Arms, and travelled with him during his visitation to his province north of the Trent. He was an acute observer and seems to have been obsessed by an urge to measure and evaluate everything. Nothing was beneath his notice, no problem too large. He provides the best estimate of the population of England for the time in *Natural and Political Observations and Conclusions upon the State and Condition of England 1696*. He edited the *Book of Roads*, designed the layout of Soho's streets and Squares (Soho Square was originally called King's Square), and drew up some of the leases on the first buildings in many parts of London.

Malthus, Thomas R. (1766–1834)

Pessimistic empiricist economist and demographer. In contrast to many economists and social commentators of his day, such as William Godwin and Rousseau, Malthus did not assume the ultimate perfectibility of either mankind or human society. He believed that, as population increased geometrically and production arithmetically, population would always outstrip production causing famine, disease and war which acted as a form of population control. This theory was set out in a pamphlet, *An Essay on the Principle of Population as it affects the Future Improvements of Society, with Remarks upon the Speculations of Mr Godwin, M. Condorcet and other Writers* (1798). This was later expanded to include his observations of conditions in Europe, into a much larger book. He stressed the need to balance production rates with the capacity for consumption which partly anticipated the theories of J. M. Keynes in the 1930s.

MARSHALL, WILLIAM (1745–1818)

Agriculturalist and philologist, he was born in Yorkshire to a farming family. He spent 14 years in the West Indies returning in 1774 to manage a 300-acre farm near Croydon, Surrey. Here he wrote *Minutes of Agriculture Made on a Farm of three hundred acres of various soils near Croydon . . . published as a sketch of the actual Business of a Farm* (1778). In 1779 he published *Experiments and observations concerning Agriculture and the Weather*. He was appointed agent on Sir Harbord Harbord's Norfolk estate in 1780 and in 1783 contributed to the Philosophical Transactions of the Royal Society, London, *An account of the Black Canker Caterpillar which Destroys Turnips in Norfolk*. In 1808 he retired to a large estate in the vale of Cleveland and was engaged in the creation of an agricultural college in Pickering when he died.

MORRIS, WILLIAM (1834–96)

Designer, craftsman, poet and socialist. He pioneered the Arts and Crafts Movement and revolutionised Victorian taste. While at Oxford he became friends with Edward Burne-Jones, who was a founder member of the Pre-Raphaelite Brotherhood, and with whom he later shared a studio. He was influenced by the writings of John Ruskin (q.v.) and subsequently worked for the Gothic Revivalist architect G. E. Street. Art included, for Morris, everything made by human hands. In 1861 he founded the firm of Morris, Marshall, Falkner and Co. (it became Morris and Co. in 1874), producing embroidery, furniture and stained glass. He also founded the Socialist League, Hammersmith Socialist Society and, in 1891, the Kelmscott Press. He is best remembered today for his designs for wallpaper, textiles and carpets.

OWEN, WILFRED (1893–1918)

A poet who, in his earliest writing, modelled himself on John Keats. His technical mastery of poetic forms was outstanding and his innovative use of assonance became very influential after his death. Owen enlisted in the Artists Rifles in 1915 and his experiences of the horrors of trench warfare prompted him to write poems full of anger, pity and powerful imagery. He was killed a week before Armistice Day and his verse was published by Siegfried Sassoon.

REITH, SIR JOHN (1889–1971)

Creator of and Director General of the British Broadcasting Company (later Corporation) from 1922 to 1938, whose influence was felt long after his departure. Originally trained in engineering, Reith had a talent for making systems work efficiently. There was, in 1922, no precedent for setting up a broadcasting service and many people believed that radio would be a nine days wonder. Reith, however, had a vision for its future and was determined to establish a status and authority for radio broadcasting. He had a somewhat dour, Calvinistic attitude and very strict standards which he sought to impose on the working and private lives of his employees. He left the BBC in 1938 and his subsequent career seemed to be an anticlimax although often demanding and various. From 1943 to 1944 he worked in the Admiralty's Combined Operations Material Department. His planning and co-ordination of the supplies and materials for the D-Day Landings was described as 'well-nigh perfect'.

RUSKIN, JOHN (1819–1900)

Writer, critic, artist and socialist, and one of the most influential men of the nineteenth century, Ruskin promoted the Gothic style in architecture as being based on natural forms in nature. He preached an aesthetic derived from Nature and the 'natural truth' to be derived from it, condemning fussy, forced, over elaborate decoration. He befriended the Pre-Raphaelite Brotherhood (less because he admired their work than because of the savage criticism they attracted) and deeply admired and defended W. M. Turner. In 1869 he became the first Slade Professor of Fine Art at Oxford. Ruskin's political philosophy, expressed in *Unto This Last* (*Cornhill Magazine* 1860, in book form 1862), had much in common with other British socialist thinkers of the time, such as William Morris (q.v.).

SPENCER, HERBERT (1820–1903)

Radical philosopher who urged the limitation of State responsibilities and the value of individualism. He championed the freedom of every individual to do whatever he wished provided it did not impinge upon the equal freedom of any other. He believed the role of the State should be confined to public security in the form of a police service and defence services only: there should be no national education, poor law or standards of sanitation. He coined the phrase 'survival of the fittest' in his *Development Hypothesis* in which he defended the theory of organic evolution seven years before Darwin's *Origin of Species* was published.

THOMAS, EDWARD (1878–1917)

Journalist, author and poet. He left home at an early age to escape parental opposition to his desire to pursue a career as a writer. He became a hack journalist at the age of 19 and lived in great poverty. Literary criticism for the *Manchester Guardian* later provided him with some slight financial security. Despite his urban upbringing he had an instinctive inclination to nature and solitude. Thomas' output was prodigious but it was not until 1912 that he began to write the poetry for which he is now famous. In 1915 Thomas joined the Artists Rifles, was transferred to the Royal Garrison Artillery and was killed at Arras in 1917.

TURNER J. M. W. (1775–1851)

Landscape painter greatly influenced by Claude Lorrain. He drew and painted from childhood and was first sent, in 1786, to Soho Academy. He first exhibited at the Royal Academy in 1790. A tour to Yorkshire and the north in 1797 greatly inspired him and developed his distinctive artistic talents. He afterwards attributed his successful career to the reception of his painting *Norham Castle on the Tweed – Summer's Morn*. He also found an enthusiastic champion in John Ruskin, but Turner's success was life long and he had been able to earn a living from art from his early years. His paintings are unequalled in range, imagination and atmosphere.

WALTON, SIR WILLIAM TURNER (1902–83)

Composer, much influenced by his own contemporaries such as Edward Elgar (q.v.) and Igor

Stravinsky. He is best known for his orchestral music, especially that composed for films, most famously a version of Shakespeare's *Henry V* (1944). He wrote two symphonies and the oratorio *Belshazzar's Feast*. For some ten years he lived with Osbert and Sacheverell Sitwell and wrote *Façade* (1923), pieces for a chamber ensemble, to accompany the strange poetry of their sister, Edith. The music was subsequently preserved in the ballet of the same name choreographed by Frederick Ashton.

YARRANTON, ANDREW (1616–84[?])

Engineer and agriculturalist, his interests ranged from canals and river navigation ways to ironworks and he was among the first to appreciate the agricultural value of clover. He published *The Great Improvement of Lands by Clover* in 1663 and *England's Improvement by Sea and Land to outdo the Dutch without Fighting* in two parts in 1677 and 1681. He travelled abroad and throughout England and acted as a consulting engineer, carrying out many river surveys and advising on the canalisation of the Dee at Chester and the Avon in Somerset.

Glossary

AGENDA 21 – the section of the agreement from the Rio Declaration of the United Nations Conference on Environmental Development in 1992, attended by 167 countries. Agenda 21 was a blueprint for the cleaning up of the environment and deals with a full range of environmental concerns It has been criticised for not promoting a fundamental change in the relationship between rich and poor nations.

AGGRADATION – building up, or raising by the addition of material. Everything from the largest rocks to the finest silts are moved by glaciers and ice sheets, by weathering and fluvial action. High ground tends to erosion, low ground to aggradation.

ANIMISM – in psychology the belief that mankind cannot be explained purely as a mechanism. In religion animism is the primitive attribution of a spiritual dimension to the material world. The human soul is understood to be a separate entity capable of independent action in life and after death. Inanimate objects, animals and natural phenomena may become possessed by a human soul, a spirit or a god.

ASSART – from the Latin ex, 'out of', and sarire, 'to hoe or weed'. To develop or reclaim land for agriculture from forests or wastes. In the medieval period it indicates rising populations and a consequent demand for land.

BIOCIDES – chemicals capable of killing living matter. These include weedkillers, insecticides and fungicides such as DDT deliberately applied to crops and subsequently found throughout the environment and in the food chain. A wide range of chemicals, such as PCBs and heavy metals are also biocides, and have entered the environment from industrial processes and by their disposal after use.

BIOMASS – the weight of living material in or as part of a living organism, population or ecosystem. Usually expressed as dry matter/unit area (kg/ha or g/m^2).

BIOTA – group of plants and animals occupying a place together.

BOD – biochemical oxygen demand. This is a measure of water purity using the amount of dissolved oxygen needed for oxidisation of organic pollutants by Aerobic Biochemical Action (ABA). The ABA is measured over a 5-day period and the results expressed as a ratio of mg/1. Drinking water standards are often expressed in BOD.

BP – before the present. The suffix 'BP' indicates that a date is expressed in years before AD 1950 (when radiocarbon dates were first published). 'Cal BP' indicates a calibrated radiocarbon date but many authors follow the convention established by the Journal *Antiquity* and express calibrated dates as 'BP' and uncalibrated dates as 'bp'. Calibrated dates are referred to tree ring sequences for greater accuracy.

BROWN EARTH – typical but not universal fertile soil of deciduous forest, also called 'forest brown earths'. A mixture of organic material and minerals from the ground rock. The organic compound decreases with depth and the soil is usually slightly acid. It has a humic A horizon and a B horizon in which minerals are released by weathering.

BSE – bovine spongiform encephalitis. A disease of the brain in cattle believed to have mutated from scrapie, a disease of sheep which crossed the species barrier when sheep proteins were included in cattle food. Some people who ate infected beef have developed Creutzfeld Jakob Disease and 90 have so far died (2001).

CAESIUM – element with the atomic number 55, its symbol is Cs. Radioactive nuclides Caesium 134 and 137 are discharged in liquid form from nuclear reprocessing plants. Caesium 137 is also part of the radioactive debris created by the atmospheric testing of nuclear weapons and is now found in the food chain. Its half-life is approximately 30 years.

CALCICOLES – plants which grow on limestone or limey soils.

CEGB – the Central Electricity Generating Board was created in the 1930s to enforce uniformity of standards in the industry and hasten the electrification of homes and businesses. It was privatised in 1990.

CLINTS AND GRYKES – features of a limestone pavement produced by the weathering of the surface. Grykes are deep gashes which follow the joint fractures of the rock and are arranged roughly at right angles so forming the flat paving stone shapes called clints.

COMMON AGRICULTURAL POLICY (CAP) of the European Union – for free trade in agricultural commodities in the EU, preference for domestic produce, control of imports from the rest of the world and common financing. Its objectives, as stated in the Treaty of Rome, include greater agricultural productivity, fair standards of living for farmers, reasonable consumer prices, market stability and secure food supplies.

CRITICAL LOADS – the threshold beyond which a pollutant is present in sufficient quantities to exert perturbations on the existing environment.

CULLET – waste glass which is to be melted for reuse.

CULTURE – stage of human social development, for example, hunter-gatherer, agricultural, industrialised, or a variation within any one stage as expressed by one aspect of the groups material culture, such as a style of hand axe or pottery. Also the sum of the social behaviour, customs, beliefs and mores of a society.

DOMESTICATED – refers to animals bred in captivity, usually for economic advantage, by a society which manages its breeding, food supply and habitat. A wild-born animal which has been tamed is not domesticated. Most domesticated animals are not tame.

ECOTONE – the narrow transitional boundary between different ecological communities, typically species rich, such as the land/water interface.

ECUMENE – the whole inhabited world.

EEZ – Exclusive Economic Zone. The zones of exclusive sea fishery extending from 30 to 200 nautical miles as designated by the first session of the Third Conference on the Law of the Sea in Caracas in 1974. They were created by unilateral acts which claimed 200 mile zones by Latin America, Africa and South Asia. The then USSR established its zone in 1976 and the USA in 1977.

ENCLOSURE – farming practice with fenced, hedged or walled fields farmed by one tenant or landowner. Where the 'open field' system was in use enclosure replaced it piecemeal from c.1300 especially when populations fell. Enclosure for sheep farming took place in the fifteenth and sixteenth centuries. Parliamentary Enclosures from c.1750 ended most open fields systems. Laxton is the sole survivor.

ENVIRONMENTAL SYSTEMS – the interaction between living organisms and the surrounding, natural or man-made environment.

EPIGRAPHY – writing on buildings or memorial stones.

FALLOW – of an arable field – left unplanted to recover its fertility.

FEUDAL – describes a hierarchical society in which each person's rights and obligations in respect of his social superiors and inferiors are clearly defined in law and reflected in rights over land and other property.

FOOD SYSTEM – a food web; a set of interconnecting food chains which comprise the pattern of routes by which food can flow through an ecosystem.

FOREST – a legal term, after the Norman Conquest, for areas preserved for the king's deer hunting. It does not necessarily mean that there were trees. One of the rights of access to the king's forest was pannage (q.v.).

FULLING – washing newly made woollen cloth to remove the natural grease in sheep wool from the material.

GEOMORPHOLOGY – the study of landforms, including those under the seas and oceans, and the forces and processes which create them.

GRANGES – a new form of landholding for profit and power, granges were created by monastic communities colonising uplands and wastes in the early Middle Ages. Fountains Abbey in Yorkshire had 26 granges, the largest number attached to any monastic house.

HEAD-DYKE – a term applied mainly in Scotland for a drystone wall built to mark off a farm's fields from the open moorland beyond.

HEAT ISLAND – a concentration of heat given off by a place of high energy consumption such as a city which reflects solar radiation and emits high levels of waste heat. The ambient temperature may be as much as 10 °C above that of the surrounding countryside. This may cause higher thunderstorm frequencies and increased rainfall downwind of the city.

HIDB – Highlands and Islands Development Board, a Government development agency established in 1965 to promote the economic and social development of the Highlands and Islands of Scotland. In 1991 the HIDB became Highlands and Islands Enterprise, a co-ordinating body for nine Local Enterprise Companies which are run under the auspices of the private sector.

HOLOCENE – the present, warm, epoch of the Quaternary period. It began c.10,000 BP, with the Flandrian interglacial. The Holocene is not really a separate epoch so much as one of a series of warm intervals in the Pleistocene (q.v.).

HYDROLOGICAL CYCLE – the movement of water between the various stores of water on the planet. The ice caps, oceans and seas are subject to evaporation, precipitation and runoff. The movement is governed by the climatic regime.

HYPSITHERMAL – the period of highest Holocene (q.v.) temperatures which occurred at different times in different parts of the world. In the Antarctic it happened between 11 and 8,000 years ago but in the Arctic between 5 and 4,000 years ago. In temperate latitudes the temperatures during the hypsithermal were typically 20 °C above those of the present day.

IDEALISM – the doctrine that in external perceptions the objects immediately known are ideas and that all of reality is of a psychic nature.

IMPROVEMENT – a development, in British agriculture, from the 1730s, with the increased use of newly invented machinery and recuperative and clearing crops. Increased animal feed meant more livestock could be kept over winter. Breed improvements became practical and the extra manure benefited the arable crops. An enclosed field system was vital. The open field system militated against innovation.

INALIENABLE – that which cannot be separated; a right protected by law or land which cannot be sold out of a family holding.

INBYE – the cultivated fields in the immediate neighbourhood of the farm buildings; the arable part of a farm which also has some hill pasture.

KELP – the *Laminaria* genus of seaweed. Traditionally on the west coasts of Europe kelp has been used for fertiliser. Now it is harvested industrially in France, UK, Ireland and Portugal for the extraction of iodine and for alginates used in cosmetics, pharmaceutical and food manufacture.

LEGUMES – also known as pulses, legumes are plants whlch produce seeds in pods: peas, beans, lentils. Legumes are valuable in crop rotation systems because they can fix atmospheric nitrogen on their roots. Subsequent crops benefit from the nitrogen thus provided in the soil.

MALTHUSIAN – economic doctrine based on the writings of Thomas R. Malthus (1766–1834). In his *Essay on the Principle of Population* of 1798 he argued that starvation was inevitable unless population growth was controlled.

MIDDEN – a refuse heap. Every society produces middens; a joy to archaeologists, the earliest are heaps of shells and fish bones associated with Mesolithic people. Agricultural communities tend to spread their middens on the fields but middens survive on many archaeological sites. In the last 50 years we have produced landfill middens of unprecedented size.

MODERNISM – a movement, which began at the end of the nineteenth century and lasted to the early or mid-1940s, and is reflected in the development of all the arts from poetry to architecture. It was marked by a breaking away from the established rules and conventions and by the use of innovative, experimental techniques and language, forms and styles.

MOR – a type of conifer-forest humus layer of organic material usually matted or compacted, it is very acid with little microbial content except for fungi. *Mor* is also found on open heath and moorland in moist, cool climates.

OESTROGENS – female sex hormones governing the development of secondary sex characteristics and fertility. They can be synthesised and are used in the promotion and control of both human and animal fertility.

OMBROGENOUS – describes a plant which derives its water and minerals entirely from precipitation. Raised bogs and blanket bogs support ombrogenous plants, typically the bog-moss *Sphagnum*.

OPEN FIELDS – a farming system originally widespread in the Midlands and elsewhere, in which a number of fields, usually three, was farmed by the allocation of strips within each to every tenant. Co-operation and interdependence was essential and individual enterprise and agricultural reform were impossible. Parliamentary Enclosures (1750–1850) put an end to most open fields; Laxton was an exception.

PALAEOBOTANY – the study of fossil plants.

PALAEOCLIMATOLOGY – the study of ancient climates.

PALAEOHYDROLOGY – the study of ancient water systems.

PANNAGE – the ancient right of turning pigs out to graze in forests. Also a term for the food thus found.

PCBs – Polychlorinated biphenyls, one of a range of compounds first synthesised in 1881 and produced from 1929 for insulating industrial electricity transformers. First detected as a pollutant in 1966, PCBs are implicated in damage to the immune system and for lowered fertility in marine birds and animals. Their use has been restricted in the USA and Europe since the 1970s.

PERIGLACIAL ZONE – at the landward limit of the great ice sheets, periglacial zones are subject to very severe conditions and are characterised by frost-shattering and other cold climate processes, with or without the presence of permafrost.

PLEISTOCENE – an epoch within the Quaternary Period. Its age estimates range from 1.6 to 2.4 million years. It covers about 0.04 per cent of the total age of the Earth.

PLUTONIUM – an element, atomic number 94, symbol Pu. Different isotopes of plutonium are produced by nuclear reactions, 13 isotopes are known of which Pu-239 is the most stable. It has a half-life of 24,000 years and is used in nuclear weapons. One kilogram yields about 10^{14} joules of energy.

PODSOLIC – refers to soils on acid sub-soils or in areas of high rainfall and drainage. The leaching effect produces an E horizon of ash grey colour due to the removal of iron and humus and a B horizon where the iron and humus accumulate sometimes forming iron pan. Podsols are characteristic of heaths and moorlands and polsolization is caused by deforestation.

POST-MILLS – wind mills constructed on a central post on which the mill can be rotated to face the prevailing wind.

PREHISTORIC – the time before surviving written records. The prehistoric period of China and the Middle East ended much earlier than the prehistoric in Western Europe.

PUMPED STORAGE – a system for generating peak-load electricity by turbines which are turned by the fall of water from a reservoir filled by pumps which lift the water from a lower level. During low demand periods the water is raised to fill the reservoir and released to drive the turbines during peak-demand periods.

RADIONUCLIDES – radioactive atoms. Such atoms have more neutrons at their nucleus than other atoms of the same element and therefore have a higher atomic number. Atmospheric carbon 12 has the isotope carbon 14, which is radioactive. The regular decay of carbon 14 at about 1 per cent every 83 years is the basis for radiocarbon dating.

RAISED BOG – a bog is an ombrogenous (q.v.) mire. Raised bogs have a low convex profile and are usually found on floodplains and in estuaries. The ombrotrophic plants usually develop over a peat of minerotrophic and/or aquatic plants and the depth of the peat raises the bog surface above the ground water influence.

REALISM – Realism asserts that (1) there is an external world, (2) that some of our beliefs about it are true, and (3) we can sometimes determine which of our beliefs are true. Classic realism asserts that we inhabit a world whose nature and existence is neither logically nor causally dependent upon any mind, that some of our beliefs about this world are true and that we have means of enquiry which enable us to discover that some of our beliefs are true.

REAVES – long parallel banks of stone and earth forming rectangular fields. Reave systems are found on Dartmoor and in parts of Yorkshire and are believed to date from the Bronze Age. They cover extensive areas and are remarkable for their regularity.

RENDZINA – a class of soils with black or brown friable surface horizons over light grey or yellow calcareous material of weathered bedrock. An 'A' horizon of rich humic material overlies a chalky 'C' horizon.

ROMANTICISM – a literary and artistic movement originating in Europe at the end of the eighteenth century. It was characterised by an interest in the natural order, Nature, primitive ways of life (the cult of the Noble Savage), scenery and a subjective association of human feelings with surroundings. In Britain it is represented in the writings of Coleridge, Wordsworth, Keats, Shelley, Byron and Sir Walter Scott.

ROUGH GRAZING – a stage in the natural progression by which ley becomes permanent pasture (meadow) then becomes rough grazing then scrub and finally forest. Rough grazing includes moors, heaths, downs and fens.

SMOG – a portmanteau word coined by Dr H. A. Voeux of the Coal Smoke Abatement Society in 1905 for a mixture of natural fog and smoke. Smog of that type became less common after the Clean Air Act of 1956. Photochemical smog develops over traffic-filled cities.

SSSI – Site of Special Scientific Interest, a designation applied to species-rich or geologically interesting areas. There are about 5000 SSSIs in the United Kingdom and they cover about 7.5 per cent of the country. SSSI status does not confer protection on a site unless owned by an official conservation agency or voluntary organisation. Some 2 per cent of SSSIs are lost completely every year.

STOICS – Followers of a Graeco-Roman philosophy which stresses the importance of duty and responsibility in the good conduct of public affairs. Stoics believe that the whole universe is governed by fate and is fundamentally rational: therefore people should conduct their lives in imitation of the calm grandeur of the universe and accept everything that happens with a stern and tranquil mind.

SUCCESSION – the regular and progressive change in the components of an ecosystem from the initial colonisation of an area to a stable state or mature ecosystem.

TECHNETIUM – the first synthetically produced element, atomic number 43, symbol Tc.

TEPHRA – a word coined in 1954 by S. Thorarinson to refer to pyroclastic material (hot rock thrown out of a volcano) which is fine enough to be transported by air. Tephra deposits can be found a long way from their volcano of origin and can be dated by potassium-argon, or fission track dating, or by radiocarbon dating of associated organic remains.

TSUNAMI – a Japanese word for a very high wave, exceptionally long which can cause devastation on coasts. Tsunamis are caused by sudden volcano or earthquake action and are also known as impulsive waves.

WILDWOOD – a term coined by Oliver Rackham for equilibrium woodland formed by the natural succession of trees colonising Britain. There are no surviving wildwoods but part of Speyside pine woods may have survived as wildwood until as late as the seventeenth century. Ancient woods are different in composition from the wildwood and all have been subject to intensive use and management.

A selected bibliography

Ball, D. F., Dale, J., Sheail, J. and Heal, O. W. (1982) *Vegetation Change in Upland Landscapes*, Bangor: ITE.

Bradley, R. (1998) *The Significance of Monuments: On the Shaping of Human Experience in Neolithic and Bronze Age Europe*, London and New York: Routledge.

Brimblecombe, P. (1987) *The Big Smoke: A History of Air Pollution in London since Medieval Times*, London and New York: Routledge.

Buchanan, M. (ed.) (1995) *St Kilda*, Edinburgh: HMSO/Glasgow Museums.

Butlin, R. A. and Roberts, N. (eds) (1995) *Ecological Relations in Historical Times*, Oxford: Blackwell.

Clapham, A. R. (ed.) (1978) *Upper Teesdale: The Area and Its Natural History*, London: Collins.

Clapp, B. W. (1994) *An Environmental History of Britain since the Industrial Revolution*, London and New York: Longman.

Collins, E. J. T. (1978) *The Economy of Upland Britain, 1750–1950: An illustrated review*, Reading: University of Reading Centre for Agricultural Strategy.

Cook, F. H. and Williamson, T. (1999) *Water Management in the English Landscape: Field, Marsh and Meadow*, Edinburgh: Edinburgh University Press.

Coull, J. R. (1996) *The Sea Fisheries of Scotland: A Historical Geography*, Edinburgh: John Donald.

Dark, K. and Dark, P. (1997) *The Landscape of Roman Britain*, Stroud: Sutton Publishing.

Environment, Department of the (1992) 'UK Environment', London: HMSO.

Environment, Department of the (1996) 'Digest of Environmental Statistics', London: HMSO.

Fleming, A. (1988) *The Dartmoor Reaves: Investigating Prehistoric Land Divisions*, London: Batsford.

Fowler, P. (1983) *The Farming of Prehistoric Britain*, Cambridge: Cambridge University Press.

Gingell, C. (1992) *The Marlborough Downs: A Later Bronze Age Landscape and Its Origins*, Devizes: Wiltshire Archaeological and Natural History Society.

Grigg, D. (1989) *English Agriculture: An Historical Perspective*, Oxford: Blackwell.

Hassan, J. (1998) *A History of Water in Modern England and Wales*, Manchester and New York: Manchester University Press.

Holderness, B. A. (1976) *Pre-Industrial England: Economy and Society from 1500–1700*, London:

Hooke, D. and Burnell, S. (eds) (1995) *Landscape and Settlement in Britain AD 400–1066*, Exeter: University of Exeter Press.

Hoskins, W. G. (1988) *The Making of the English Landscape*, London: Hodder and Stoughton [with an Introduction and Commentary by C. Taylor].

Hulme, M. and Barrow, E. (eds) (1997) *Climates of the British Isles: Present, Past and Future*, London and New York: Routledge.

James, N. D. G. (1981) *A History of English Forestry*, Oxford: Blackwell.

Mackay, D. (1995) *Scotland's Rural Land Use Agencies: The History and Effectiveness in Scotland of the Forestry Commission, Nature Conservancy Council and Countryside Commission for Scotland*, Aberdeen: Scottish Cultural Press.

McLaren, D., Bullock, S. and Yousuf, N. (1998) *Tomorrow's World: Britain's Share in a Sustainable Future*, London: Earthscan.

Mellers, W. (1989) *Vaughan Williams and the Vision of Albion*, London: Barrie and Jenkins.

Muir, R. (1999) *Approaches to Landscape*, Basingstoke: Macmillan.

Palmer, M. and Neaverson, P. (1994) *Industry in the Landscape, 1700–1900*, London and New York: Routledge.

Parry, M. L. and Slater, T. R. (1980) *The Making of the Scottish Countryside*, London: Croom Helm.

Pearsall, W. H., revised Pennington, W. (1968) *Mountains and Moorlands*, Fontana Library, London: Collins.

Perry, A. H. (1981) *Environmental Hazards in the British Isles*, London: Allen and Unwin.

Rackham, O. (1997) *The History of the Countryside*, London: Phoenix.

Sheail, J. (1976) *Nature in Trust: The History of Nature Conservation in Britain*, Glasgow and London: Blackie.

Sheail, J. (1981) *Rural Conservation in Inter-War Britain*, Oxford: Clarendon Press.

Smout, T. C. (1997) *Scottish Woodland History*, Edinburgh: Scottish Cultural Press.

Smout, T. C. (2000) *Nature Contested: Environmental History in Scotland and Northern England, 1600 to the present*, Edinburgh: Edinburgh University Press.

Smout, T. C. (ed.) (1993) *Scotland since Prehistory: Natural Change and Human Impact*, Aberdeen: Scottish Cultural Press.

Stamp, L. D. (1962) *The Land of Britain: Its Use and Misuse*, London: Longman Green/Geographical Publications Ltd, 3rd edn.

Stamp, L. D. and Hoskins, W. G. (1963) *The Common Lands of England and Wales*, London: Collins.

Stroud, D. A., Reed, T. M., Pienkowski, M. W. and Lindsay, P. A. (1987) *Birds, Bogs and Forestry: The Peatlands of Caithness and Sutherland*, Peterborough: NCC.

Thirsk, J. (1997) *Alternative Agriculture: A History from the Black Death to the Present Day*, Oxford: Oxford University Press.

Thirsk, J. (ed.) (2000) *The English Rural Landscape*, Oxford: Oxford University Press.

Thompson, D. B. A., Hester, A. and Usher, M. B. (eds) (1995) *Heaths and Moorland: Cultural Landscapes*, Edinburgh: HMSO.

Trinder. B. (1982) *The Making of the Industrial Landscape*, London: J. M. Dent.

Turnock, D. (1995) *The Making of the Scottish Landscape*, Aldershot: Scolar Press.

Williams, M. (1970) *The Draining of the Somerset Levels*, Cambridge: Cambridge University Press.

Williamson, T. (1995) *Polite Landscapes. Gardens and Society in Eighteenth-Century England*, Stroud: Sutton Publishing.

Williamson, T. (1997) *The Norfolk Broads. A Landscape History*, Manchester and New York: Manchester University Press.

Winter, J. (1999) *Secure from Rash Assault: Sustaining the Victorian Environment*, Berkeley, Los Angeles and London: University of California Press.

Wright, P. (1985) *On Living in an Old Country: The Past in Contemporary Britain*, London: Verso.

Index

Note: Refs to captions of Plates and Figures are in **bold**.

Nature Contested
Environmental History in Scotland and Northern England since 1600

T. C. Smout,
Director of the Institute for Environmental History at the University of St Andrews

July 2000   210pp   Illustrated
Hardback   0 7486 1410 9   £45.00
Paperback   0 7486 1411 7   £14.99

This book is about how we have treated nature in some of the most valued landscapes in Europe. Combining social and cultural history with ecology and geography T.C. Smout has written an environmental history that is both profound and accessible.

The contest between two views of nature – conservation versus development; use versus delight – is at the centre of this book.

The author begins by taking a hard look at our encounters with the natural world. He shows how the Scots and the northern English never shared the southerner's view of their environment as intimidating, and describes how conflict between using and enjoying the land gradually arose and gave birth to modern conservation ideas. He reveals how the history of the woods – especially the 'Great Wood of Caledon' – is quite different from popular myth, and examines the history and fate of the soil and the fields; of the rivers, lakes and lochs; of the hills and mountains; and of the modern quarrel over the countryside.

'By the end,' the author writes, 'I hope to have presented on my theatre a dramatic tale that tells us a fair amount not only of northern Britain, but something about the globe and the European west as a whole over the last four hundred years.'

The Harvest of the Hills
Rural Life in Northern England and the Scottish Borders, 1400-1700

Angus J. L. Winchester,
Senior Lecturer in History at Lancaster University

August 2000   194pp   c. 50 illustrations
Paperback   1 85331 239 8   £19.95

This illustrated environmental history of rural life in Northern England and the Scottish Borders in the late medieval and early modern periods explores the relationship between society and the environment – the ways in which humans responded to and used the environment in which they lived. The author uses the orders and byelaws made by manorial courts to build up a picture of how pastoral society in the Pennine, Lake District and Border hills husbanded the resources of the uplands. It offers an upland, pastoral paradigm of land use, the management of common land, and the transition from medieval to early-modern farming systems to balance the extensive literature on lowland culture.

Through a lively text and carefully selected illustrations the author captures the distinctive local culture of traditional pastoral communities in Britain.

- The first book to focus on agrarian byelaws from the manorial records of upland communities
- Provides a detailed picture of traditional hill farming practices
- Offers a model of the development of pastoral farming systems from c. 1400-1700
- Full texts of six compilations of byelaws from upland manors included as an appendix
- Generously illustrated

Order from
Marston Book Services, PO Box 269, Abingdon, Oxon OX14 4YN
Tel 01235 465500 • Fax 01235 465555
Email: direct.order@marston.co.uk

Visit our website www.eup.ed.ac.uk

All details correct at time of printing but subject to change without notice